Progress in Mathematics

SADLIER-OXFORD

Rose Anita McDonnell

Catherine D. LeTourneau

Anne Veronica Burrows

M. Winifred Kelly

Mary Grace Fertal

with
Dr. Elinor R. Ford

Series Consultants

Christine D. Thomas, Ph.D.
Assistant Professor of Mathematics Education
Georgia State University
Atlanta, GA

Margaret Mary Bell, S.H.C.J., Ph.D.
Director, Teacher Certification
Rosemont College
Rosemont, PA

Ana Maria Rodriguez
Math Specialist
St. Agatha School
Miami, FL

Re McClung, Ed.D.
Administrator of Instructional Services
and Staff Development
Centralia School District
Buena Park, CA

Alan Amundsen
Math Specialist
Los Angeles, CA

Dennis W. Nelson, Ed.D.
Director of Basic Skills
Mesa Public Schools
Mesa, AZ

Tim Mason
Math Specialist
Palm Beach County School District
West Palm Beach, FL

Sadlier-Oxford
A Division of William H. Sadlier, Inc.

Series Reviewers
The publisher wishes to thank the following teachers and administrators who read portions of the series prior to publication for their comments and suggestions.

Maria Bono Whitestone, NY	Sr. Lynn Roebert San Dimas, CA	Sr. Ruthanne Gypalo East Rockaway, NY	Madonna Atwood Creve Coeur, MO
Jennifer Fife Yardley, PA	Anna Cano-Amato Brooklyn, NY	Sr. Anita O'Dwyer North Arlington, NJ	Marlene Kitrosser Bronx, NY
Donna Violi Melbourne, FL	Galen Chappelle Los Angeles, CA		

Acknowledgments

Every good faith effort has been made to locate the owners of copyrighted material to arrange permission to reprint selections. In several cases this has proved impossible. The publisher will be pleased to consider necessary adjustments in future printings.

Thanks to the following for permission to reprint the copyrighted materials listed below.

"High on the Wall" (text only) by Charles Causley, copyright © 1975 by Charles Causley. From COLLECTED POEMS, Macmillan Publishers, Ltd. Reprinted by permission of David Higham Associates Ltd.

"Houses" (text only) by Aileen Fisher. From UP THE WINDY HILL by Aileen Fisher. Copyright 1958 Aileen Fisher. © Renewed 1986 Aileen Fisher. Originally published by Aberlard Schuman. Used by permission of Marian Reiner for the author.

"How Many Seconds in a Minute?" (text only) by Christina G. Rossetti.

"The Inchworm's Trip" (text only) by Sandra Liatsos. From POEMS TO COUNT ON by Sandra Liatsos. Copyright © 1995 by Sandra Liatsos (Scholastic Professional Books). Used by permission of Marian Reiner for the author.

"Maps" by Goldie Capers Smith.

"Mysteries of the Deep" (text only) by Christine Barrett. Printed with permission of the author.

"On with the Show" (text only) by Marie A. Cooper. Printed with permission of the author.

"One Little Kitten" (text only) by Carolyn Graham. Copyright © 1996 by Carolyn Graham. Used by permission of the author.

"The Race is On" (text only) by Rebecca Kai Dotlich. Copyright © 2000 by Rebecca Kai Dotlich. Reprinted by permission of Curtis Brown, Ltd.

"A Speedy Young Driver" (text only) by Sandra Liatsos. From POEMS TO COUNT ON by Sandra Liatsos. Copyright © 1995 by Sandra Liatsos (Scholastic Professional Books). Used by permission of Marian Reiner for the author.

Excerpt from "Toothy Crocodile" (text only) from RIDICHOLAS NICHOLAS by J. Patrick Lewis. Copyright © 1995 by J. Patrick Lewis. Used by permission of Dial Books for Young Readers, a division of Penguin Putnam Inc.

"Twos" from THE BUTTERFLY JAR by Jeff Moss. Copyright © 1989 by Jeff Moss. Used by permission of Bantam Books, a division of Random House, Inc.

Anastasia Suen, Literature Consultant

All manipulative products generously provided by ETA, Vernon Hills, IL.

Series Covers
Design and Production by Design Five.

Photo Credits

Janette Beckman: 348, 389, 395, 405, 447, 456, 458.
Jane Bernard: 29, 36, 101, 107, 109, 115, 117, 133, 183, 191, 197, 227, 235, 261, 266, 271, 301, 303, 306, 321, 331, 411.
FPG International/ Bill Losh: 319 top;
Arthur Tilley: 319 bottom right.

Ken Karp: 87, 111, 145, 229(a), 229(b), 277, 366.
Clay Patrick McBride: 61, 63, 66, 67, 69, 72, 75, 77, 80, 83, 95, 118, 139, 141, 143, 147, 149, 151, 153, 161, 175, 205, 210, 215, 245, 254, 255, 273, 299, 305, 309, 310, 329, 330, 337, 339, 341, 343, 347, 349, 357, 367, 373, 376, 377, 383, 387, 413, 417, 419.

Tony Stone Images/ Daryl Balfour: 20;
Ian Shaw: 168; *Lori Adamski Peek*: 320 top left and right, 319 bottom left;
George Kamper: 320 bottom left; *Joe McBride*: 320 bottom right; *A. Witte/ C.Mahaney*: 446.

Illustrators

Bernard Adnet
Joe Boddy
Lisa Blackshear
John Corkery
Deborah Drummond
Rob Dunlavey
Cameron Eagle

Dave Garbot
Tim Haggerty
Steve Henry
Nathan Young Jarvis
Dave Jonason
Andy Levine
Lori Osiecki

Bart Rivers
Stephen Schudlich
Matt Straub
Blake Thornton
Dirk Wunderlich

Contents

* Algebraic Reasoning

CHAPTER 3
Addition and Subtraction: Facts to 18

CHAPTER 4
Money and Time

✳ Algebraic Reasoning

CHAPTER 5
Addition of Two-Digit Numbers

*Algebraic Reasoning

CHAPTER 6
Subtraction of Two-Digit Numbers

CHAPTER 7
Geometry, Fractions, and Probability

CHAPTER 8
Measurement

✱Algebraic Reasoning

* Algebraic Reasoning

2 groups of 4 equals 8 in all.

CHAPTER 11
Place Value to 1000 and Subtraction

CHAPTER 12
Moving On in Math

End-of-Book Materials

∗ **Algebraic Reasoning**

Dear Family:

We are pleased that your child will be learning mathematics with us.

Progress in Mathematics, now in its fifth decade of user-proven success, is a complete basal mathematics program. Written by experienced teacher-authors, the program combines the best of past educational methods with the most up-to-date approaches to meet the mathematical needs of *all* children.

The scope and sequence of *Progress in Mathematics* meets the new Standards. Using *Progress in Mathematics*, children may progress as quickly as they can or as slowly as they must.

Second-grade topics to be studied include: addition/subtraction, place value, multiplication/division, time, money, geometry, measurement, fractions, graphing, and introductory concepts in statistics and probability. Special attention is given to critical thinking, problem solving, and the uses of.

But overall success in achieving the goals of this program is dependent upon active teacher-family-student interaction. You can help your child achieve her/his maximum learning level in mathematics by: (*a*) talking to your child about mathematics in everyday situations, (*b*) encouraging your child so that she/he will like mathematics, (*c*) providing quiet space and time for homework, and (*d*) instilling the idea that by practicing concepts she/he can have fun while learning mathematics.

Throughout the year, your child will bring home *Math Alive at Home* pages for you and your child to complete together. These pages include fun activities that will help you relate the mathematics your child is learning to everyday life.

We know that by using *Progress in Mathematics* your child will learn to value math, become confident in her/his ability to do math and solve problems, and learn to reason and communicate mathematically.

Sincerely,

The Sadlier-Oxford Family

Skills Update

A Review of Mathematical Skills from Grade 1

CONTENTS

∗ Algebraic Reasoning

This magazine belongs to _____.

I can count to 20.

1 2 3 4 5 6 7 8 9 10
11 12 13 14 15 16 17 18 19 20

Ring how many.

1. 4   5   6

2. 10   11   12

3. 14   15   16

Write how many.

4. ____

5. ____

6. ____

7. ____

_____ can skip count to twenty.

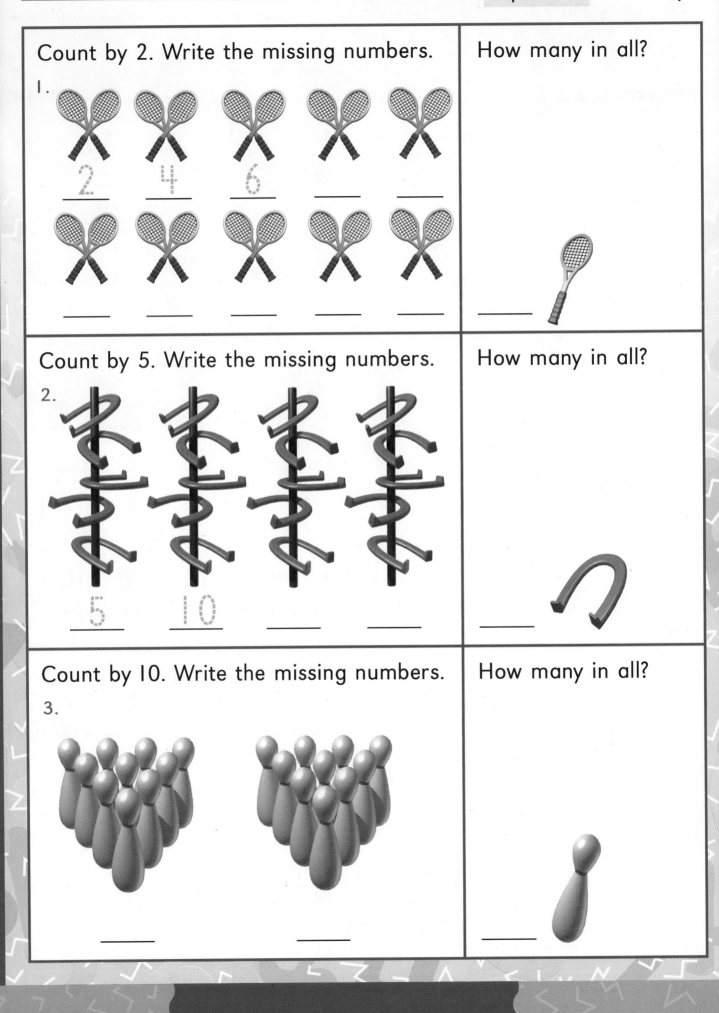

Count by 2. Write the missing numbers.

1.

2 4 6 ___ ___

___ ___ ___ ___ ___

How many in all?

___

Count by 5. Write the missing numbers.

2.

5 10 ___ ___

How many in all?

___

Count by 10. Write the missing numbers.

3.

___ ___

How many in all?

___

Skip Counting by 2, 5, 10

_____ can **compare** numbers.

Compare 19 ☆ to 16 ★.

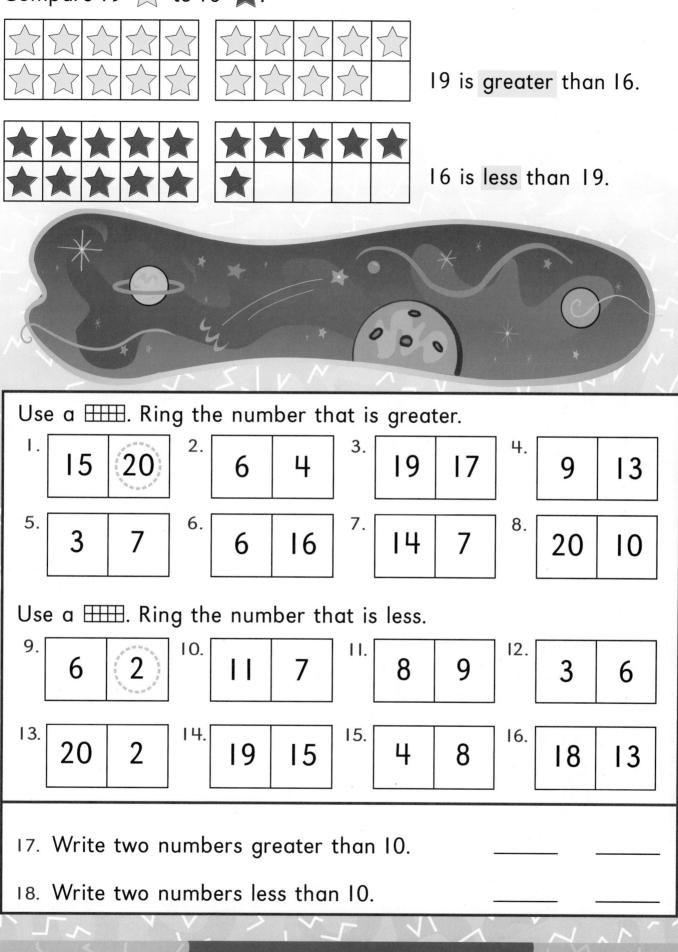

19 is **greater** than 16.

16 is **less** than 19.

Use a ▦. Ring the number that is greater.

1. 15　(20)　　2. 6　4　　3. 19　17　　4. 9　13

5. 3　7　　6. 6　16　　7. 14　7　　8. 20　10

Use a ▦. Ring the number that is less.

9. 6　(2)　　10. 11　7　　11. 8　9　　12. 3　6

13. 20　2　　14. 19　15　　15. 4　8　　16. 18　13

17. Write two numbers greater than 10. _____ _____

18. Write two numbers less than 10. _____ _____

I can **subtract to separate**.

$4 + 2 = 6$ sum

$6 - 2 = 4$ difference

Add to find how many in all. You may use 🎲 to check.

1.
$\begin{array}{r} 2 \\ +3 \\ \hline 5 \end{array}$
$\begin{array}{r} 1 \\ +1 \\ \hline \end{array}$
$\begin{array}{r} 3 \\ +1 \\ \hline \end{array}$
$\begin{array}{r} 3 \\ +2 \\ \hline \end{array}$
$\begin{array}{r} 3 \\ +3 \\ \hline \end{array}$
$\begin{array}{r} 5 \\ +1 \\ \hline \end{array}$

2. $5 + 3 = $ ___ $2 + 2 = $ ___ $4 + 1 = $ ___

3. $6 + 0 = $ ___ $4 + 4 = $ ___ $2 + 1 = $ ___

4. $2 + 5 = $ ___ $5 + 0 = $ ___ $3 + 4 = $ ___

Subtract to find how many are left. You may use 🎲 to check.

5.
$\begin{array}{r} 4 \\ -1 \\ \hline 3 \end{array}$
$\begin{array}{r} 3 \\ -2 \\ \hline \end{array}$
$\begin{array}{r} 2 \\ -1 \\ \hline \end{array}$
$\begin{array}{r} 5 \\ -2 \\ \hline \end{array}$
$\begin{array}{r} 8 \\ -1 \\ \hline \end{array}$
$\begin{array}{r} 3 \\ -3 \\ \hline \end{array}$

6. $7 - 3 = $ ___ $8 - 4 = $ ___ $2 - 0 = $ ___

7. $8 - 0 = $ ___ $4 - 2 = $ ___ $6 - 3 = $ ___

8. $5 - 5 = $ ___ $6 - 1 = $ ___ $7 - 2 = $ ___

MATH JOURNAL

9. Draw a joining picture.

10. Draw a separating picture.

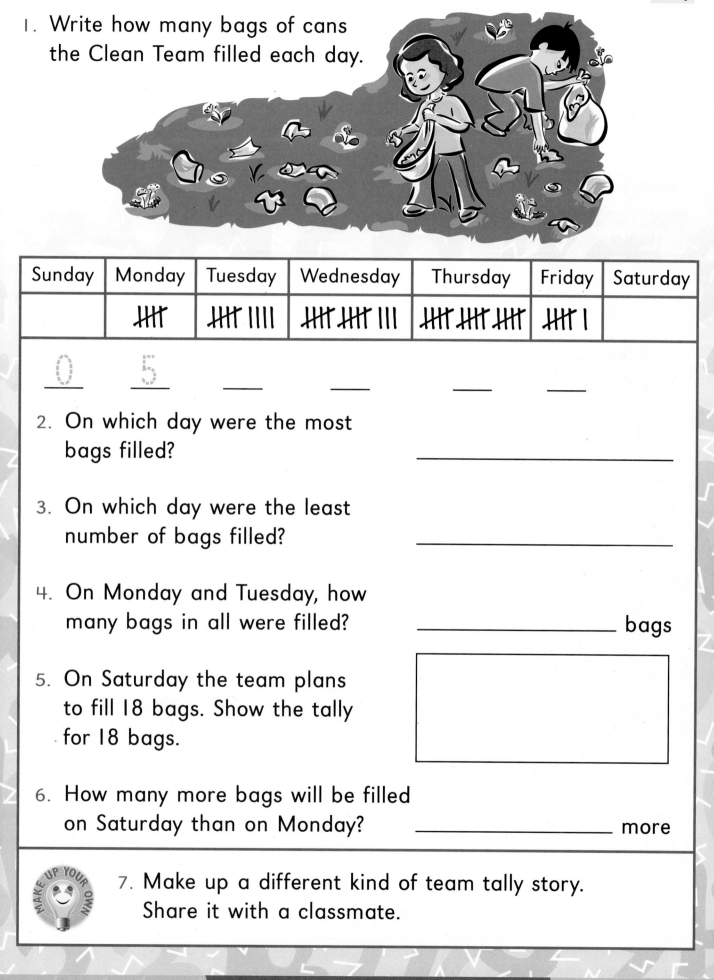

_____ can tally.

1. Write how many bags of cans the Clean Team filled each day.

Sunday	Monday	Tuesday	Wednesday	Thursday	Friday	Saturday
	卌	卌 llll	卌 卌 lll	卌 卌 卌	卌 l	
___0___	___5___	___	___	___	___	___

2. On which day were the most bags filled?

_____

3. On which day were the least number of bags filled?

_____

4. On Monday and Tuesday, how many bags in all were filled?

_____ bags

5. On Saturday the team plans to fill 18 bags. Show the tally for 18 bags.

6. How many more bags will be filled on Saturday than on Monday?

_____ more

7. Make up a different kind of team tally story. Share it with a classmate.

_____ can count and write tens.

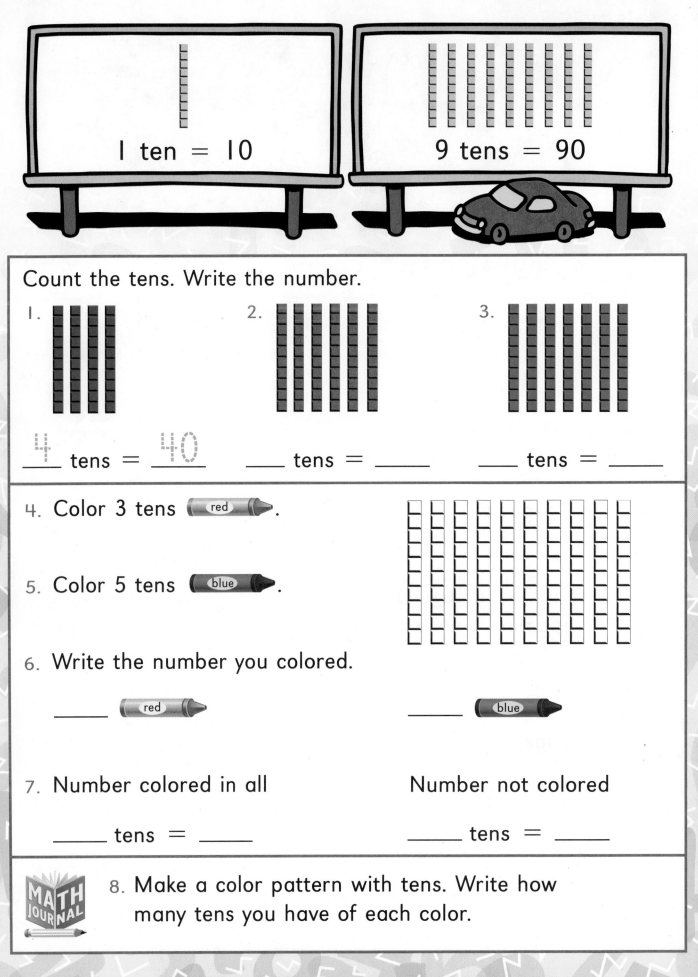

1 ten = 10

9 tens = 90

Count the tens. Write the number.

1. __4__ tens = __40__

2. _____ tens = _____

3. _____ tens = _____

4. Color 3 tens red.

5. Color 5 tens blue.

6. Write the number you colored.

_____ red

_____ blue

7. Number colored in all

_____ tens = _____

Number not colored

_____ tens = _____

8. Make a color pattern with tens. Write how many tens you have of each color.

Tens

_____ can count on with pennies, nickels, and dimes.

I penny  I nickel   I dime

I cent I¢   5 cents 5¢   10 cents 10¢

1. Count on by one. Write how much.

_____ ¢

2. Count on by five. Write how much.

_____ ¢

3. Count on by ten. Write how much.
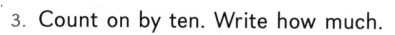
_____ ¢

Write how much.

4.
 _____

5.
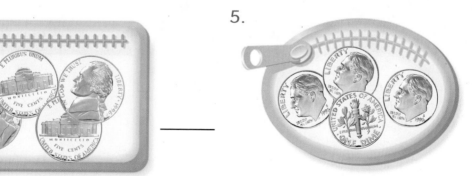 _____

6. Use 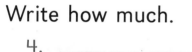 to color the purse with more money.

Penny, Nickel, Dime

REVIEW OF GRADE 1 SKILLS

_____ can tell time.

I can name the hour.

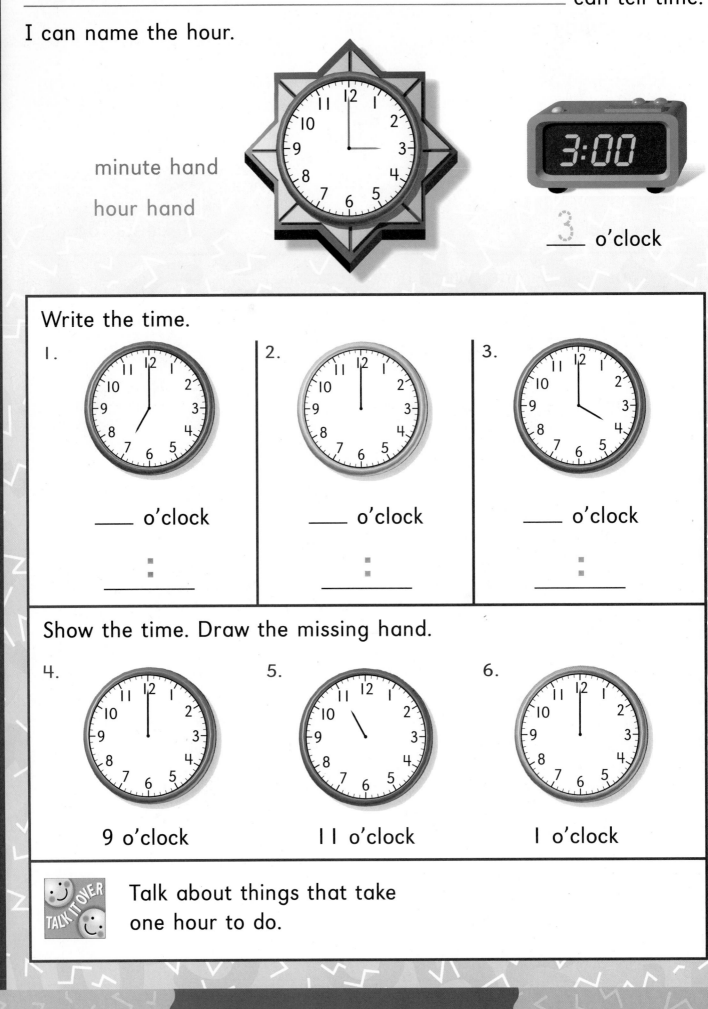

minute hand

hour hand

3:00

___ o'clock

Write the time.

1.
___ o'clock

_____

2.
___ o'clock

_____

3.
___ o'clock

_____

Show the time. Draw the missing hand.

4.
9 o'clock

5.
11 o'clock

6.
1 o'clock

TALK IT OVER

Talk about things that take one hour to do.

Clock Sense: Hours

_____ can add tens.

3 ones + 5 ones = 8 ones

$$\begin{array}{r} 3 \\ +5 \\ \hline 8 \end{array}$$

3 + 5 = 8

3 tens + 5 tens = 8 tens

tens	ones
3	0
+ 5	0
8	0

30 + 50 = 80

Write how many in all. First add ones.

1.

tens	ones
3	0
+ 4	0
7	0

2.

tens	ones
2	0
+ 3	0

3.

tens	ones
1	0
+ 4	0

Add to find the sum.

4.
$$\begin{array}{r} 50 \\ +20 \\ \hline 70 \end{array}$$
$$\begin{array}{r} 10 \\ +70 \\ \hline \end{array}$$
$$\begin{array}{r} 20 \\ +20 \\ \hline \end{array}$$
$$\begin{array}{r} 70 \\ +20 \\ \hline \end{array}$$
$$\begin{array}{r} 30 \\ +30 \\ \hline \end{array}$$

5.
$$\begin{array}{r} 20 \\ +60 \\ \hline \end{array}$$
$$\begin{array}{r} 80 \\ +10 \\ \hline \end{array}$$
$$\begin{array}{r} 10 \\ +50 \\ \hline \end{array}$$
$$\begin{array}{r} 50 \\ +40 \\ \hline \end{array}$$
$$\begin{array}{r} 30 \\ +60 \\ \hline \end{array}$$

6.
$$\begin{array}{r} 50 \\ +30 \\ \hline \end{array}$$
$$\begin{array}{r} 60 \\ +10 \\ \hline \end{array}$$
$$\begin{array}{r} 20 \\ +50 \\ \hline \end{array}$$
$$\begin{array}{r} 10 \\ +80 \\ \hline \end{array}$$
$$\begin{array}{r} 40 \\ +40 \\ \hline \end{array}$$

FINDING TOGETHER

7. Join tens to show 60 in all
in three different ways.

Add Tens

_____ can subtract tens.

6 ones − 2 ones = 4 ones

6 tens − 2 tens = 4 tens

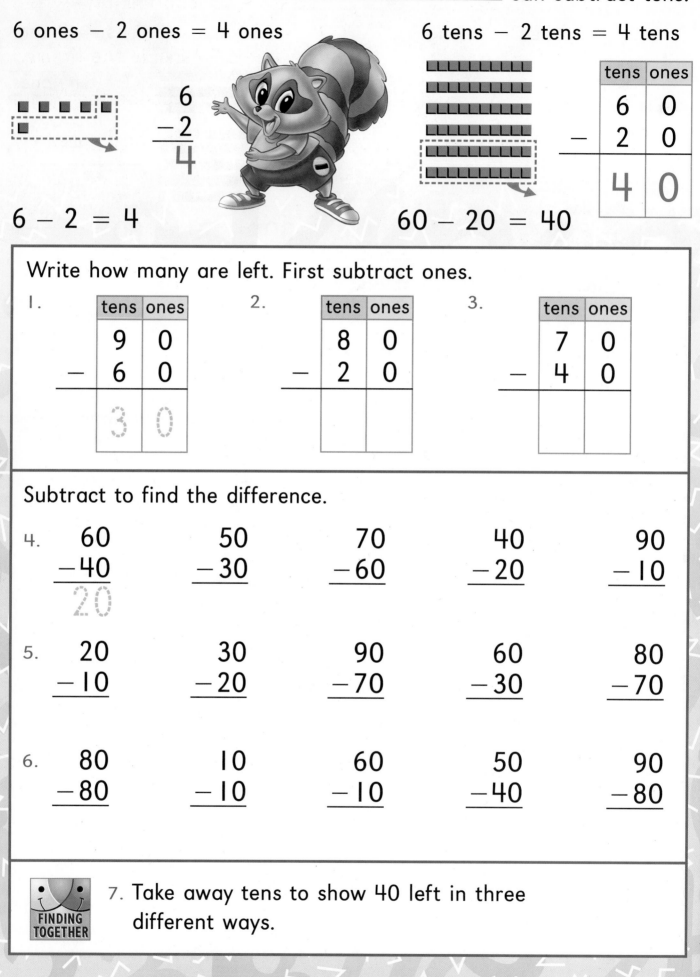

$$\begin{array}{r} 6 \\ -2 \\ \hline 4 \end{array}$$

6 − 2 = 4

tens	ones
6	0
− 2	0
4	0

60 − 20 = 40

Write how many are left. First subtract ones.

1.

tens	ones
9	0
− 6	0
3	0

2.

tens	ones
8	0
− 2	0

3.

tens	ones
7	0
− 4	0

Subtract to find the difference.

4.
$$\begin{array}{r} 60 \\ -40 \\ \hline 20 \end{array}$$
$$\begin{array}{r} 50 \\ -30 \\ \hline \end{array}$$
$$\begin{array}{r} 70 \\ -60 \\ \hline \end{array}$$
$$\begin{array}{r} 40 \\ -20 \\ \hline \end{array}$$
$$\begin{array}{r} 90 \\ -10 \\ \hline \end{array}$$

5.
$$\begin{array}{r} 20 \\ -10 \\ \hline \end{array}$$
$$\begin{array}{r} 30 \\ -20 \\ \hline \end{array}$$
$$\begin{array}{r} 90 \\ -70 \\ \hline \end{array}$$
$$\begin{array}{r} 60 \\ -30 \\ \hline \end{array}$$
$$\begin{array}{r} 80 \\ -70 \\ \hline \end{array}$$

6.
$$\begin{array}{r} 80 \\ -80 \\ \hline \end{array}$$
$$\begin{array}{r} 10 \\ -10 \\ \hline \end{array}$$
$$\begin{array}{r} 60 \\ -10 \\ \hline \end{array}$$
$$\begin{array}{r} 50 \\ -40 \\ \hline \end{array}$$
$$\begin{array}{r} 90 \\ -80 \\ \hline \end{array}$$

FINDING TOGETHER

7. Take away tens to show 40 left in three different ways.

Subtract Tens

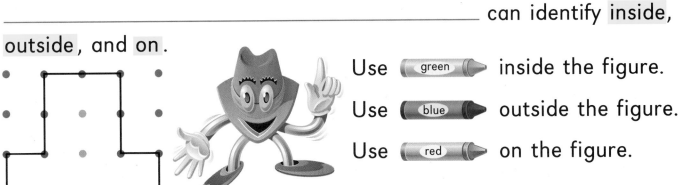

_____ can identify inside, outside, and on.

Use 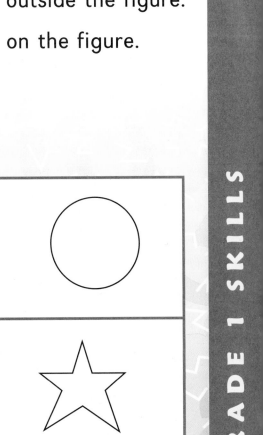 green inside the figure.

Use blue outside the figure.

Use red on the figure.

Help the Hatter make hats.
Use the color code to color each figure.

1.

2.

3.

Count the dots inside, outside, and on.
Write the number.

4.

inside ___

outside ___

on ___

Now use dot paper to make up your own.

Position: Inside, Outside, On

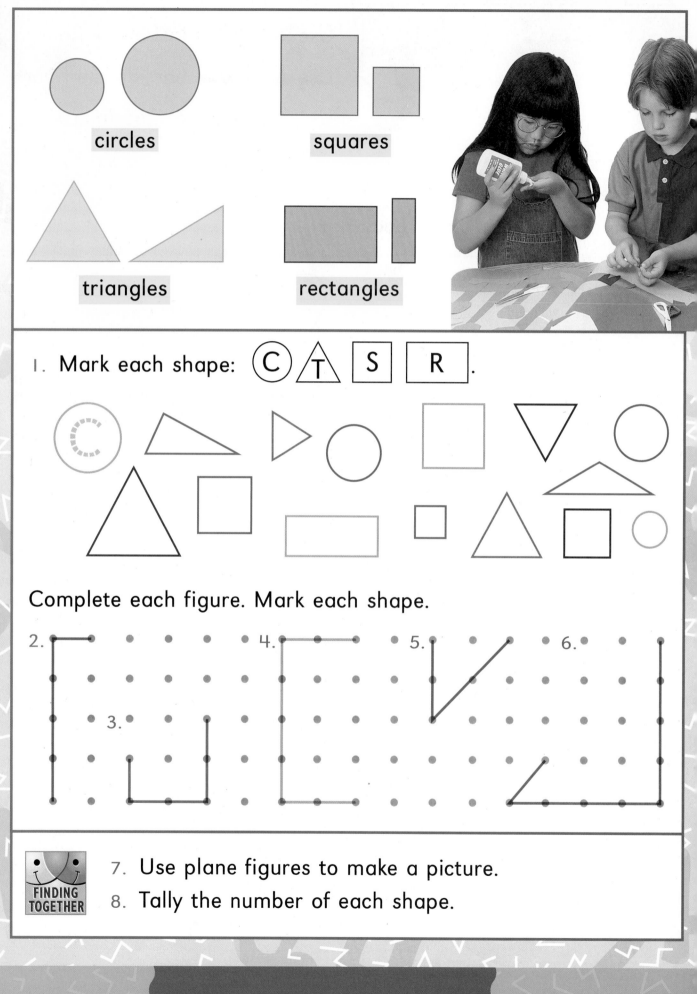

circles

squares

triangles

rectangles

1. Mark each shape: Ⓒ △T □S □R .

Complete each figure. Mark each shape.

2.

3.

4.

5.

6.

FINDING TOGETHER

7. Use plane figures to make a picture.

8. Tally the number of each shape.

Plane Figures

_____ can identify equal parts of a whole.

My circle has 5 equal parts.

Color Code

purple for 2 equal parts
blue for 3 equal parts
green for 4 equal parts
red for 5 equal parts

Color each figure with equal parts. Use Amal's color code.

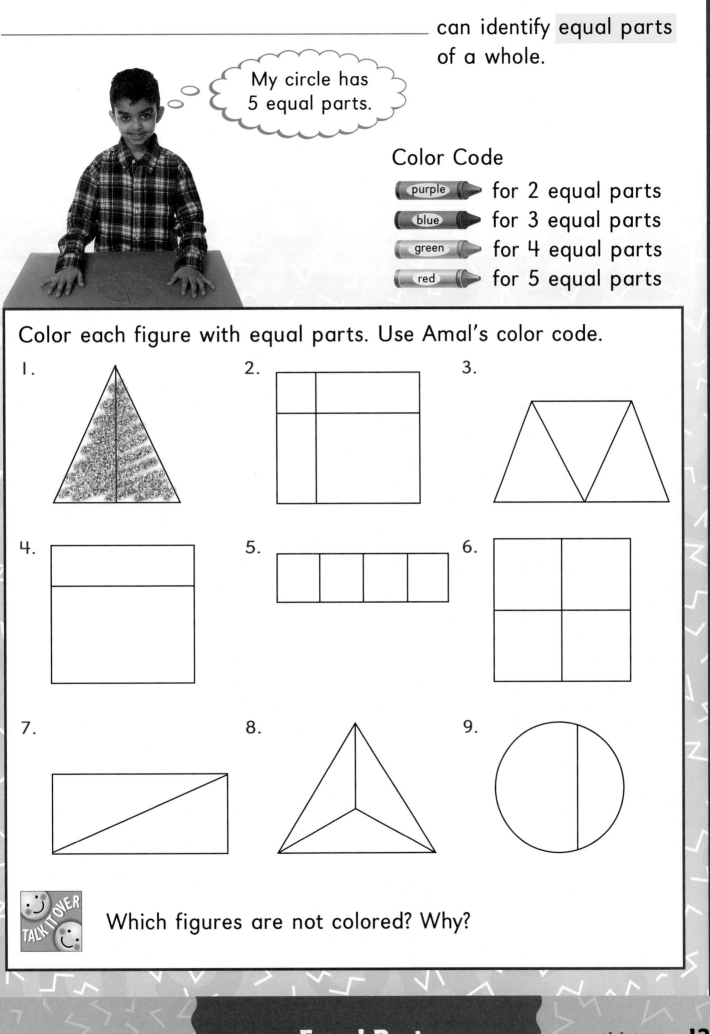

1.

2.

3.

4.

5.

6.

7.

8.

9.

Which figures are not colored? Why?

_____ can measure length .

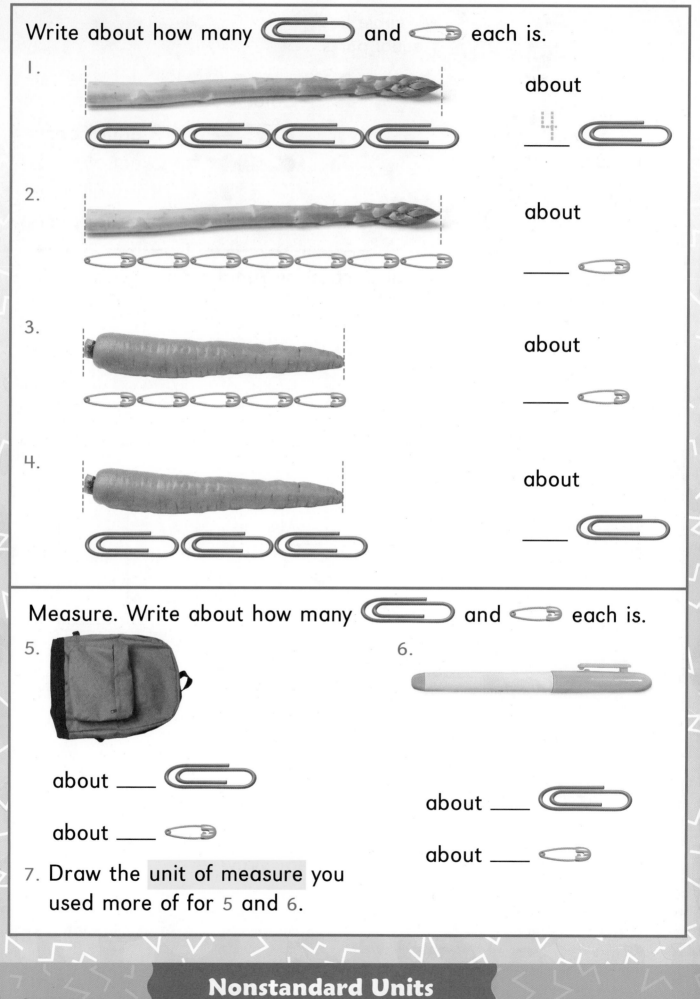

Write about how many 🖇 and 🧷 each is.

1.
about
___ 4 🖇

2.
about
___ 🧷

3.
about
___ 🧷

4.
about
___ 🖇

Measure. Write about how many 🖇 and 🧷 each is.

5.
about ___ 🖇
about ___ 🧷

6.
about ___ 🖇
about ___ 🧷

7. Draw the unit of measure you used more of for 5 and 6.

Nonstandard Units of Length

Problem Solving

This problem-solving magazine belongs to _____

To be a super problem solver, use these steps.

Read → Put yourself in the problem. Picture what is happening. Be sure to study the facts. Know what the question asks.

Draw →

Think → Plan what you will do to solve the problem.

Write → Work your plan. Ring your choice. Sometimes you will add or subtract.

Check → List and label your answer. Be sure it makes sense.

A

fold

Use this strategy: Draw a Picture.

Read → Essay saw 6 on Monday. She saw 3 today. How many did Essay find in all?

Draw → Draw found each day.

Monday

Tuesday

Think → Add to find how many in all.

Write → Write the number sentence.

6 + 3 = 9

Check → Essay saw ___ aliens in all.

Use to act it out.

Here are some problem-solving strategies.

Act It Out

Draw a Picture

Find a Pattern

Write a Number Sentence

Choose the Operation

Use Information from a Table

Use a Model

Logical Reasoning

Make a Table

Use a Graph

Use a Map

Ask a Question

Hidden Information

Two-Step Problems

Guess and Test

Extra Information

Make an Organized List

Missing Information

fold

Use this strategy: Logical Reasoning.

Read In a race the 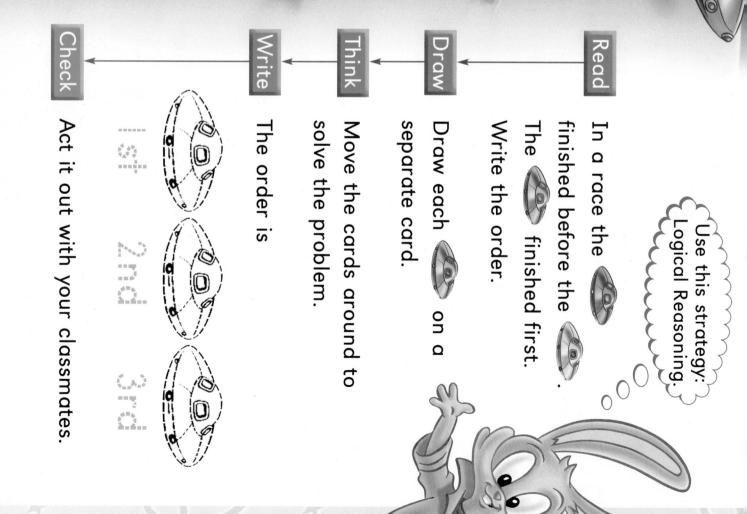 finished before the . The finished first. Write the order.

Draw Draw each on a separate card.

Think Move the cards around to solve the problem.

Write The order is

1st 2nd 3rd

Check Act it out with your classmates.

C

CRITICAL THINKING

Make a chart to show how many animals of each kind there are in the theater. Are there more or less than 20 in all?

Dear Family,

Today your child began Chapter 1. As she/he studies number and operation sense, you may want to read the poem below, which was read in class. Encourage your child to talk about some of the math ideas shown on page 17.

Look for the 🏛 at the bottom of each skills lesson. The suggestion on the page gives you an opportunity to improve your child's understanding of math. You may want to have countables available for your child to use throughout the chapter.

Home Activity

Name Numbers

Try this activity with your child. Write the names of three of your child's favorite animals or three unusual ones. Have your child count the letters in each name and write the number and the number word for each. Then have your child add a pet name for each animal (see below). After you finish each lesson in this chapter, change the name of the animal and adjust the skills.

Porcupine   Patsy Porcupine
9 nine   14 fourteen

Home Reading Connection

One Little Kitten

One little kitten
Two big cats
Three baby butterflies
Four big rats
Five fat fishes
Six sad seals
Seven silly seagulls
Eight happy eels;

Nine nervous lizards
Ten brave bees
Eleven smelly elephants
Twelve fat fleas
Thirteen alligators
Fourteen whales
Fifteen donkeys
with fifteen tails.

Carolyn Graham

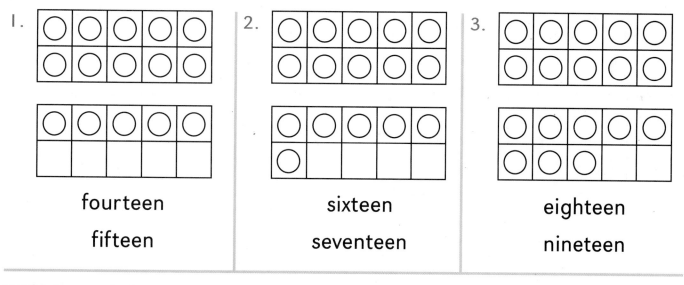

zero	one	two	three	four	five	six

seven	eight	nine	ten	eleven	twelve	thirteen

Color the counters. Ring the number word.

1.

fourteen

fifteen

2.

sixteen

seventeen

3.

eighteen

nineteen

 Which number words have a number word at the beginning of them?

Two groups of 10 are twenty.
Tell how to model twenty.

 1-1 Help your child model each number word from zero to twenty.

twenty-one **21**

Count the group. Write the number word.

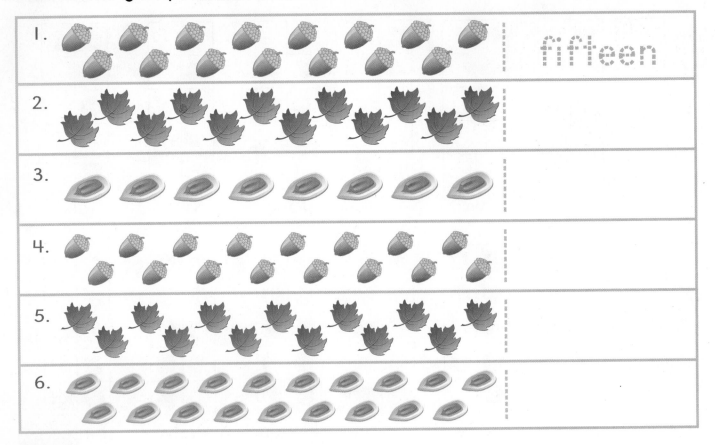

1. *fifteen*

2. _____

3. _____

4. _____

5. _____

6. _____

 PROBLEM SOLVING Write the number word.

7. The 🦌 ate fifteen 🫐.
   On Monday it ate one fewer.
   How many 🫐 did it eat?   _____

8. My sister and I saw ten 🐰.
   Then we saw one more.
   How many 🐰 did we see?   _____

9. Marco counted 20 🐸.
   I counted as many as he did.
   How many 🐸 did I count?   _____

 10. Draw a picture and write a problem in
     your Math Journal about a woodland
     animal. Use number words to solve.

Name _____

Count on for just after.

4 is just after 3.

Count back for just before.

17 is just before 18.

← | | | | | | | | | | | | | | | | | | | | | →
0 1 2 3 4 5 6 7 8 9 10 11 12 13 14 15 16 17 18 19 20

TALK IT OVER Explain how to count on or to count back to find the number that comes between.

Write the number that comes just after.

1. 6, _7_   2, ___   5, ___   13, ___

2. 11, ___   17, ___   16, ___   19, ___

Write the number that comes just before.

3. _7_,8   ___,11   ___,13   ___,18

4. ___,15   ___,16   ___,1   ___,10

Write the number that comes between.

5. 4, _5_, 6   7, ___, 9   3, ___, 5

6. 12, ___, 14   14, ___, 16   9, ___, 11

7. 18, ___, 20   10, ___, 12   17, ___, 19

CHALLENGE Write the numbers that come between.

8. 4, ___, ___, 7   9. 16, ___, ___, 19

Write the numbers in order on each animal's footprint.

1.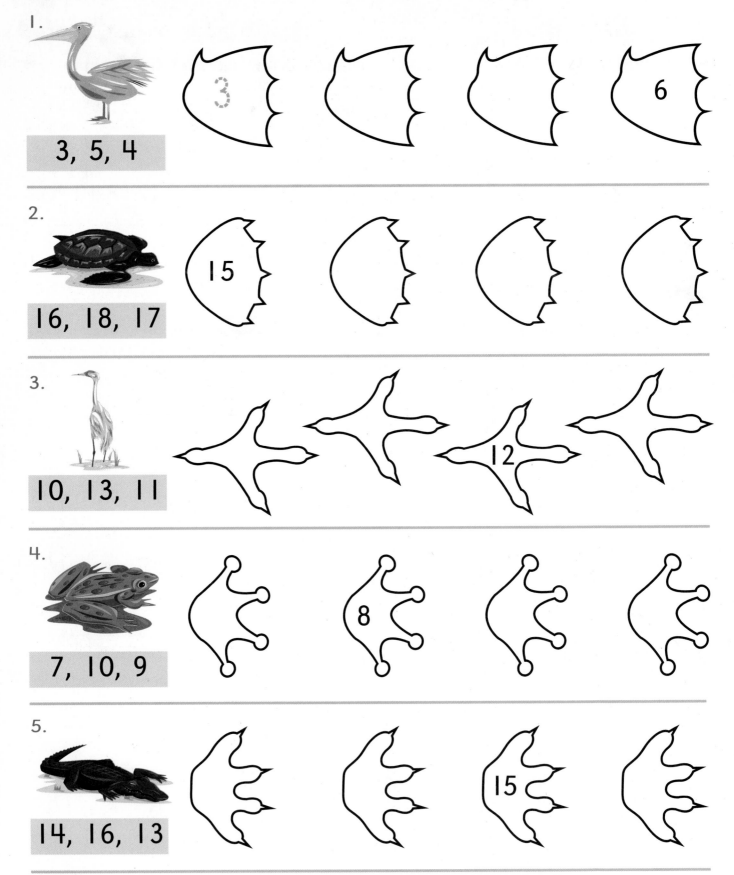

3, 5, 4

3 _ _ 6

2.

16, 18, 17

15 _ _ _

3.

10, 13, 11

_ _ 12 _

4.

7, 10, 9

_ 8 _ _

5.

14, 16, 13

_ _ 15 _

6. Draw a picture in your Math Journal to show a number order story about a marsh animal.

Name

I saw 15 horses.
Rico saw 12.

Compare to tell
the greater number.

15 is greater than 12.
15 > 12

12 is less than 15.
12 < 15

TALK IT OVER Why is it easy to tell which is greater, 7 or 17?

Model and compare. Write less or greater.
Then write < or >.

1. 8 is _greater_ than 7.
   8 ⬤> 7

2. 14 is _____ than 15.
   14 ◯ 15

3. 16 is _____ than 19.
   16 ◯ 19

4. 1 is _____ than 0.
   1 ◯ 0

5. 13 is _____ than 18.
   13 ◯ 18

6. 12 is _____ than 11.
   12 ◯ 11

7. 9 is _____ than 19.
   9 ◯ 19

8. 17 is _____ than 20.
   17 ◯ 20

1-3 Write two numbers less than
twenty. Help your child model
and compare them.

twenty-five **25**

You can compare numbers on a number line.

←——+——→
  0 1 2 3 4 5 6 7 8 9 10 11 12 13 14 15 16 17 18 19 20

$$8 > 5 \text{ and } 8 < 10$$

Compare. Write $<$ or $>$.

1. 4 $<$ 5   1 ◯ 9   7 ◯ 5

2. 5 ◯ 10   12 ◯ 6   13 ◯ 15

3. 6 ◯ 2   2 ◯ 12   10 ◯ 3

4. 10 ◯ 15   20 ◯ 17   0 ◯ 10

5. 7 ◯ 11   14 ◯ 11   6 ◯ 9

6. 15 ◯ 14   19 ◯ 20   4 ◯ 1

7. 13 ◯ 16   9 ◯ 8   16 ◯ 15

PROBLEM SOLVING

8. Mark saw greater than 10 🐷 but less than 16. How many 🐷 might he have seen?

_____ 🐷

9. Keesha saw greater than 15 🦆 but less than 20. How many 🦆 might she have seen?

_____ 🦆

10. Write a problem in your Math Journal like 8 or 9. Have a classmate solve it.

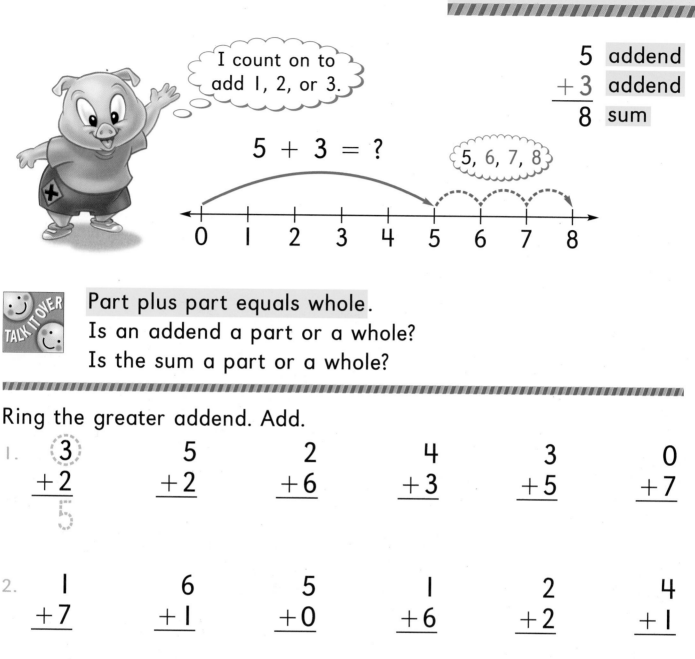

I count on to add 1, 2, or 3.

$$5 + 3 = ?$$

5, 6, 7, 8

$$
\begin{array}{r}
5 \\
+3 \\
\hline
8
\end{array}
\quad
\begin{array}{l}
\text{addend} \\
\text{addend} \\
\text{sum}
\end{array}
$$

TALK IT OVER

Part plus part equals whole.
Is an addend a part or a whole?
Is the sum a part or a whole?

Ring the greater addend. Add.

1. $\begin{array}{r} 3 \\ +2 \\ \hline 5 \end{array}$ $\qquad$ $\begin{array}{r} 5 \\ +2 \\ \hline \end{array}$ $\qquad$ $\begin{array}{r} 2 \\ +6 \\ \hline \end{array}$ $\qquad$ $\begin{array}{r} 4 \\ +3 \\ \hline \end{array}$ $\qquad$ $\begin{array}{r} 3 \\ +5 \\ \hline \end{array}$ $\qquad$ $\begin{array}{r} 0 \\ +7 \\ \hline \end{array}$

2. $\begin{array}{r} 1 \\ +7 \\ \hline \end{array}$ $\qquad$ $\begin{array}{r} 6 \\ +1 \\ \hline \end{array}$ $\qquad$ $\begin{array}{r} 5 \\ +0 \\ \hline \end{array}$ $\qquad$ $\begin{array}{r} 1 \\ +6 \\ \hline \end{array}$ $\qquad$ $\begin{array}{r} 2 \\ +2 \\ \hline \end{array}$ $\qquad$ $\begin{array}{r} 4 \\ +1 \\ \hline \end{array}$

3. $\begin{array}{r} 4 \\ +2 \\ \hline \end{array}$ $\qquad$ $\begin{array}{r} 4 \\ +4 \\ \hline \end{array}$ $\qquad$ $\begin{array}{r} 2 \\ +5 \\ \hline \end{array}$ $\qquad$ $\begin{array}{r} 5 \\ +1 \\ \hline \end{array}$ $\qquad$ $\begin{array}{r} 8 \\ +0 \\ \hline \end{array}$ $\qquad$ $\begin{array}{r} 3 \\ +3 \\ \hline \end{array}$

4. $\begin{array}{r} 5 \\ +3 \\ \hline \end{array}$ $\qquad$ $\begin{array}{r} 3 \\ +4 \\ \hline \end{array}$ $\qquad$ $\begin{array}{r} 2 \\ +4 \\ \hline \end{array}$ $\qquad$ $\begin{array}{r} 6 \\ +2 \\ \hline \end{array}$ $\qquad$ $\begin{array}{r} 7 \\ +1 \\ \hline \end{array}$ $\qquad$ $\begin{array}{r} 0 \\ +8 \\ \hline \end{array}$

1-4 Have your child count on to add 6 + 1, 5 + 2, and 4 + 3.

twenty-seven **27**

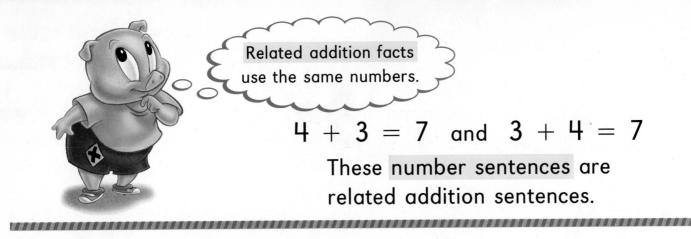

Related addition facts use the same numbers.

$4 + 3 = 7$ and $3 + 4 = 7$

These number sentences are related addition sentences.

Find the sum. Write the related addition fact.

1. $6 + 0 = \underline{6}$

$\underline{0} + \underline{6} = \underline{6}$

$5 + 2 = \underline{}$

$\underline{} + \underline{} = \underline{}$

$2 + 3 = \underline{}$

$\underline{} + \underline{} = \underline{}$

2. $4 + 2 = \underline{}$

$\underline{} + \underline{} = \underline{}$

$2 + 6 = \underline{}$

$\underline{} + \underline{} = \underline{}$

$6 + 1 = \underline{}$

$\underline{} + \underline{} = \underline{}$

3. $0 + 4 = \underline{}$

$\underline{} + \underline{} = \underline{}$

$3 + 5 = \underline{}$

$\underline{} + \underline{} = \underline{}$

$1 + 5 = \underline{}$

$\underline{} + \underline{} = \underline{}$

Find the sum.

4. $3 + 3 = \underline{}$ $2 + 2 = \underline{}$ $4 + 4 = \underline{}$

SHARE YOUR THINKING

5. The number sentences in 4 are doubles.
    Doubles do not have related addition facts. Why not?

PROBLEM SOLVING

6. Tina saw 2 🐄.
    Rhea saw 3 more than Tina.
    How many 🐄 did Rhea see?
    How many did they see in all?

    Rhea saw ____ 🐄.   They saw ____ 🐄 in all.

28 twenty-eight

Name

Speech bubble (girl): 4 + 5 is I more than 4 + 4.

Speech bubble (boy): 5 + 4 is I less than 5 + 5.

$$\begin{array}{r} 4 \\ +5 \\ \hline 9 \end{array} \qquad \begin{array}{r} 5 \\ +4 \\ \hline 9 \end{array}$$

These are **related addition facts**.

Add.

1.
$$\begin{array}{r} 7 \\ +3 \\ \hline 10 \end{array} \quad \begin{array}{r} 8 \\ +1 \\ \hline \end{array} \quad \begin{array}{r} 4 \\ +4 \\ \hline \end{array} \quad \begin{array}{r} 3 \\ +7 \\ \hline \end{array} \quad \begin{array}{r} 0 \\ +9 \\ \hline \end{array} \quad \begin{array}{r} 3 \\ +5 \\ \hline \end{array} \quad \begin{array}{r} 7 \\ +1 \\ \hline \end{array}$$

2.
$$\begin{array}{r} 5 \\ +5 \\ \hline \end{array} \quad \begin{array}{r} 3 \\ +6 \\ \hline \end{array} \quad \begin{array}{r} 8 \\ +2 \\ \hline \end{array} \quad \begin{array}{r} 2 \\ +7 \\ \hline \end{array} \quad \begin{array}{r} 1 \\ +8 \\ \hline \end{array} \quad \begin{array}{r} 4 \\ +3 \\ \hline \end{array} \quad \begin{array}{r} 2 \\ +8 \\ \hline \end{array}$$

3.
$$\begin{array}{r} 9 \\ +0 \\ \hline \end{array} \quad \begin{array}{r} 5 \\ +2 \\ \hline \end{array} \quad \begin{array}{r} 1 \\ +9 \\ \hline \end{array} \quad \begin{array}{r} 4 \\ +5 \\ \hline \end{array} \quad \begin{array}{r} 6 \\ +3 \\ \hline \end{array} \quad \begin{array}{r} 2 \\ +6 \\ \hline \end{array} \quad \begin{array}{r} 9 \\ +1 \\ \hline \end{array}$$

4.
$$\begin{array}{r} 7 \\ +2 \\ \hline \end{array} \quad \begin{array}{r} 6 \\ +4 \\ \hline \end{array} \quad \begin{array}{r} 3 \\ +3 \\ \hline \end{array} \quad \begin{array}{r} 5 \\ +4 \\ \hline \end{array} \quad \begin{array}{r} 4 \\ +6 \\ \hline \end{array} \quad \begin{array}{r} 2 \\ +4 \\ \hline \end{array} \quad \begin{array}{r} 1 \\ +6 \\ \hline \end{array}$$

 SECOND LOOK

✔ the related facts in each row above.

 TALK IT OVER

How are the ✔ facts alike? Name the related addition facts for sums of 9 in 1–4.

1-5 Have your child use a number line to show related addition facts such as 6 + 3 and 3 + 6.

twenty-nine **29**

These number lines show related facts.

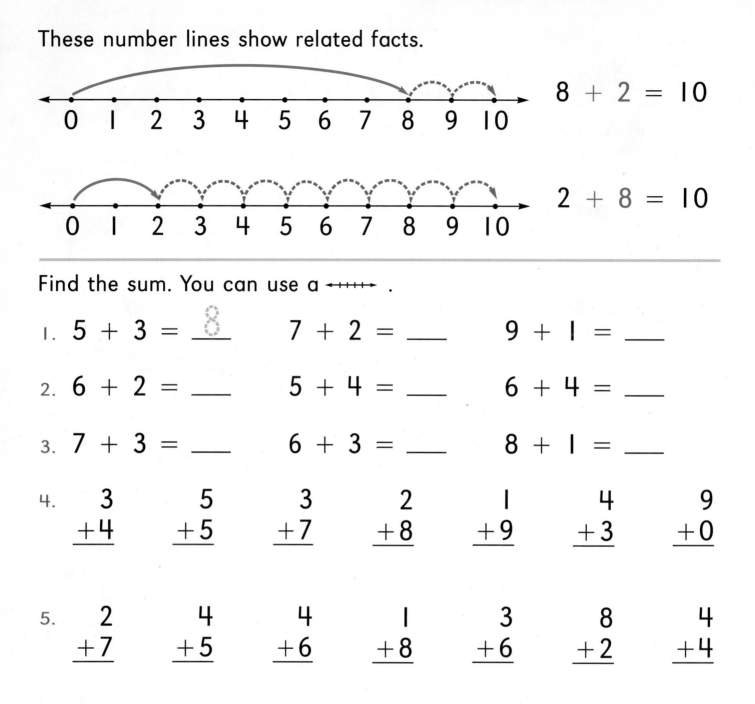

$8 + 2 = 10$

$2 + 8 = 10$

Find the sum. You can use a ⟷.

1. $5 + 3 = \underline{8}$ $7 + 2 = \underline{\hphantom{0}}$ $9 + 1 = \underline{\hphantom{0}}$

2. $6 + 2 = \underline{\hphantom{0}}$ $5 + 4 = \underline{\hphantom{0}}$ $6 + 4 = \underline{\hphantom{0}}$

3. $7 + 3 = \underline{\hphantom{0}}$ $6 + 3 = \underline{\hphantom{0}}$ $8 + 1 = \underline{\hphantom{0}}$

4.
$$\begin{array}{ccccccc} 3 & 5 & 3 & 2 & 1 & 4 & 9 \\ +4 & +5 & +7 & +8 & +9 & +3 & +0 \\ \hline \end{array}$$

5.
$$\begin{array}{ccccccc} 2 & 4 & 4 & 1 & 3 & 8 & 4 \\ +7 & +5 & +6 & +8 & +6 & +2 & +4 \\ \hline \end{array}$$

FINDING TOGETHER

6. Martin can show 10¢ in 9 different ways with heads and tails. Finish his table.

Heads	9¢	8¢	¢	¢	¢	¢	¢	¢	¢
Tails	1¢	2¢	¢	¢	¢	¢	¢	¢	¢
¢ in all	10¢	10¢	¢	¢	¢	¢	¢	¢	¢

Name _____

Reese sorts addition facts into three groups.

Doubles	Near Doubles	
$4 + 4 = 8$	$4 + 5 = 9$	$4 + 3 = 7$
	doubles + 1	doubles − 1

 Which near double is 1 less than a double?

Which near double is 1 more than a double?

Complete the doubles and near doubles facts.
Tally to check.

Doubles	Doubles + 1	Doubles − 1
1. $3 + \underline{3} = \underline{6}$	$3 + \underline{4} = \underline{}$	$3 + \underline{2} = \underline{}$
2. $1 + \underline{} = \underline{}$	$1 + \underline{} = \underline{}$	$1 + \underline{} = \underline{}$
3. $5 + \underline{} = \underline{}$	$5 + \underline{} = \underline{}$	$5 + \underline{} = \underline{}$
4. $2 + \underline{} = \underline{}$	$2 + \underline{} = \underline{}$	$2 + \underline{} = \underline{}$

CHALLENGE

5. $6 + \underline{} = \underline{}$ $6 + \underline{} = \underline{}$ $6 + \underline{} = \underline{}$

1-6 Have your child use 12 objects to retell each addition story shown above.

thirty-one **31**

Write the missing addends.

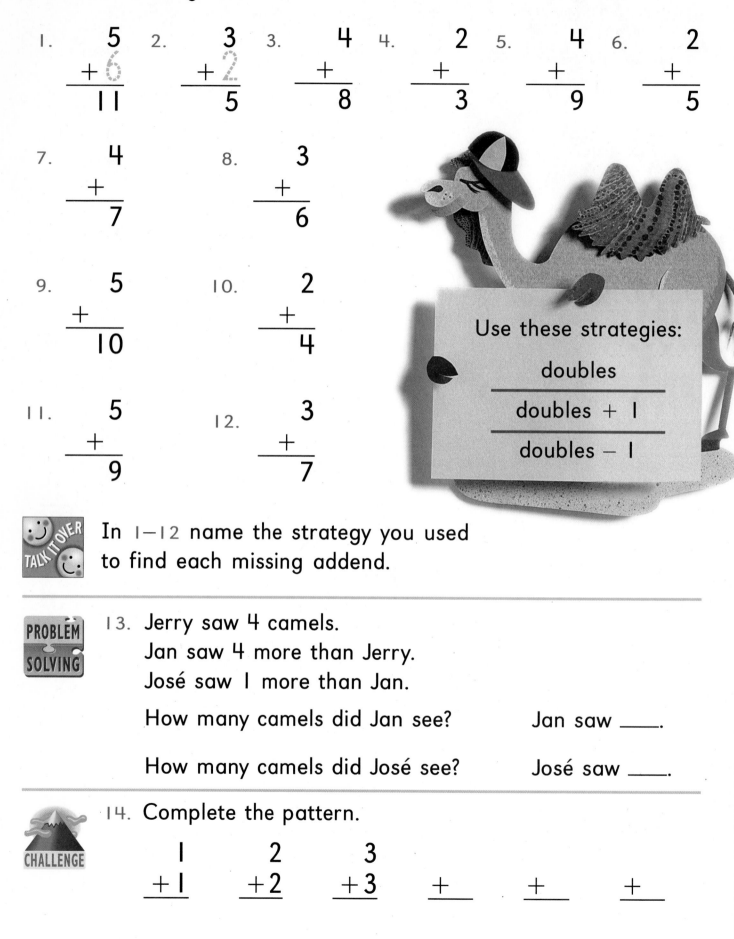

1.
$$\begin{array}{r} 5 \\ +6 \\ \hline 11 \end{array}$$

2.
$$\begin{array}{r} 3 \\ +2 \\ \hline 5 \end{array}$$

3.
$$\begin{array}{r} 4 \\ + \\ \hline 8 \end{array}$$

4.
$$\begin{array}{r} 2 \\ + \\ \hline 3 \end{array}$$

5.
$$\begin{array}{r} 4 \\ + \\ \hline 9 \end{array}$$

6.
$$\begin{array}{r} 2 \\ + \\ \hline 5 \end{array}$$

7.
$$\begin{array}{r} 4 \\ + \\ \hline 7 \end{array}$$

8.
$$\begin{array}{r} 3 \\ + \\ \hline 6 \end{array}$$

9.
$$\begin{array}{r} 5 \\ + \\ \hline 10 \end{array}$$

10.
$$\begin{array}{r} 2 \\ + \\ \hline 4 \end{array}$$

11.
$$\begin{array}{r} 5 \\ + \\ \hline 9 \end{array}$$

12.
$$\begin{array}{r} 3 \\ + \\ \hline 7 \end{array}$$

Use these strategies:

doubles
_____
doubles + 1
_____
doubles − 1

TALK IT OVER In 1–12 name the strategy you used to find each missing addend.

PROBLEM SOLVING

13. Jerry saw 4 camels.
Jan saw 4 more than Jerry.
José saw 1 more than Jan.

How many camels did Jan see?   Jan saw _____.

How many camels did José see?   José saw _____.

CHALLENGE

14. Complete the pattern.

$$\begin{array}{r} 1 \\ +1 \\ \hline \end{array} \qquad \begin{array}{r} 2 \\ +2 \\ \hline \end{array} \qquad \begin{array}{r} 3 \\ +3 \\ \hline \end{array} \qquad \begin{array}{r} \\ + \\ \hline \end{array} \qquad \begin{array}{r} \\ + \\ \hline \end{array} \qquad \begin{array}{r} \\ + \\ \hline \end{array}$$

Name _____

I make tens to add 9 + 2.

$$\begin{array}{r} 9 \\ +2 \\ \hline \end{array}$$

9 + 2 equals 9 + 1 + 1.

Find the sum. Draw ○ and ● to make ten.

1. $$\begin{array}{r} 8 \\ +4 \\ \hline 12 \end{array}$$

2. $$\begin{array}{r} 6 \\ +5 \\ \hline \end{array}$$

3. $$\begin{array}{r} 4 \\ +7 \\ \hline \end{array}$$

4. $$\begin{array}{r} 3 \\ +8 \\ \hline \end{array}$$

5. $$\begin{array}{r} 6 \\ +6 \\ \hline \end{array}$$

6. $$\begin{array}{r} 7 \\ +5 \\ \hline \end{array}$$

7. $$\begin{array}{r} 9 \\ +3 \\ \hline \end{array}$$

8. $$\begin{array}{r} 4 \\ +8 \\ \hline \end{array}$$

9. $$\begin{array}{r} 7 \\ +4 \\ \hline \end{array}$$

10. $$\begin{array}{r} 5 \\ +7 \\ \hline \end{array}$$

11. $$\begin{array}{r} 8 \\ +3 \\ \hline \end{array}$$

12. $$\begin{array}{r} 5 \\ +6 \\ \hline \end{array}$$

TALK IT OVER

How does knowing 9 + 2 = 11 help you add 2 + 9?

Name the related facts in 1–12.

1-7 Ask your child to add 9 to 2 and then to 3 by making ten.

thirty-three **33**

Count on to add on the number line.

$3 + 9 = 12$

$9 + 3 = 12$

0 1 2 3 4 5 6 7 8 9 10 11 12

Find the sum.

1. $6 + 6 = 12$ $9 + 2 = \rule{1cm}{0.4pt}$ $7 + 4 = \rule{1cm}{0.4pt}$

2. $8 + 2 = \rule{1cm}{0.4pt}$ $7 + 5 = \rule{1cm}{0.4pt}$ $9 + 3 = \rule{1cm}{0.4pt}$

3. $8 + 4 = \rule{1cm}{0.4pt}$ $6 + 5 = \rule{1cm}{0.4pt}$ $8 + 3 = \rule{1cm}{0.4pt}$

4.
$$\begin{array}{ccccccc} 5 & 7 & 6 & 2 & 4 & 5 & 7 \\ +6 & +3 & +6 & +9 & +8 & +4 & +4 \end{array}$$

5.
$$\begin{array}{ccccccc} 5 & 3 & 6 & 6 & 3 & 4 & 9 \\ +7 & +9 & +5 & +4 & +8 & +7 & +2 \end{array}$$

SECOND LOOK ✔ the even sums in 1–5.

PROBLEM SOLVING

6. Dani saw 8 🦀.
Teri saw 3 more.
How many 🦀 did
they see in all?

$\rule{1cm}{0.4pt} \bigcirc \rule{1cm}{0.4pt} = \rule{1cm}{0.4pt}$

$\rule{1cm}{0.4pt}$ 🦀

7. Liam counted 5 🐬.
Rafe counted 7 more.
How many 🐬 did
the boys count altogether?

$\rule{1cm}{0.4pt} \bigcirc \rule{1cm}{0.4pt} = \rule{1cm}{0.4pt}$

$\rule{1cm}{0.4pt}$ 🐬

CHALLENGE Color (yellow) the sums that are one less than thirteen.

8. | nine plus three | | seven plus four | | six plus six |

Name _____

You can group addends in different ways.

I look for doubles and sums of 10 first.

I count on first.

```
 ②
 7  ⟩4
+②
 11
```

```
 2
 ⑥  ⟩10
+④
 12
```

```
 ⑥
 ②  6, 7, 8
+ 3
 11
```

Ring the addends you group first. Then add.

1.
```
 ⑤     8     6     5     4     7     3
 ②     2     1     5     0     3     3
+3    +2    +4    +1    +4    +2    +3
 10
```

2.
```
 3     5     3     4     4     4     1
 5     2     3     3     1     4     3
+3    +4    +4    +2    +7    +3    +6
```

3. 6 + 2 + 4 = ____

4. 1 + 2 + 9 = ____

5. 2 + 2 + 5 = ____

6. 5 + 3 + 4 = ____

Use doubles or sums of ten to complete each addition sentence. Show how to solve each using both strategies.

7. 2 + ____ + ____ = ____

8. 1 + ____ + ____ = ____

1-8 Have your child explain how she/he grouped addends to solve exercises 1–6.

thirty-five **35**

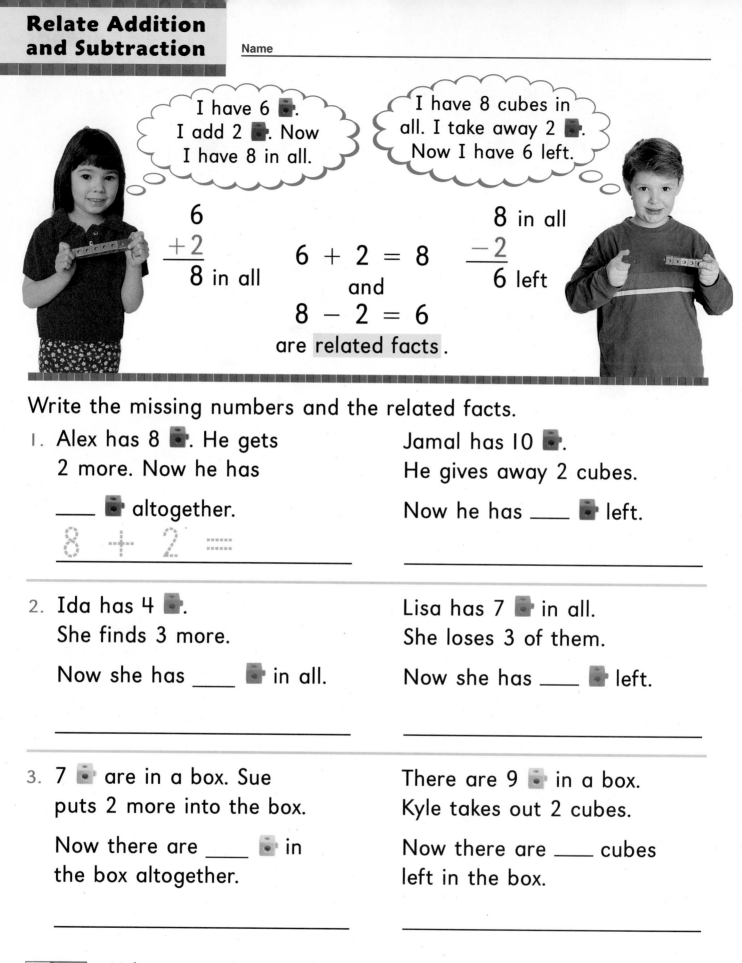

I have 6 🔲.
I add 2 🔲. Now
I have 8 in all.

I have 8 cubes in
all. I take away 2 🔲.
Now I have 6 left.

$$\begin{array}{r} 6 \\ +2 \\ \hline 8 \text{ in all} \end{array}$$

$$6 + 2 = 8$$
and
$$8 - 2 = 6$$
are related facts.

$$\begin{array}{r} 8 \text{ in all} \\ -2 \\ \hline 6 \text{ left} \end{array}$$

Write the missing numbers and the related facts.

1. Alex has 8 🔲. He gets
2 more. Now he has

____ 🔲 altogether.

8 + 2 = _____
_____

Jamal has 10 🔲.
He gives away 2 cubes.

Now he has ____ 🔲 left.

_____

2. Ida has 4 🔲.
She finds 3 more.

Now she has ____ 🔲 in all.

_____

Lisa has 7 🔲 in all.
She loses 3 of them.

Now she has ____ 🔲 left.

_____

3. 7 🔲 are in a box. Sue
puts 2 more into the box.

Now there are ____ 🔲 in
the box altogether.

_____

There are 9 🔲 in a box.
Kyle takes out 2 cubes.

Now there are ____ cubes
left in the box.

_____

TALK IT OVER Where are the addends and sum
in a subtraction sentence?

Have your child tell problems
for 4 + 1 = 5 and 5 − 1 = 4.
1-9

Subtraction Strategies

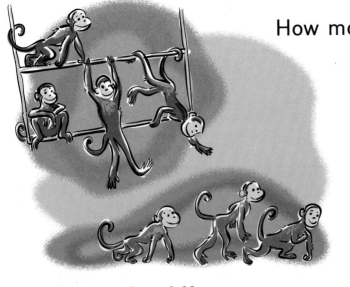

How many monkeys are on the bars?

Whole minus part equals part.

$$7 - 3 = 4$$

$$\begin{array}{r} 7 \text{ in all} \\ -\,3 \\ \hline 4 \text{ difference} \end{array}$$

There are 4 monkeys on the bars.

Is the difference a part or the whole?
Name the subtraction sentence for finding
how many monkeys are on the ground.

Subtract. Then write the whole minus part a different way.

1. $8 - 2 = \underline{6}$ $5 - 2 = \underline{}$ $8 - 8 = \underline{}$

$8 - 6 = 2$ _____   _____

2. $5 - 4 = \underline{}$ $7 - 0 = \underline{}$ $6 - 4 = \underline{}$

_____   _____   _____

3. $8 - 5 = \underline{}$ $6 - 1 = \underline{}$ $7 - 5 = \underline{}$

_____   _____   _____

Subtract.

4. $6 - 3 = \underline{}$ $8 - 4 = \underline{}$ $4 - 2 = \underline{}$

The number sentences in 1–3 are
related subtraction facts. Doubles do
not have related subtraction facts. Why not?

**Have your child take 7 objects to
model subtraction for whole minus
part equals part.**

I count back
to subtract
1, 2, or 3.

$$7 - 2 = ?$$

7 ← Start with the
−2   number in all.
5

7, 6, 5

```
  +---+---+---+---+---+---+---+
  0   1   2   3   4   5   6   7
```

Subtract.

1.
$$\begin{array}{r} 8 \\ -2 \\ \hline 6 \end{array}$$
$$\begin{array}{r} 8 \\ -1 \\ \hline \end{array}$$
$$\begin{array}{r} 6 \\ -5 \\ \hline \end{array}$$
$$\begin{array}{r} 7 \\ -6 \\ \hline \end{array}$$
$$\begin{array}{r} 6 \\ -2 \\ \hline \end{array}$$
$$\begin{array}{r} 8 \\ -3 \\ \hline \end{array}$$
$$\begin{array}{r} 5 \\ -4 \\ \hline \end{array}$$

2.
$$\begin{array}{r} 8 \\ -5 \\ \hline \end{array}$$
$$\begin{array}{r} 7 \\ -7 \\ \hline \end{array}$$
$$\begin{array}{r} 6 \\ -0 \\ \hline \end{array}$$
$$\begin{array}{r} 7 \\ -0 \\ \hline \end{array}$$
$$\begin{array}{r} 8 \\ -6 \\ \hline \end{array}$$
$$\begin{array}{r} 7 \\ -5 \\ \hline \end{array}$$
$$\begin{array}{r} 5 \\ -2 \\ \hline \end{array}$$

3.
$$\begin{array}{r} 7 \\ -3 \\ \hline \end{array}$$
$$\begin{array}{r} 8 \\ -4 \\ \hline \end{array}$$
$$\begin{array}{r} 6 \\ -4 \\ \hline \end{array}$$
$$\begin{array}{r} 5 \\ -3 \\ \hline \end{array}$$
$$\begin{array}{r} 8 \\ -8 \\ \hline \end{array}$$
$$\begin{array}{r} 6 \\ -1 \\ \hline \end{array}$$
$$\begin{array}{r} 5 \\ -1 \\ \hline \end{array}$$

4.
$$\begin{array}{r} 7 \\ -4 \\ \hline \end{array}$$
$$\begin{array}{r} 7 \\ -2 \\ \hline \end{array}$$
$$\begin{array}{r} 8 \\ -7 \\ \hline \end{array}$$
$$\begin{array}{r} 7 \\ -1 \\ \hline \end{array}$$
$$\begin{array}{r} 6 \\ -3 \\ \hline \end{array}$$
$$\begin{array}{r} 8 \\ -0 \\ \hline \end{array}$$
$$\begin{array}{r} 6 \\ -6 \\ \hline \end{array}$$

SECOND LOOK Ring the differences you found
by counting back in 1–4.

CRITICAL THINKING What number am I?

5. I am greater than 7 − 5.
   I am less than 7 − 3.

   I am ____.

6. I am greater than 8 − 8.
   I am less than 8 − 6.

   I am ____.

Name _____

How many more 🪙 does each need
to have 10¢ to spend?

Jay has 6 🪙.   Lee has 4 🪙.

In related subtraction facts,
the whole and
parts are the same.

10¢ − 6¢ = 4¢   10¢ − 4¢ = 6¢

Jay needs 4 🪙.   Lee needs 6 🪙.

$$\begin{array}{r} 10¢ \\ -6¢ \\ \hline 4¢ \end{array} \qquad \begin{array}{r} 10¢ \\ -4¢ \\ \hline 6¢ \end{array}$$

Subtract.

1.
$$\begin{array}{r} 9 \\ -4 \\ \hline 5 \end{array} \quad \begin{array}{r} 10 \\ -1 \\ \hline \end{array} \quad \begin{array}{r} 8 \\ -7 \\ \hline \end{array} \quad \begin{array}{r} 9 \\ -9 \\ \hline \end{array} \quad \begin{array}{r} 9 \\ -5 \\ \hline \end{array} \quad \begin{array}{r} 10 \\ -5 \\ \hline \end{array} \quad \begin{array}{r} 7 \\ -4 \\ \hline \end{array}$$

2.
$$\begin{array}{r} 9 \\ -1 \\ \hline \end{array} \quad \begin{array}{r} 10 \\ -7 \\ \hline \end{array} \quad \begin{array}{r} 8 \\ -3 \\ \hline \end{array} \quad \begin{array}{r} 9 \\ -0 \\ \hline \end{array} \quad \begin{array}{r} 7 \\ -5 \\ \hline \end{array} \quad \begin{array}{r} 10 \\ -3 \\ \hline \end{array} \quad \begin{array}{r} 10 \\ -9 \\ \hline \end{array}$$

3.
$$\begin{array}{r} 9 \\ -7 \\ \hline \end{array} \quad \begin{array}{r} 10 \\ -2 \\ \hline \end{array} \quad \begin{array}{r} 8 \\ -4 \\ \hline \end{array} \quad \begin{array}{r} 9 \\ -2 \\ \hline \end{array} \quad \begin{array}{r} 8 \\ -5 \\ \hline \end{array} \quad \begin{array}{r} 10 \\ -4 \\ \hline \end{array} \quad \begin{array}{r} 7 \\ -6 \\ \hline \end{array}$$

4.
$$\begin{array}{r} 8¢ \\ -6¢ \\ \hline \end{array} \quad \begin{array}{r} 10¢ \\ -8¢ \\ \hline \end{array} \quad \begin{array}{r} 9¢ \\ -3¢ \\ \hline \end{array} \quad \begin{array}{r} 9¢ \\ -6¢ \\ \hline \end{array} \quad \begin{array}{r} 9¢ \\ -8¢ \\ \hline \end{array} \quad \begin{array}{r} 10¢ \\ -6¢ \\ \hline \end{array} \quad \begin{array}{r} 8¢ \\ -8¢ \\ \hline \end{array}$$

 ✔ the related facts in each row above.

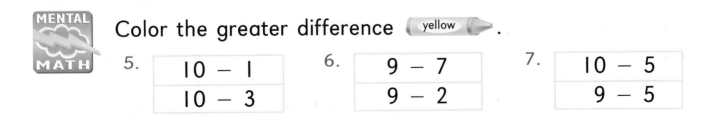

Color the greater difference 🖍 yellow .

5.
| 10 − 1 |
| 10 − 3 |

6.
| 9 − 7 |
| 9 − 2 |

7.
| 10 − 5 |
| 9 − 5 |

These number lines show related facts.

$9 - 2 = 7$

$9 - 7 = 2$

Find the difference. You can use a ⟵┼┼┼⟶.

1. $10 - 5 = \underline{5}$ $9 - 4 = \underline{\ \ }$ $10 - 4 = \underline{\ \ }$

2. $9 - 6 = \underline{\ \ }$ $10 - 7 = \underline{\ \ }$ $10 - 9 = \underline{\ \ }$

3.
$\begin{array}{r} 10 \\ -\ 2 \\ \hline \end{array}$
$\begin{array}{r} 9 \\ -7 \\ \hline \end{array}$
$\begin{array}{r} 10 \\ -\ 5 \\ \hline \end{array}$
$\begin{array}{r} 7 \\ -5 \\ \hline \end{array}$
$\begin{array}{r} 8 \\ -4 \\ \hline \end{array}$
$\begin{array}{r} 9 \\ -8 \\ \hline \end{array}$
$\begin{array}{r} 8 \\ -2 \\ \hline \end{array}$

4.
$\begin{array}{r} 10 \\ -\ 3 \\ \hline \end{array}$
$\begin{array}{r} 9 \\ -2 \\ \hline \end{array}$
$\begin{array}{r} 10 \\ -\ 6 \\ \hline \end{array}$
$\begin{array}{r} 8 \\ -5 \\ \hline \end{array}$
$\begin{array}{r} 8 \\ -6 \\ \hline \end{array}$
$\begin{array}{r} 9 \\ -9 \\ \hline \end{array}$
$\begin{array}{r} 8 \\ -3 \\ \hline \end{array}$

5.
$\begin{array}{r} 10 \\ -\ 1 \\ \hline \end{array}$
$\begin{array}{r} 9 \\ -3 \\ \hline \end{array}$
$\begin{array}{r} 10 \\ -\ 7 \\ \hline \end{array}$
$\begin{array}{r} 10 \\ -\ 8 \\ \hline \end{array}$
$\begin{array}{r} 9 \\ -5 \\ \hline \end{array}$
$\begin{array}{r} 9 \\ -1 \\ \hline \end{array}$
$\begin{array}{r} 9 \\ -0 \\ \hline \end{array}$

PROBLEM SOLVING

6. Jessica is 9 years old. Her sister is 5 years old. How much older is Jessica than her sister?

7. Joy has 10 books. She read 5 of them. How many books are left for her to read?

_____ years older

_____ books left to read

Name _____

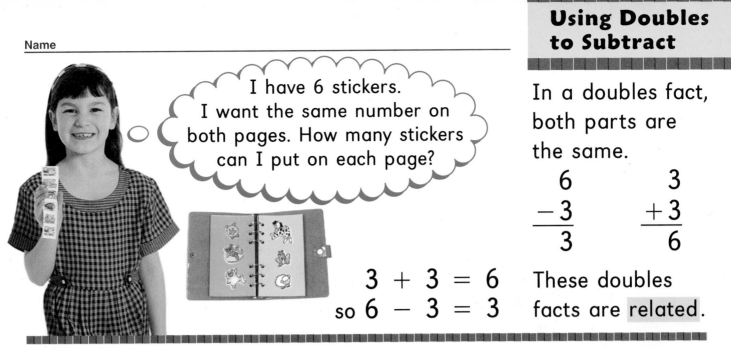

I have 6 stickers. I want the same number on both pages. How many stickers can I put on each page?

In a doubles fact, both parts are the same.

$$\begin{array}{r} 6 \\ -3 \\ \hline 3 \end{array} \qquad \begin{array}{r} 3 \\ +3 \\ \hline 6 \end{array}$$

$3 + 3 = 6$
so $6 - 3 = 3$

These doubles facts are related.

Draw stickers. Put the same number on both pages.
Write the related doubles facts.

1. 4 stickers in all

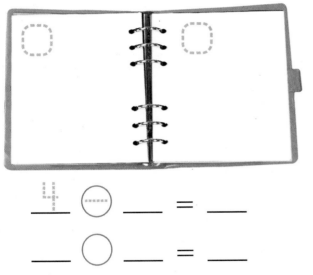

___4___ ◯------ ___ = ___

___ ◯ ___ = ___

2. 8 stickers in all

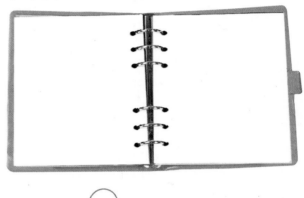

___ ◯ ___ = ___

___ ◯ ___ = ___

3. 10 stickers in all

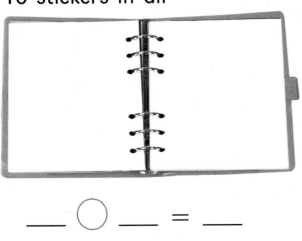

___ ◯ ___ = ___

___ ◯ ___ = ___

4. 12 stickers in all

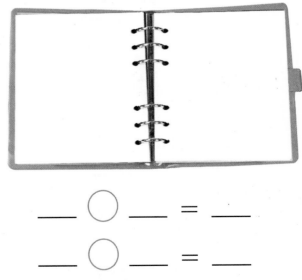

___ ◯ ___ = ___

___ ◯ ___ = ___

1-12   **Fold a sheet of paper in half, then have your child draw the same number of animals on each side to show doubles and tell the related doubles fact.**

forty-one   **41**

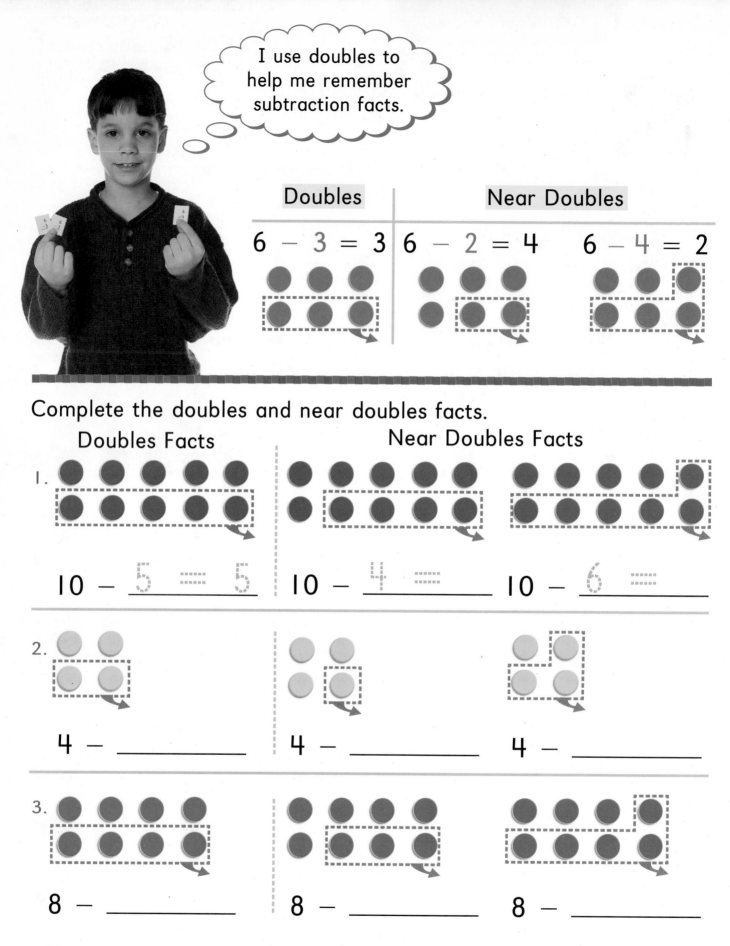

I use doubles to help me remember subtraction facts.

Doubles	Near Doubles	
$6 - 3 = 3$	$6 - 2 = 4$	$6 - 4 = 2$

Complete the doubles and near doubles facts.

Doubles Facts

Near Doubles Facts

1. $10 - 5 = 5$ $10 - 4 =$ _____ $10 - 6 =$ _____

2. $4 -$ _____ $4 -$ _____ $4 -$ _____

3. $8 -$ _____ $8 -$ _____ $8 -$ _____

Look at the near doubles in each row above.
What do you notice about them?

Name _____

Seeing a pattern can help you subtract.

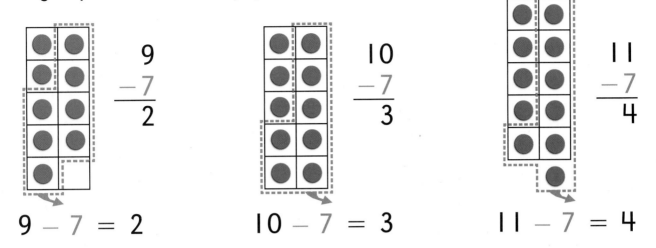

$$\begin{array}{r} 9 \\ -7 \\ \hline 2 \end{array} \qquad \begin{array}{r} 10 \\ -7 \\ \hline 3 \end{array} \qquad \begin{array}{r} 11 \\ -7 \\ \hline 4 \end{array}$$

$$9 - 7 = 2 \qquad 10 - 7 = 3 \qquad 11 - 7 = 4$$

In the pattern above the part taken away stays the same.
How does the whole change in the pattern?
Name and model the next fact.

Find the difference.

1.
$$\begin{array}{r} 12 \\ -5 \\ \hline 7 \end{array} \qquad \begin{array}{r} 12 \\ -6 \\ \hline \end{array} \qquad \begin{array}{r} 11 \\ -4 \\ \hline \end{array} \qquad \begin{array}{r} 11 \\ -5 \\ \hline \end{array} \qquad \begin{array}{r} 10 \\ -3 \\ \hline \end{array} \qquad \begin{array}{r} 10 \\ -4 \\ \hline \end{array}$$

2.
$$\begin{array}{r} 12 \\ -4 \\ \hline \end{array} \qquad \begin{array}{r} 11 \\ -3 \\ \hline \end{array} \qquad \begin{array}{r} 10 \\ -2 \\ \hline \end{array} \qquad \begin{array}{r} 12 \\ -3 \\ \hline \end{array} \qquad \begin{array}{r} 11 \\ -2 \\ \hline \end{array} \qquad \begin{array}{r} 10 \\ -1 \\ \hline \end{array}$$

3.
$$\begin{array}{r} 11 \\ -9 \\ \hline \end{array} \qquad \begin{array}{r} 12 \\ -8 \\ \hline \end{array} \qquad \begin{array}{r} 11 \\ -6 \\ \hline \end{array} \qquad \begin{array}{r} 12 \\ -9 \\ \hline \end{array} \qquad \begin{array}{r} 11 \\ -7 \\ \hline \end{array} \qquad \begin{array}{r} 11 \\ -8 \\ \hline \end{array}$$

SECOND LOOK ✔ rows that make a pattern in 1–3.

MATH JOURNAL Write the next three facts in the pattern below.

4. $10 - 9 \quad 11 - 9 \quad 12 - 9 \quad 10 - 8 \quad 11 - 8 \quad 12 - 8$

1-13 Have your child draw a ten-frame to show how to solve 2 exercises on this page.

forty-three **43**

Count back from the whole to subtract on a number line.

$$12 - 4 = 8$$
$$12 - 8 = 4$$

Find the difference. You may use the number line.

1. $12 - 6 = \underline{6}$ $12 - 3 = \underline{}$ $11 - 5 = \underline{}$

2. $11 - 8 = \underline{}$ $12 - 9 = \underline{}$ $12 - 7 = \underline{}$

3. $11 - 6 = \underline{}$ $11 - 3 = \underline{}$ $10 - 6 = \underline{}$

4.
12	11	12	11	11	11
− 5	− 4	− 9	− 7	− 9	− 5

5.
12	12	11	12	11	11
− 7	− 4	− 6	− 8	− 8	− 2

MENTAL MATH

Color the difference that is
6. least (yellow)
7. greatest (blue)

| 11 − 3 |
| 11 − 5 | 11 − 7 |

CHALLENGE

8. Find the missing numbers. The whole is the middle number.

3
7   9
4   12   5
8   6

9   5
6   11   4
8   7

5   2
8
6   10   1
3   8

Name _____

This is a **fact family** for my model.

6, 3, 9

$$\begin{array}{r} 6 \\ +3 \\ \hline 9 \end{array} \qquad \begin{array}{r} 9 \\ -3 \\ \hline 6 \end{array} \qquad \begin{array}{r} 3 \\ +6 \\ \hline 9 \end{array} \qquad \begin{array}{r} 9 \\ -6 \\ \hline 3 \end{array}$$

TALK IT OVER Why do 6, 3, and 9 make a fact family?

Write each fact family.

1.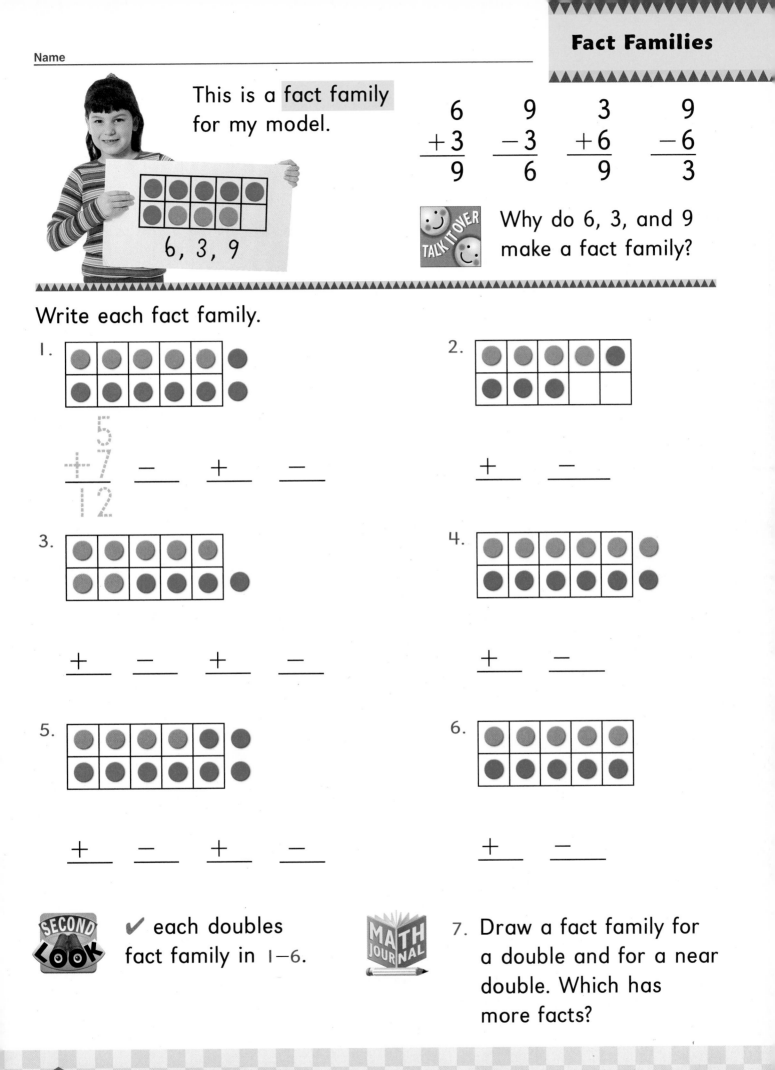

   $\dfrac{+7}{12}$ — ＿ + ＿ — ＿

2. ＿ + ＿ — ＿

3. ＿ + ＿ — ＿ + ＿ — ＿

4. ＿ + ＿ — ＿

5. ＿ + ＿ — ＿ + ＿ — ＿

6. ＿ + ＿ — ＿

SECOND LOOK ✔ each doubles fact family in 1—6.

MATH JOURNAL 7. Draw a fact family for a double and for a near double. Which has more facts?

1-14 Have your child find which of the numbers 6, 4, 5, 9 she/he can use to make a fact family.

forty-five **45**

X the number that does not belong to the fact family. Write the facts.

Part	Whole
+ Part	− Part
Whole	Part

1.

8	~~4~~
3	11

8 + 3 = 11

11 − 3 = ____

_____   _____

2.

3	7
8	5

_____   _____

_____   _____

3.

8	2
10	9

_____   _____

_____   _____

4.

5	7
11	6

_____   _____

_____   _____

5.

9	12
4	3

_____   _____

_____   _____

6. Write two fact families in your Math Journal. Use 8, 4, and 12.

You can find a missing addend on a number line.

7 + ☐ = 10
addend addend sum

0 1 2 3 4 5 6 7 8 9 10

7 + 3 = 10

Find the missing addend. Use a number line.

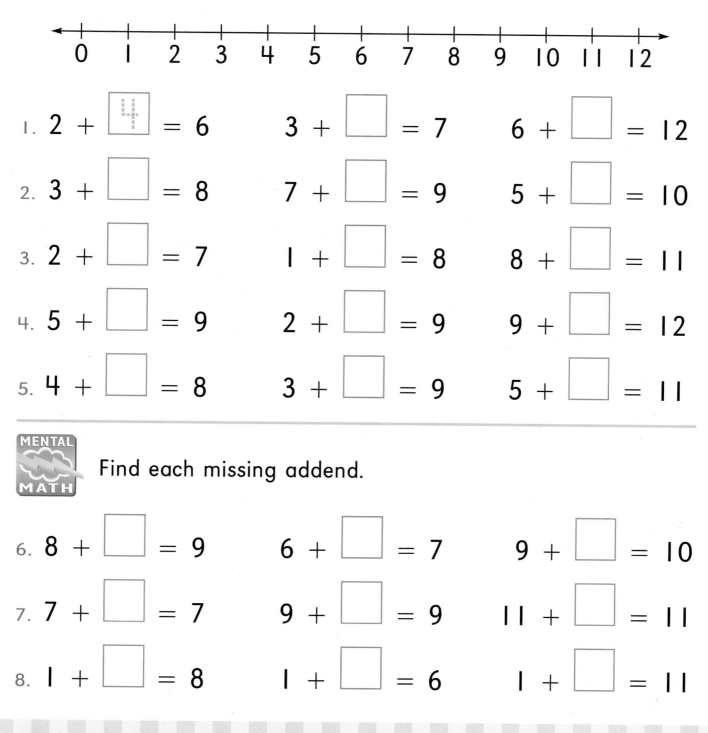

0 1 2 3 4 5 6 7 8 9 10 11 12

1. 2 + ☐4 = 6   3 + ☐ = 7   6 + ☐ = 12

2. 3 + ☐ = 8   7 + ☐ = 9   5 + ☐ = 10

3. 2 + ☐ = 7   1 + ☐ = 8   8 + ☐ = 11

4. 5 + ☐ = 9   2 + ☐ = 9   9 + ☐ = 12

5. 4 + ☐ = 8   3 + ☐ = 9   5 + ☐ = 11

MENTAL MATH Find each missing addend.

6. 8 + ☐ = 9   6 + ☐ = 7   9 + ☐ = 10

7. 7 + ☐ = 7   9 + ☐ = 9   11 + ☐ = 11

8. 1 + ☐ = 8   1 + ☐ = 6   1 + ☐ = 11

1-15 Start with 12 pennies. Have your child turn away as you hide some pennies under a cup. Then ask her/him to look and tell how many pennies are under the cup.

forty-seven **47**

Use a ▦. Draw to find the missing addend.

1.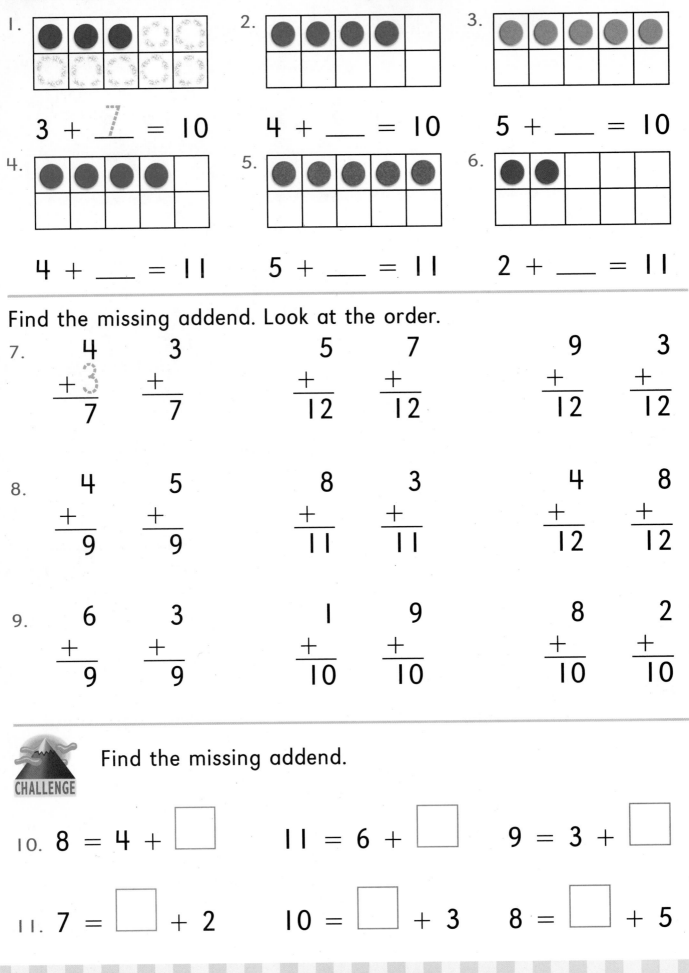

    3 + _7_ = 10

2. 4 + ___ = 10

3. 5 + ___ = 10

4. 4 + ___ = 11

5. 5 + ___ = 11

6. 2 + ___ = 11

Find the missing addend. Look at the order.

7.
$$\begin{array}{r} 4 \\ +3 \\ \hline 7 \end{array}$$
$$\begin{array}{r} 3 \\ + \\ \hline 7 \end{array}$$
$$\begin{array}{r} 5 \\ + \\ \hline 12 \end{array}$$
$$\begin{array}{r} 7 \\ + \\ \hline 12 \end{array}$$
$$\begin{array}{r} 9 \\ + \\ \hline 12 \end{array}$$
$$\begin{array}{r} 3 \\ + \\ \hline 12 \end{array}$$

8.
$$\begin{array}{r} 4 \\ + \\ \hline 9 \end{array}$$
$$\begin{array}{r} 5 \\ + \\ \hline 9 \end{array}$$
$$\begin{array}{r} 8 \\ + \\ \hline 11 \end{array}$$
$$\begin{array}{r} 3 \\ + \\ \hline 11 \end{array}$$
$$\begin{array}{r} 4 \\ + \\ \hline 12 \end{array}$$
$$\begin{array}{r} 8 \\ + \\ \hline 12 \end{array}$$

9.
$$\begin{array}{r} 6 \\ + \\ \hline 9 \end{array}$$
$$\begin{array}{r} 3 \\ + \\ \hline 9 \end{array}$$
$$\begin{array}{r} 1 \\ + \\ \hline 10 \end{array}$$
$$\begin{array}{r} 9 \\ + \\ \hline 10 \end{array}$$
$$\begin{array}{r} 8 \\ + \\ \hline 10 \end{array}$$
$$\begin{array}{r} 2 \\ + \\ \hline 10 \end{array}$$

CHALLENGE Find the missing addend.

10. 8 = 4 + ☐   11 = 6 + ☐   9 = 3 + ☐

11. 7 = ☐ + 2   10 = ☐ + 3   8 = ☐ + 5

Read ⟶ Think ⟶ Write ⟶ Check

List the numbers you guess. Test each sum.

1. It is between 10 and 16.
   It is an even number.
   It is the sum of 6 + 6.
   What is the number?

   List: 11, 12, 13, 14, 15

   Even: 12, 14

   Sum: 6 + 6 = 12   The number is 12.

2. It is less than 8.
   It is an odd number.
   It is the missing addend
   in 4 + □ = 9.
   What is the number?

   List: 1, 2, 3, 4, 5, 6, 7

   Odd: 1,

   4 +   The number is ___.

3. It is less than 15 and
   greater than 9.
   It is an even number.
   It belongs to the
   fact family of 4 and 6.
   What is the number?

   List: _____

   Even: _____

   The number is ___.

4. It is between 10 and
   15. It is an odd number.
   It is the missing whole
   in □ − 5 = 6.
   What is the number?

   List: _____

   Odd: _____

   The number is ___.

PROBLEM SOLVING

1-16 Have your child explain how she/he used the *Guess and Test* problem-solving strategy on this page.

forty-nine **49**

5. It is between 2 and 8.   List: _____
   It is the sum of a double.
   It is greater than 5.
   What is the number?   The number is _____.

6. It is between 9 and   List: _____
   14. It is the sum of
   two even numbers.
   It is not before 11.
   What is the number?   The number is _____.

7. It is less than 3   List: _____
   doubled. It is greater
   than 6 − 3. It is an
   odd number. What
   is the number?   The number is _____.

8. It is greater than   List: _____
   10 more than 5.
   It is an even number.
   It is less than the number
   that comes just before 19.
   What is the number?   The number is _____.

MAKE UP YOUR OWN

9. It is between 10 and _____.

   It is the sum of ___ and ___.
   What is the number?

Name _____

Read ➝ Think ➝ Write ➝ Check

I use 🎲 to check my answer.

Add to join.

Subtract to separate or to find how many more are needed.

1. There are 10 🐴.
4 of them are running.
The rest are eating.
How many 🐴 are eating?

add
or
subtract

$\frac{10}{-\ 4}$

_____ 🐴 are eating.

2. Nadia saw 3 big 🐷.
She saw 5 small 🐷.
How many 🐷 did she see in all?

add
or
subtract

$\frac{3}{+\ 5}$

Nadia saw _____ 🐷 in all.

3. 9 🐦 are in the yard.
2 🐦 fly away.
How many 🐦 are left?

add
or
subtract

_____ 🐦 are left.

4. Carlos saw 8 🐄.
Then he saw 3 more.
How many 🐄 did he see altogether?

add
or
subtract

Carlos saw _____ 🐄 altogether.

1-17 In this lesson your child solved problems by using the *Choose the Operation* strategy.

fifty-one **51**

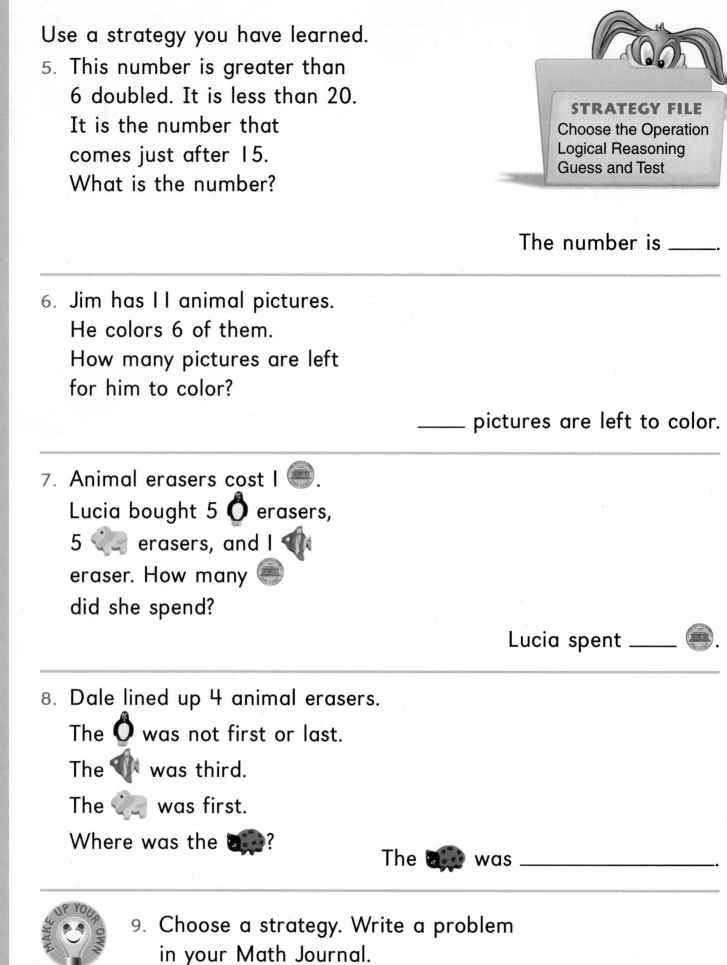

Use a strategy you have learned.

STRATEGY FILE
Choose the Operation
Logical Reasoning
Guess and Test

5. This number is greater than
   6 doubled. It is less than 20.
   It is the number that
   comes just after 15.
   What is the number?

   The number is _____.

6. Jim has 11 animal pictures.
   He colors 6 of them.
   How many pictures are left
   for him to color?

   _____ pictures are left to color.

7. Animal erasers cost 1 🪙.
   Lucia bought 5 🐧 erasers,
   5 🐘 erasers, and 1 🐟
   eraser. How many 🪙
   did she spend?

   Lucia spent _____ 🪙.

8. Dale lined up 4 animal erasers.
   The 🐧 was not first or last.
   The 🐟 was third.
   The 🐘 was first.
   Where was the 🐞?

   The 🐞 was _____.

9. Choose a strategy. Write a problem
   in your Math Journal.

Name _____

Count the group. Ring the number word.

1.		eleven	twelve	thirteen
2.		nine	seven	five
3.		twelve	twenty	ten

Write the missing numbers.

4. ___, 17   0, ___   12, ___ 14

Compare. Write < or >. Tally to check.

5. 9 ◯ 11   13 ◯ 16   20 ◯ 10

Add or subtract.

6.
$$\begin{array}{r}3\\+7\\\hline\end{array}\qquad\begin{array}{r}6\\+3\\\hline\end{array}\qquad\begin{array}{r}5\\+5\\\hline\end{array}\qquad\begin{array}{r}7\\+4\\\hline\end{array}\qquad\begin{array}{r}4\\+8\\\hline\end{array}\qquad\begin{array}{r}6\\+5\\\hline\end{array}\qquad\begin{array}{r}8\\+3\\\hline\end{array}$$

7.
$$\begin{array}{r}9\\-2\\\hline\end{array}\qquad\begin{array}{r}9\\-5\\\hline\end{array}\qquad\begin{array}{r}8\\-4\\\hline\end{array}\qquad\begin{array}{r}9\\-9\\\hline\end{array}\qquad\begin{array}{r}12\\-6\\\hline\end{array}\qquad\begin{array}{r}11\\-2\\\hline\end{array}\qquad\begin{array}{r}10\\-8\\\hline\end{array}$$

8. Ring doubles facts in 6 and 7.

9. Write the fact family for 3, 9, and 12.

_____   _____

_____   _____

PROBLEM SOLVING   10. T.J. fed 6 🐿, 4 🐱, and 2 🐶.
How many animals did T.J. feed?   ____ animals

REINFORCEMENT

Help Dulcé sort the numbers from 1 to 20.

1-Digit Numbers

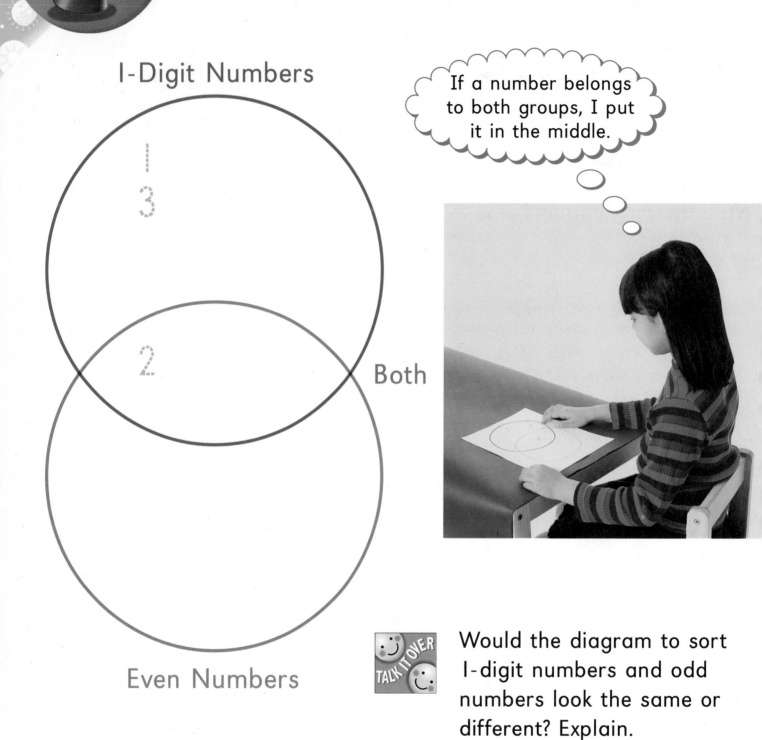

1

3

2

Both

Even Numbers

If a number belongs to both groups, I put it in the middle.

Would the diagram to sort 1-digit numbers and odd numbers look the same or different? Explain.

Draw 2 circles that overlap. Sort 2-digit numbers from odd numbers.

This page extends your child's understanding of odd and even numbers.

Name _____

1. Model and complete these addition and subtraction sentences.

Doubles $4 +$ _____ $=$ _____

$10 -$ _____ $=$ _____

Near Doubles $4 +$ _____ $=$ _____

$12 -$ _____ $=$ _____

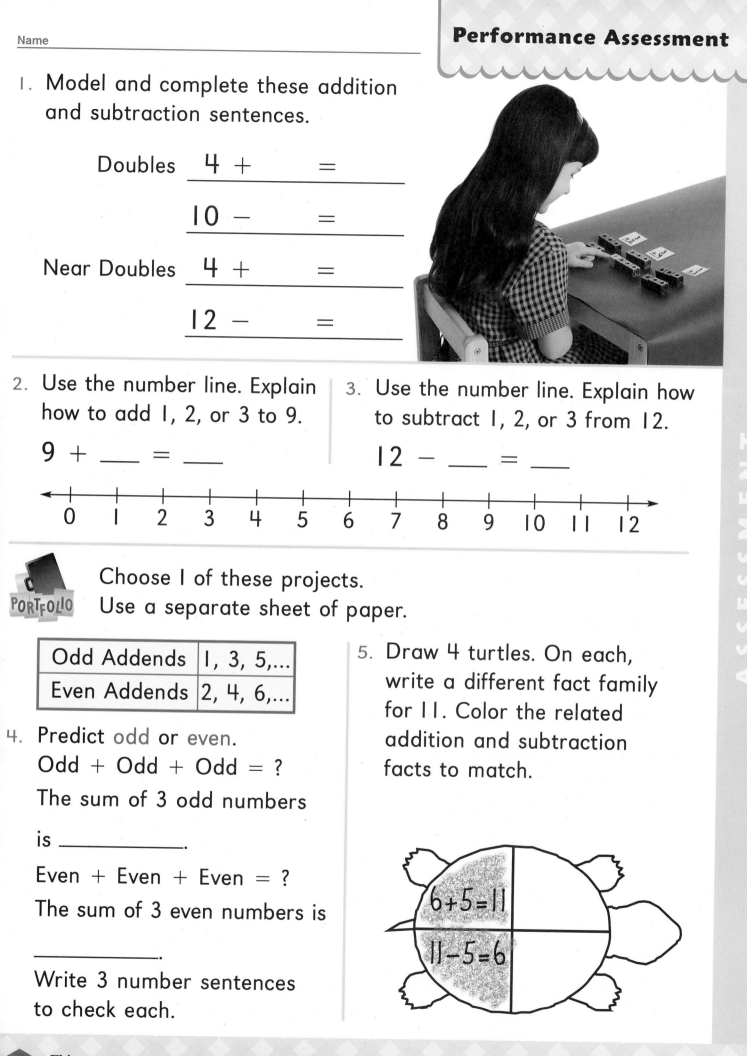

2. Use the number line. Explain how to add 1, 2, or 3 to 9.

$9 +$ ___ $=$ ___

3. Use the number line. Explain how to subtract 1, 2, or 3 from 12.

$12 -$ ___ $=$ ___

```
←——+——+——+——+——+——+——+——+——+——+——+——+——+——→
   0   1   2   3   4   5   6   7   8   9  10  11  12
```

PORTFOLIO Choose 1 of these projects. Use a separate sheet of paper.

Odd Addends	1, 3, 5,...
Even Addends	2, 4, 6,...

4. Predict odd or even.
Odd + Odd + Odd = ?
The sum of 3 odd numbers

is _____.

Even + Even + Even = ?
The sum of 3 even numbers is

_____.

Write 3 number sentences to check each.

5. Draw 4 turtles. On each, write a different fact family for 11. Color the related addition and subtraction facts to match.

$6+5=11$

$11-5=6$

This page provides a variety of informal assessment opportunities in order to measure your child's understanding of the skills taught in Chapter 1.

fifty-five **55**

Name _____

Write the missing number words.

1. _____, twenty eight, _____ ten, _____, twelve

Compare. Write < or >.

2. 7 ◯ 9   15 ◯ 12   17 ◯ 8

Add or subtract.

3. 7 + 3 = ___   6 + 6 = ___   5 + 4 = ___

4. 8 − 5 = ___   11 − 7 = ___   10 − 4 = ___

5.
2	8	7	3	9	5	2
+6	+4	+2	+8	+1	+6	+7

6.
12	9	10	8	12	11	10
− 5	−8	− 2	−2	− 7	− 9	− 5

SECOND LOOK ✔ related facts in 5 and 6.

Find the missing addend.

7. 6 + ☐ = 9

8. 8 + ☐ = 12

Find the sum.

9. 1 + 2 + 9 = ___

10. 3 + 3 + 4 = ___

PROBLEM SOLVING 11. 12 🐟 are in the lake.
3 of them swim away.
How many are left?

___ are left

This page is a formal assessment of your child's understanding of the content presented in Chapter 1.

1

2

CRITICAL THINKING

My invention has 10 ⬤ in the front, 10 ⬤ in the back, 20 on the right side and 20 on the left side. How many wheels do I need to add to have 100 altogether?

For more information about Chapter 2, visit the Family Information Center at **www.sadlier-oxford.com**

Internet

Dear Family,

Today your child began Chapter 2. As he/she studies place value to 100, you may want to read the poem below, which was read in class. Encourage your child to talk about some of the math ideas shown on page 57.

Look for the 🏢 at the bottom of each skills lesson. The suggestion on the page gives you an opportunity to improve your child's understanding of math. You may want to have strips for tens rods and squares for ones units available for your child to use throughout the chapter.

Home Activity

Sticker Maker

Try this activity with your child. Draw the following tens rods and ones units (see diagram). Ask your child to write how many tens and ones and the number. Continue the activity by drawing a different number of tens and ones to 100.

__3__ tens __4__ ones = __34__

Home Reading Connection

A Speedy Young Driver

A speedy young driver from Gar
always added more wheels to his car.
    "I once thought ten or twenty,"
    he said, "were just plenty,
but a hundred are better by far."

That daring young driver from Gar
added eighty more wheels to his car.
    With a hundred in place
    those wheels ran out of space,
so he added more floor to his car.

Sandra Liatsos

Name _____

In 6 races, the red car always came in first. The purple car always came in last.

Color the cars to show the 6 different ways the speedy cars might have finished the race.

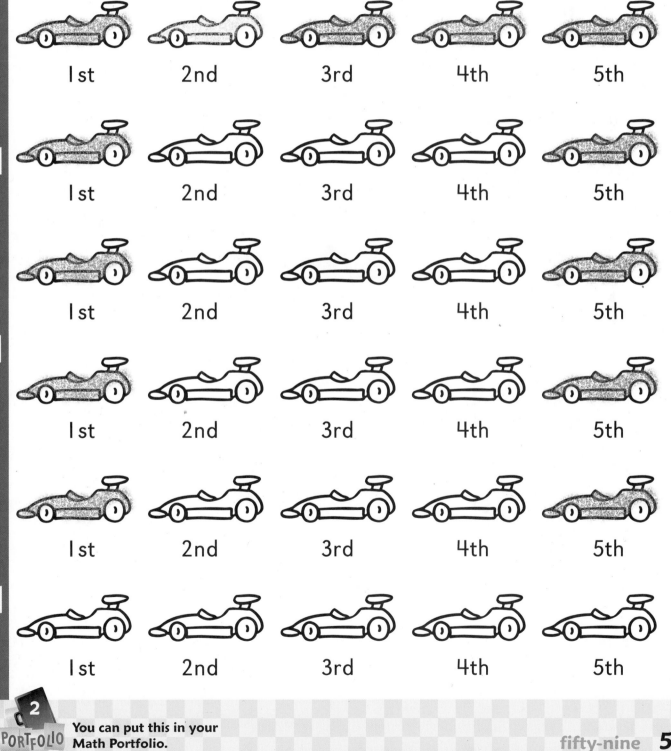

	1st	2nd	3rd	4th	5th

C O N N E C T I O N S

Who invented the lightbulb?

1st Estimate the number of bulbs in each bin. Use these bins as a guide.

2nd Match the letters on the bins with the estimates below.

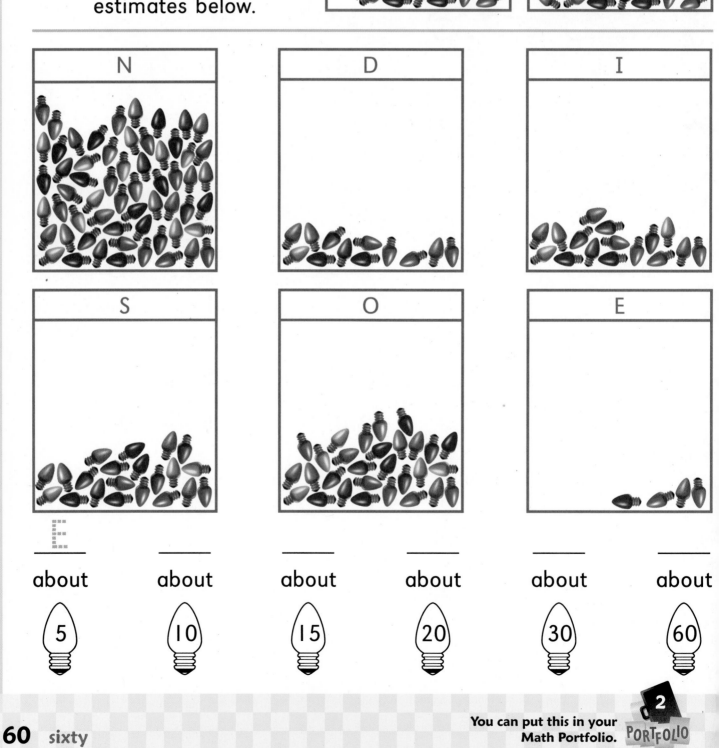

about 5

about 10

about 15

about 20

about 30

about 60

You can put this in your Math Portfolio. PORTFOLIO

Name _____

Marcel made 2 numbers. He can put the digits in the tens or ones place.

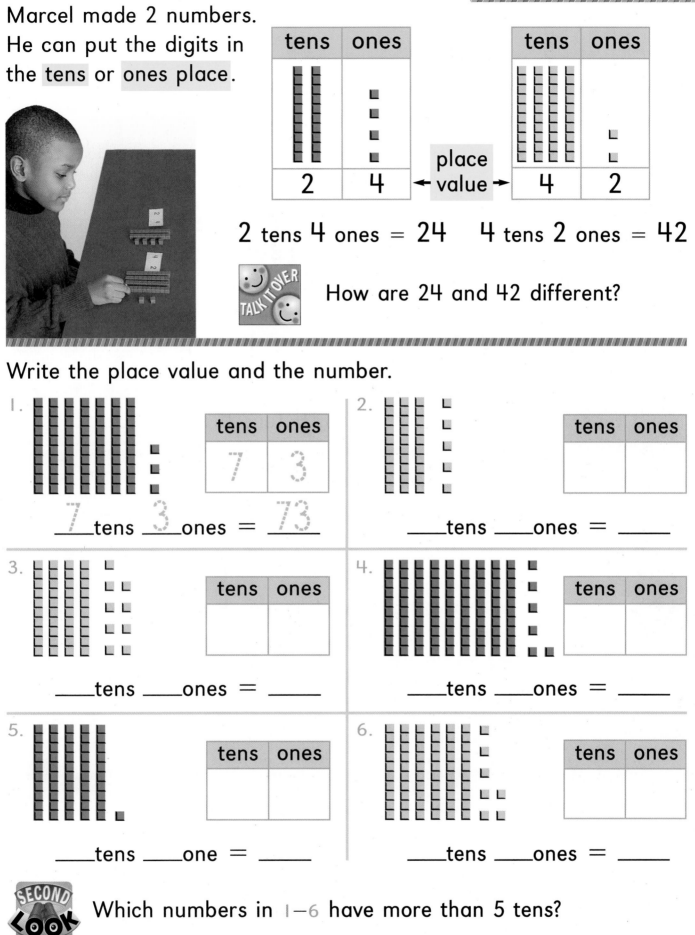

tens	ones
2	4

place ← value →

tens	ones
4	2

2 tens 4 ones = 24   4 tens 2 ones = 42

TALK IT OVER How are 24 and 42 different?

Write the place value and the number.

1.

tens	ones
7	3

___7___ tens ___3___ ones = ___73___

2.

tens	ones

_____ tens _____ ones = _____

3.

tens	ones

_____ tens _____ ones = _____

4.

tens	ones

_____ tens _____ ones = _____

5.

tens	ones

_____ tens _____ one = _____

6.

tens	ones

_____ tens _____ ones = _____

SECOND LOOK Which numbers in 1–6 have more than 5 tens?

 2-1 Ask your child to model 2 two-digit numbers that have a 6 in the ones place.

Write the place value and the number.

1.

tens	ones

____tens ____ones = _____

2.

tens	ones

____tens ____ones = _____

3.

tens	ones

____tens ____ones = _____

4.

tens	ones

____tens ____ones = _____

Write the number.

5. 5 tens 2 ones = _____

6. 4 tens 5 ones = _____

7. 7 tens 0 ones = _____

8. 3 tens 8 ones = _____

9. 9 tens 3 ones = _____

10. 6 tens 1 one = _____

11. 8 tens 4 ones = _____

12. 2 tens 6 ones = _____

SECOND LOOK In 5–12 ✔ those that have between 4 and 7 ones.

PROBLEM SOLVING Write the place value of the next number in this pattern. Model the pattern.

13.

tens	ones		tens	ones		tens	ones		tens	ones
9	8		8	7		7	6			

MATH JOURNAL 14. List in order the 2-digit numbers that have the same number of tens and ones. Describe the pattern you see.

= 11

I make groups of tens to write the place value.

36 ones = 3 tens 6 ones

tens	ones
3	6

place-value chart

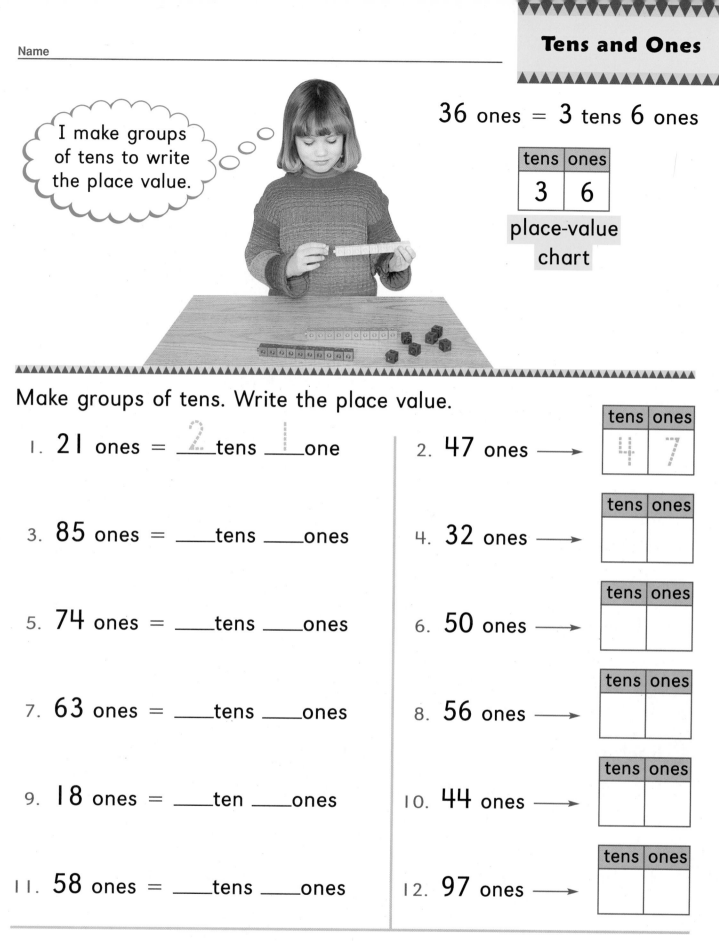

Make groups of tens. Write the place value.

1. 21 ones = __2__ tens __1__ one

2. 47 ones ⟶

tens	ones
4	7

3. 85 ones = ____ tens ____ ones

4. 32 ones ⟶

tens	ones

5. 74 ones = ____ tens ____ ones

6. 50 ones ⟶

tens	ones

7. 63 ones = ____ tens ____ ones

8. 56 ones ⟶

tens	ones

9. 18 ones = ____ ten ____ ones

10. 44 ones ⟶

tens	ones

11. 58 ones = ____ tens ____ ones

12. 97 ones ⟶

tens	ones

SHARE YOUR THINKING

How is grouping 28 ones in tens and ones like trading 28 pennies for dimes and pennies?

2-2 **Have your child use countables to group 32 ones in tens and ones.**

Here are 4 ways to describe a number.
✗ the one that does not belong.

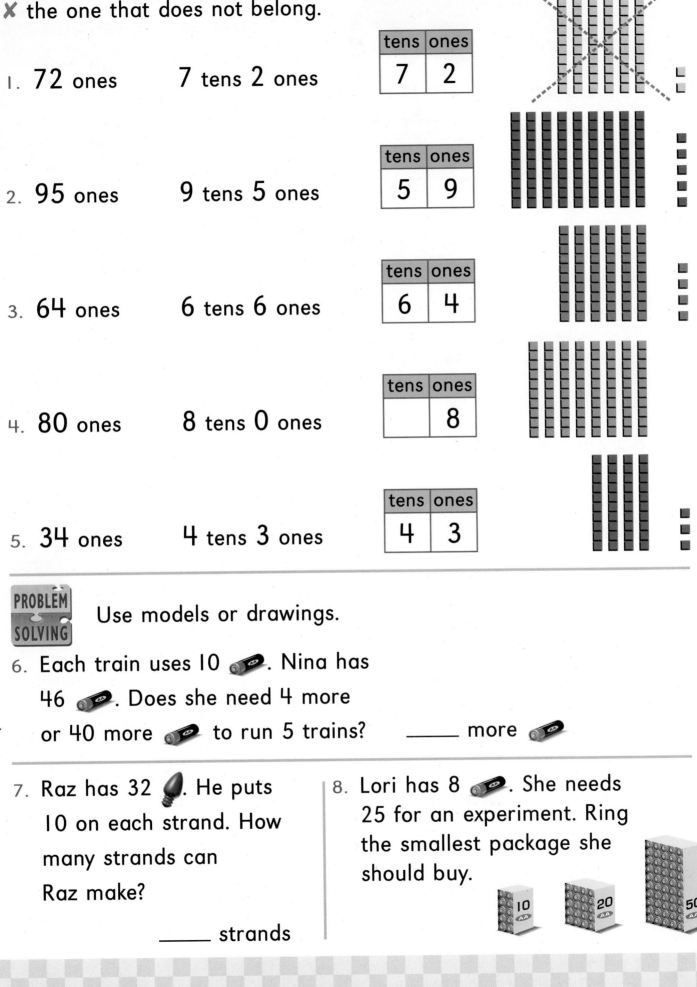

1. 72 ones   7 tens 2 ones

tens	ones
7	2

2. 95 ones   9 tens 5 ones

tens	ones
5	9

3. 64 ones   6 tens 6 ones

tens	ones
6	4

4. 80 ones   8 tens 0 ones

tens	ones
	8

5. 34 ones   4 tens 3 ones

tens	ones
4	3

PROBLEM SOLVING   Use models or drawings.

6. Each train uses 10 🔋. Nina has
   46 🔋. Does she need 4 more
   or 40 more 🔋 to run 5 trains?   _____ more 🔋

7. Raz has 32 💡. He puts
   10 on each strand. How
   many strands can
   Raz make?

   _____ strands

8. Lori has 8 🔋. She needs
   25 for an experiment. Ring
   the smallest package she
   should buy.

Name

Use this table to write number words.

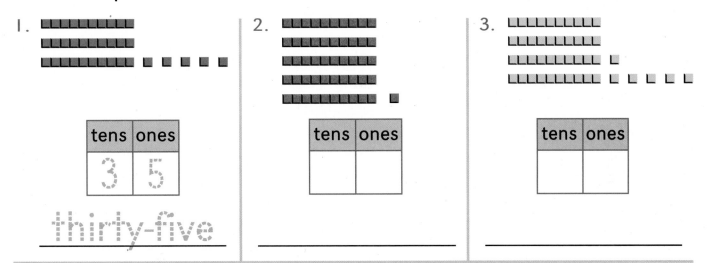

2 tens 7 ones

twenty-seven

2 tens	twenty	6 tens	sixty
3 tens	thirty	7 tens	seventy
4 tens	forty	8 tens	eighty
5 tens	fifty	9 tens	ninety

Write the place value and the number word.

1.

tens	ones
3	5

thirty-five _____

2.

tens	ones

_____

3.

tens	ones

_____

Write the number for each number word.

4. fifty-four

54

5. sixty-five

_____

6. seventy-six

_____

7. twenty-two

_____

8. thirty-three

_____

9. forty-four

_____

10. ninety-seven

_____

11. eighty-six

_____

12. seventy-five

_____

13. fifty-eight

_____

14. forty-eight

_____

15. thirty-eight

_____

SHARE YOUR THINKING

Model each number for 4–15 row by row.
Describe any pattern you see.

Write the number word.

1. 8 tens 9 ones

eighty-nine

2. 7 tens 1 one

3. 6 tens 0 ones

4. 3 tens 2 ones

5. 2 tens 9 ones

6. 9 tens 4 ones

1st Color the number.

2nd Count how many are not colored.

3rd Write the place value and the number word
     for how many are not colored.

7. forty (yellow)

____tens ____ones or

_____ not colored

8. twenty-five (blue)

____tens ____ones or

_____ not colored

9. eighty-five (red)

____ten ____ones or

_____ not colored

PROBLEM SOLVING

10. Taisha colored 4 tens 5 ones.
    Raul colored 4 tens and 5 tens.
    Dina colored four ⌴⌴⌴⌴⌴ five ⌴.
    Who did not color forty-five? _____

Here are 4 ways to write a number.

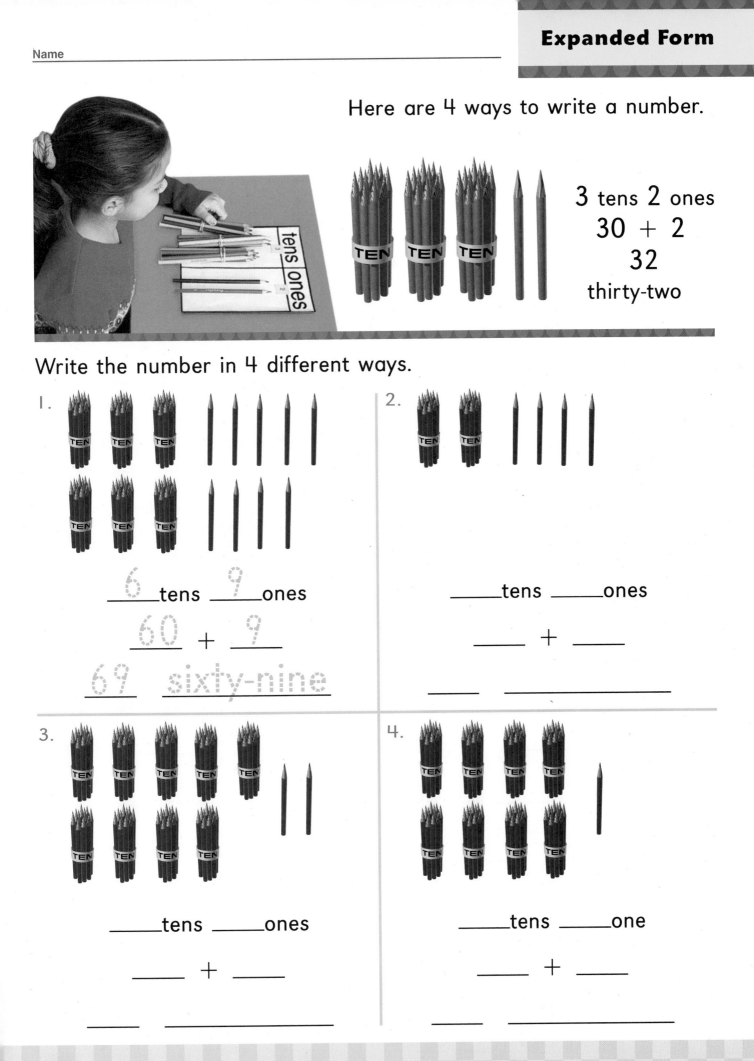

3 tens 2 ones
30 + 2
32
thirty-two

Write the number in 4 different ways.

1.

_____6_____tens _____9_____ones

_____60_____ + _____9_____

_____69_____ sixty-nine

2.

_____tens _____ones

_____ + _____

_____ _____

3.

_____tens _____ones

_____ + _____

_____ _____

4.

_____tens _____one

_____ + _____

_____ _____

Complete. Write each number in expanded form.

1. 6 tens 4 ones = 64
   60 + 4 = 64

2. 5 tens 7 ones = ____
   ____ + ____ = ____

3. 8 tens 2 ones = ____
   ____ + ____ = ____

4. 9 tens 5 ones = ____
   ____ + ____ = ____

5. 4 tens 6 ones = ____
   ____ + ____ = ____

6. 9 tens 0 ones = ____
   ____ + ____ = ____

7. 2 tens 3 ones = ____
   ____ + ____ = ____

8. 7 tens 8 ones = ____
   ____ + ____ = ____

SECOND LOOK In 1–8 ring numbers greater than 60.

Write each number in a different way.

9. 80 + 3 = 83

tens	ones
8	3	83

10. 30 + 7 = ____

tens	ones
		____

11. 90 + 2 = ____

tens	ones
		____

12. 40 + 1 = ____
    forty-one

13. 50 + 4 = ____
    ____

14. 60 + 9 = ____
    ____

15. 70 + 5 = ____
    ____

16. 20 + 1 = ____
    ____

17. 90 + 6 = ____
    ____

FINDING TOGETHER 18. Make up a number pattern like 91, 81, 71.
Write each number in 4 different ways.

Name

Compare 37 and 32.

| 1st | Compare tens. |
| 2nd | When the tens are equal, compare the ones. |

32

Both numbers have 3 tens.

37

37 is greater than 32.

$$37 > 32$$

32 is less than 37.

$$32 < 37$$

Ring the number that is greater.

| 1. | 42 | (63) | 20 | 54 | 64 | 46 |
| 2. | 71 | 77 | 49 | 45 | 93 | 91 |

Ring the number that is less.

| 3. | (65) | 72 | 36 | 22 | 67 | 97 |
| 4. | 38 | 35 | 47 | 49 | 56 | 59 |

Compare. Write $<$ or $>$.

5. 32 $<$ 42   57 ◯ 56   89 ◯ 98

6. 82 ◯ 84   79 ◯ 74   63 ◯ 93

7. 40 ◯ 30   97 ◯ 87   25 ◯ 52

TALK IT OVER When you compare 2-digit numbers, should you begin at the left or at the right? Why?

2-5 Have your child use the < and > symbols to compare the number 10 less than 62 with the number 10 more than 26.

sixty-nine **69**

Compare numbers
on a hundred chart.

1	2	3	4	5	6	7	8	9	10
11	12	13	14	15	16	17	18	19	20
21	22	23	24	25	26	27	28	29	30
31	32	33	34	35	36	37	38	39	40
41	42	43	44	45	46	47	48	49	50
51	52	53	54	55	56	57	58	59	60

32 is 10 less than 42.

52 is 10 more than 42.

Write the number 10 more than each.

1.

8	22	46	35	50	63	81
18						

Write the number 10 less than each.

2.

26						
36	19	68	40	54	85	97

What stays the same and what changes in 1–2?
Make a table to show 1 more and 1 less than each.
Describe the pattern you see.

3. Kim scored 78 points.
Daryl scored 87 points.
Who scored fewer
points?

78 ◯ 87

 _____ scored fewer points.

4. The girls collected more
than the boys. How many
leaves did the girls collect?

Leaves Collected

57	75

 The girls collected _____ .

Name _____

Use a number line to order numbers.

←—+—+—+—+—+—+—+—+—+—+—+—→
80 81 82 83 84 85 86 87 88 89 90

80 is just before 81.

82 is just after 81.

81 is between 80 and 82.

80, 81, 82 are in order from least to greatest.

Write the missing number.

1. Just Before	2. Just After	3. Between
_37_, 38	20, _21_	57, _58_, 59
___, 50	29, ___	19, ___, 21
___, 96	65, ___	98, ___, 100
___, 87	46, ___	23, ___, 25
___, 100	99, ___	86, ___, 88

Complete the number line.

4. 32, 33, 30

←—+———+———+———+———→
_30_   31   ___   ___

5. 70, 67, 68

←—+———+———+———+———→
___   ___   69   ___

6. 52, 55, 53

←—+———+———+———+———→
___   ___   54   ___

7. 77, 78, 76

←—+———+———+———+———→
___   ___   ___   79

TALK IT OVER Which number is the least and which number is the greatest in 4–7?

2-6 Have your child list the ages of each family member in order from greatest to least.

seventy-one 71

I use models to order numbers.

Compare tens first. If tens are equal, compare ones.

(3)4   (2)5   (2)7

Least to greatest → 25, 27, 34

Model the numbers. Then write them in order from least to greatest.

1. (2)3, (1)3, (3)1   13   23   ____

2. 41, 21, 12   ____   ____   ____

3. 15, 45, 51   ____   ____   ____

4. 34, 40, 30   ____   ____   ____

PROBLEM SOLVING

5. I sold 73 🌰 on Monday, 58 🌰 on Tuesday, and 69 🌰 on Wednesday.

When did I sell the most?

_____

When did I sell the least?

_____

6. Tara read 1 more 📖 than Pablo. Pablo read 10 less than Deven. Check who read the fewest.

Deven read 35 📖.

Pablo read ____ 📖.

Tara read ____ 📖.

18	19	20	13	12	8	5	10	14	11	0	9
A	B	C	D	E	F	I	M	N	O	P	R

LISTEN

A 1 ten 8 ones __18__

6 + 4 = ___

11, ___, 13

1 more than 8 ___

___ is just before 6

10 more than 10 ___

___ is just after 17

Date Pilgrims Landed

$$\begin{array}{cccc} 9 & 2 & 11 & 5 \\ -8 & +4 & -9 & -5 \end{array}$$

5 doubled ___

___ is just before 12

6 + 5 = ___

8 + 2 + 4 = ___

Year Explored

$$\begin{array}{cccc} 7 & 4 & 12 & 2 \\ -6 & +5 & -6 & +7 \end{array}$$

2

This page reviews the mathematical content presented in Chapter 1.

seventy-three **73**

Odd and Even Patterns

Name _____

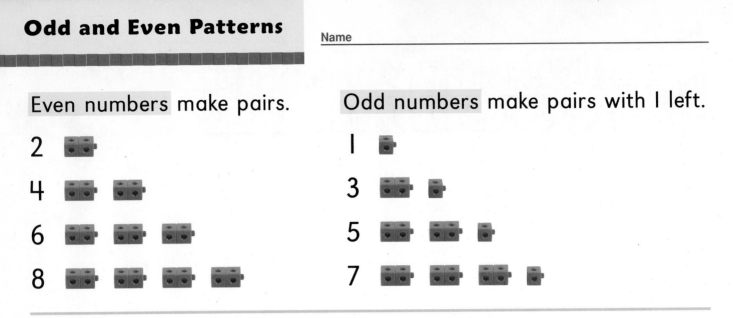

Even numbers make pairs.

2

4

6

8

Odd numbers make pairs with 1 left.

1

3

5

7

Write the missing numbers.

1	2	3	4	5	6	7	8	9	10
11	12	13							20
21		23	24						30
31			34						40
41				45					50
51					56				60
61						67			70
71	72	73	74	75	76	77	78	79	80
81								89	90
91									100

Color even numbers (yellow) and odd numbers (blue).

What 5 digits are in the ones place of even numbers?
What 5 digits are in the ones place of odd numbers?
How is saying even numbers like counting by 2s?

Ask your child to show why 23 is an odd number and 24 is an even number by pairing objects or by linking pairs of paper clips.

2-7

Name _____

Mario made 4 patterns
by counting on or
by counting back.

52, 62, 72, 82

36, 35, 34, 33

49, 39, 29, 19

79, 80, 81, 82

1	2	3	4	5	6	7	8	9	10
11	12	13	14	15	16	17	18	19	20
21	22	23	24	25	26	27	28	29	30
31	32	33	34	35	36	37	38	39	40
41	42	43	44	45	46	47	48	49	50
51	52	53	54	5		7	58	59	60
61	62	63	64				68	69	70
71	72	73	74	7		7	78	79	80
81	82	83	84	85			88	89	90
91	92		94	9			98	99	100

How did Mario count by 10s
or 1s to make each pattern?

Count by 1s. Write the missing number.
✔ when you count back.

1. 65, 66, 67, 68, ____, ____, ____

2. 90, ____, 88, 87, ____, ____, 84

3. 43, ____, 41, ____, ____, 38, 37

4. ____, 95, ____, 97, ____, 99, 100

Count by 10s. Write the missing number.
✔ when you count back.

5. 100, 90, ____, ____, 60, ____, 40

6. 35, ____, ____, 65, 75, 85, ____

7. 4, ____, 24, ____, ____, 54, 64

8. ____, ____, 57, 47, ____, 27, 17

2-8 Ask your child to name the next number
in the 4 patterns Mario made at the top
of this page.

Count by 2s. Write the missing numbers.

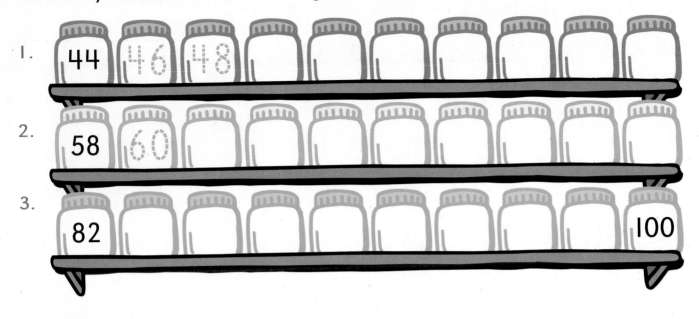

1. 44 46 48 ___ ___ ___ ___ ___ ___ ___

2. 58 60 ___ ___ ___ ___ ___ ___ ___ ___

3. 82 ___ ___ ___ ___ ___ ___ ___ ___ 100

Count by 5s. Write the missing numbers.

4. 5, 10, 15, ___, ___, ___, ___, 40

5. 40, 45, ___, ___, ___, ___, ___, ___

6. 100, 95, 90, ___, ___, ___, ___, ___

7. I am an odd number between 51 and 60. I have the same number of tens and ones.

I am _____.

8. I am an even number between 30 and 50. I have the same number of tens and ones.

I am _____.

9. I am an odd number less than 30. The digit in the tens place is one more than the digit in the ones place.

I am _____.

Odd?

Even?

Cara skip counts by 3s on a hundred chart.

3, 6, 9, ...

Brian skip counts by 4s on a hundred chart.

4, 8, 12, ...

1	2	3	4	5	6	7	8	9	10
11	12	13	14	15	16	17	18	19	20
21	22	23	24	25	26	27	28	29	30
31	32	33	34	35	36	37	38	39	40
41	42	43	44	45	46	47	48	49	50

1	2	3	4	5	6	7	8	9	10
11	12	13	14	15	16	17	18	19	20
21	22	23	24	25	26	27	28	29	30
31	32	33	34	35	36	37	38	39	40
41	42	43	44	45	46	47	48	49	50

Count by 3s. Write the missing numbers.

1.

51	52	53		55	56		58	59	
61	62		64	65		67	68		70
71		73	74		76	77		79	80
	82	83		85	86		88	89	
91	92		94	95		97	98		100

Count by 4s. Write the missing numbers.

2.

51	52	53	54	55		57	58	59	
61	62	63		65	66	67		69	70
71		73	74	75		77	78	79	
81	82	83		85	86	87		89	90
91		93	94	95		97	98	99	

2-9 Draw 10 △ and ask your child to count by 3s to tell how many sides in all.

seventy-seven **77**

Skip count to find how many in all.

1. How many wheels in all?   24 wheels

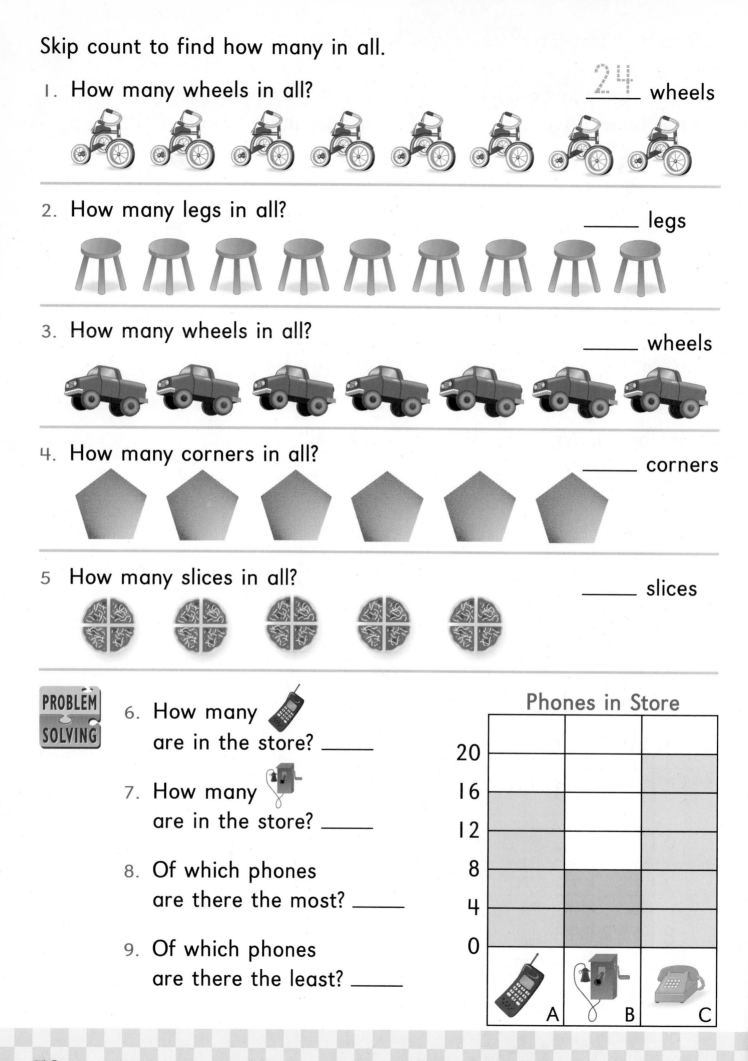

2. How many legs in all?   _____ legs

3. How many wheels in all?   _____ wheels

4. How many corners in all?   _____ corners

5 How many slices in all?   _____ slices

PROBLEM SOLVING

6. How many are in the store? _____

7. How many are in the store? _____

8. Of which phones are there the most? _____

9. Of which phones are there the least? _____

Phones in Store

20
16
12
8
4
0

A   B   C

Name _____

Each week the students read books about inventions. This tally table shows how many books they read.

Use the tally table to make a pictograph. A pictograph shows data with pictures.

Books about Inventions

Week	Tally	Number
First	ᵾᵾᵾ ᵾᵾᵾ	10
Second	ᵾᵾᵾ ᵾᵾᵾ IIII	14
Third	ᵾᵾᵾ I	
Fourth	ᵾᵾᵾ ᵾᵾᵾ ᵾᵾᵾ I	

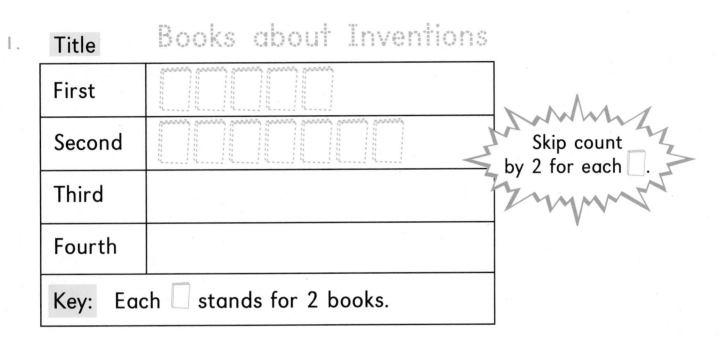

1. Title   Books about Inventions

First	▯▯▯▯▯
Second	▯▯▯▯▯▯▯
Third	
Fourth	
Key:	Each ▯ stands for 2 books.

Skip count by 2 for each ▯.

PROBLEM SOLVING

2. How many more books were read in the fourth week than in the third week? _____ more books

3. In which week were the most books read? _____ week

4. In which weeks were between 10 and 20 books read? _____ and _____ weeks

5. Were more or less than 40 books read altogether? _____ than 40 books

2-10   Have your child count aloud by 2s to tell how many books were read each week.

seventy-nine **79**

This tally table shows how many were collected each day.

 Name the days that have the same number of dimes. How many dimes were collected on those days?

Use the tally table to make a bar graph.

Money for the Homeless

Day	Dimes
Monday	~~IIII~~ ~~IIII~~ ~~IIII~~
Tuesday	~~IIII~~ ~~IIII~~
Wednesday	~~IIII~~ ~~IIII~~ ~~IIII~~
Thursday	~~IIII~~ ~~IIII~~ ~~IIII~~
Friday	~~IIII~~ ~~IIII~~ ~~IIII~~ ~~IIII~~

1. Title   **Money for the Homeless**

	0	5	10	15	20	25
Monday						
Tuesday						
Wednesday						
Thursday						
Friday						

Number of Dimes

2. On which day did the group collect the most dimes?   _____

3. On which day did it collect the fewest?   _____

4. How many more dimes did the group collect on Friday than on Tuesday?   _____ more

5. How many dimes were collected in all?   _____ dimes in all

6. How many more dimes will it take to reach 100?   _____ more

Name _____

thirty-first	31st
thirtieth	30th
twenty-ninth	29th
twenty-eighth	28th
twenty-seventh	27th
twenty-sixth	26th
twenty-fifth	25th
twenty-fourth	24th
twenty-third	23rd
twenty-second	22nd
twenty-first	21st
twentieth	20th
nineteenth	19th
eighteenth	18th
seventeenth	17th
sixteenth	16th
fifteenth	15th
fourteenth	14th
thirteenth	13th
twelfth	12th
eleventh	11th

Use the numbers first to thirty-first to tell the position.

Color these floors red.

1. twenty-fifth   twenty-sixth
   twenty-seventh   twenty-second
   twenty-ninth   thirty-first

Color these floors blue.

2. 24th   20th   19th
   30th   21st   16th
   28th   23rd   18th

PROBLEM SOLVING

3. Color the floors between the 11th and fifteenth yellow.

4. Which floor comes just before the 20th floor?

   _____

5. Which floor comes just after the 29th floor?

   _____

6. How many floors are above the twentieth floor?

   _____

7. How many floors are between the fifteenth and the thirtieth floors? _____

8. How many floors are between the ninth and the nineteenth floors? _____

2-11   Using the building above, have your child name 2 floors between the nineteenth and twenty-third floors.

eighty-one **81**

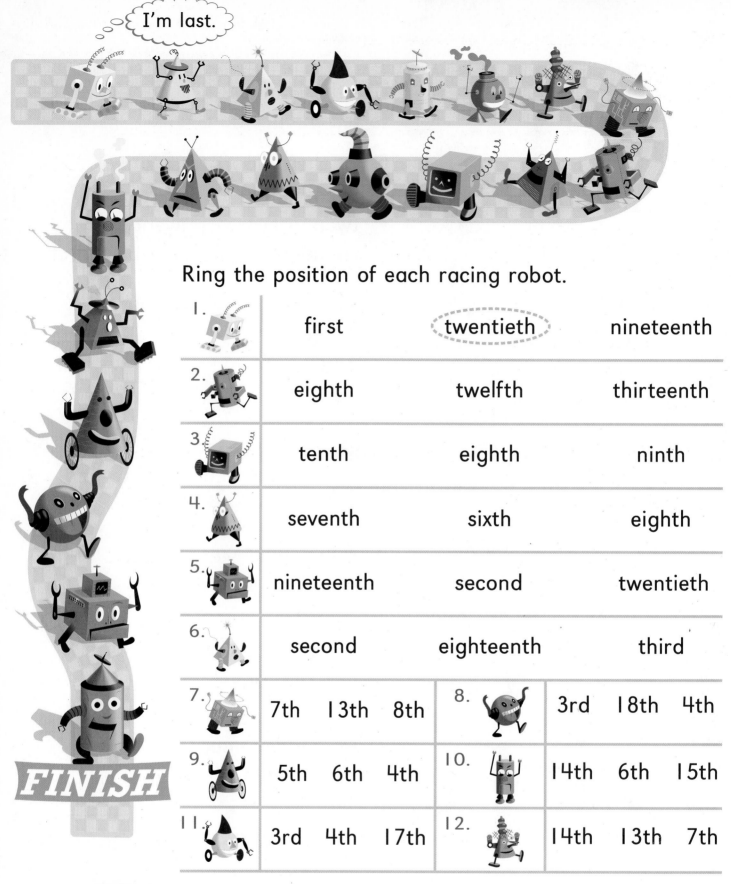

I'm last.

Ring the position of each racing robot.

1.	first	(twentieth)	nineteenth
2.	eighth	twelfth	thirteenth
3.	tenth	eighth	ninth
4.	seventh	sixth	eighth
5.	nineteenth	second	twentieth
6.	second	eighteenth	third

7.	7th 13th 8th	8.	3rd 18th 4th
9.	5th 6th 4th	10.	14th 6th 15th
11.	3rd 4th 17th	12.	14th 13th 7th

13. I can roll. I am between seventeenth and twentieth. Draw me.

14. I am not red. I cannot roll. I am before 10th place. Draw me.

Name

Dwayne has 32 hats. Does he have about 30 or about 40 hats?

To find about how many, round to the nearest ten.

Is 32 closer to 30 or to 40? Count the spaces on the number line.

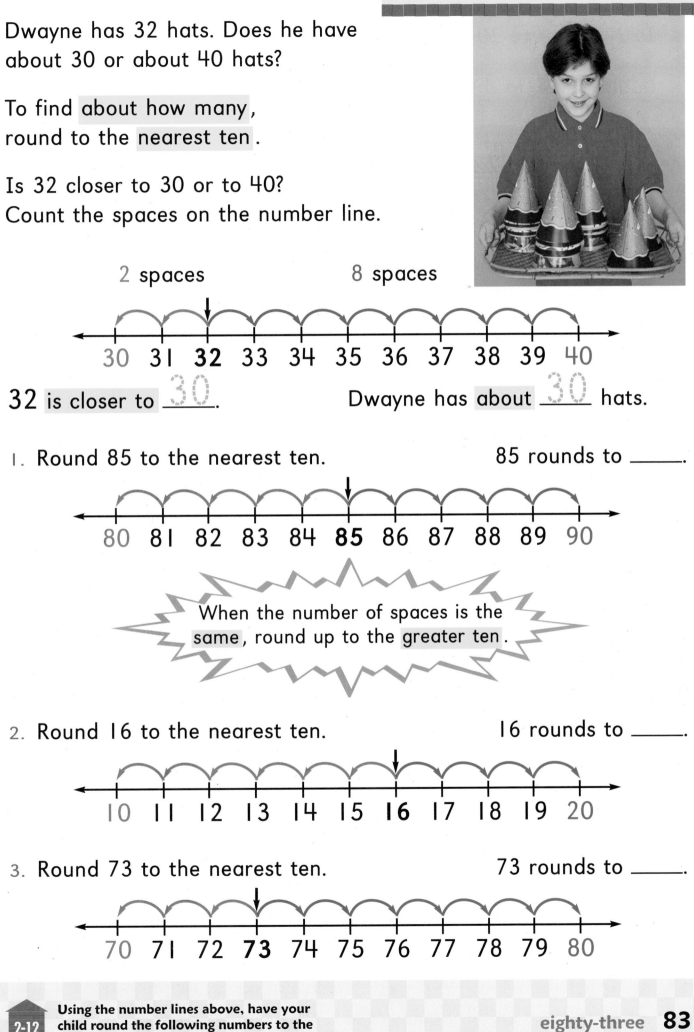

2 spaces   8 spaces

30 31 **32** 33 34 35 36 37 38 39 40

32 is closer to _30_.   Dwayne has about _30_ hats.

1. Round 85 to the nearest ten.   85 rounds to _____.

80 81 82 83 84 **85** 86 87 88 89 90

When the number of spaces is the same, round up to the greater ten.

2. Round 16 to the nearest ten.   16 rounds to _____.

10 11 12 13 14 15 **16** 17 18 19 20

3. Round 73 to the nearest ten.   73 rounds to _____.

70 71 72 **73** 74 75 76 77 78 79 80

2-12 Using the number lines above, have your child round the following numbers to the nearest ten: 38, 81, 12, 77.

Draw the hops to estimate.

1. Is 24 closer to 20 or to 30?

   Round __24__ to the nearest ten.   __24__ rounds to __20__.

   20 21 22 23 **24** 25 26 27 28 29 30

2. Is 6 tens 8 ones closer to 60 or to 70?

   Round _____ to the nearest ten.   _____ rounds to _____.

   60 61 62 63 64 65 66 67 **68** 69 70

3. Is fifty-one closer to 50 or to 60?

   Round _____ to the nearest ten.   _____ rounds to _____.

   50 **51** 52 53 54 55 56 57 58 59 60

4. Is 40 + 7 closer to 40 or to 50?

   Round _____ to the nearest ten.   _____ rounds to _____.

   40 41 42 43 44 45 46 **47** 48 49 50

5. Is 95 closer to 90 or to 100?
   Round 95 to the nearest ten.   _____ rounds to _____.

   90 91 92 93 94 **95** 96 97 98 99 100

FINDING TOGETHER

6. Show on a number line at least 4 different numbers that round to 50.

84 eighty-four

Name

Sometimes when numbers are in order, they make a pattern.

ADMIT 10   ADMIT 20   ADMIT 5   ADMIT 16

| 1st | Order the numbers. |

7   10   4   13

4, 7, 10, 13

5, 10, 16, 20

| 2nd | Check for a pattern. |

There is a pattern.

No pattern

| 3rd | Write the rule. |

Rule: + 3

No rule

Write the numbers in order.

1.

3   5   7

9   3   5   11   7

___ ___ ___ ___ ___

Is there a pattern? __Yes__   Rule: __+2__

2.

0   12   14   4   8

___ ___ ___ ___ ___

Is there a pattern? _____   Rule: _____

3.

13   3   33   23   43

___ ___ ___ ___ ___

Is there a pattern? _____   Rule: _____

SHARE YOUR THINKING

In 1—3 what numbers do not belong where there is no pattern? How can you make each into a pattern?

2-13 Have your child create several number patterns, each one illustrating a different rule.

eighty-five **85**

Sometimes the rule for a pattern uses subtraction.

95	85	75	65	

The missing number is **55**.

Rule: -10

Write the missing numbers to complete the pattern.
Write the rule.

1.

35	30	25	20	15		

Rule: _____

2.

30	33	36		42		48

Rule: _____

3.

24	20		12		4	

Rule: _____

4.

	86	76		56	46	

Rule: _____

5.

	44	48	52			64

Rule: _____

6.

	24	21	18		12	

Rule: _____

CHALLENGE Use each rule and starting number to make a pattern.

7. Rule:
$+ 1, -10$

64	65	55	56				

8. Rule:
$- 2, +10$

6	4	14	12				

9. Rule:
$- 10, +5$

90	80	85					

Skip Counting Patterns

A calculator has 3 kinds of keys.

| ON/AC | CE/C | = |

| + | − | × | ÷ |

| 0 | 1 | 2 | 3 | 4 |

| 5 | 6 | 7 | 8 | 9 | • |

Ring the correct key or keys for each.

1. Turn on and clear.

| + | = | (ON/AC) |

2. Clear only.

| ON/AC | = | CE/C |

---

3. Enter the number 17.

| 7 | 1 | + | 1 | 7 | 1 | 7 |

4. Subtract 6 − 4.

| 4 | − | 6 | 6 | − | 4 |

Press the keys. Write the number you see.
Ring odd or even.

5.

ON/AC	+		0
+	2	=	2
		=	
		=	

odd   even

6.

ON/AC	1		
+	2	=	
		=	
		=	

odd   even

This lesson teaches the functions of the control, operations, and number keys, as well as how to skip count on a calculator.

TECHNOLOGY

Complete the pattern.

1. Count on by 10s from 60.

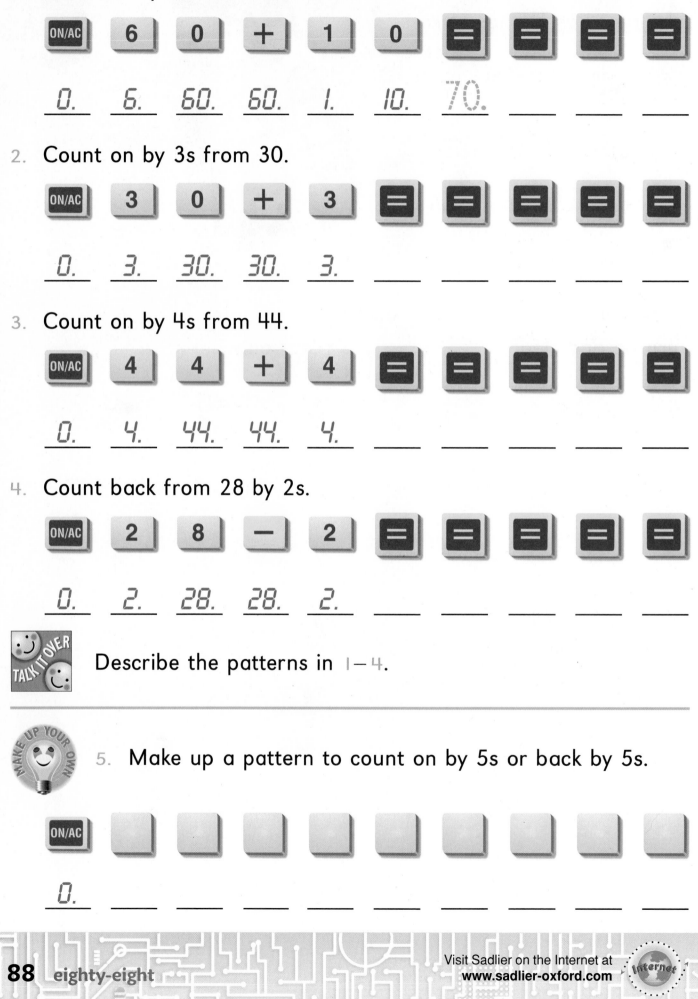

0.   6.   60.   60.   1.   10.   70. ___ ___ ___ ___

2. Count on by 3s from 30.

0.   3.   30.   30.   3. ___ ___ ___ ___ ___

3. Count on by 4s from 44.

0.   4.   44.   44.   4. ___ ___ ___ ___ ___

4. Count back from 28 by 2s.

0.   2.   28.   28.   2. ___ ___ ___ ___ ___

TALK IT OVER Describe the patterns in 1–4.

MAKE UP YOUR OWN 5. Make up a pattern to count on by 5s or back by 5s.

0. ___ ___ ___ ___ ___ ___ ___ ___ ___

1. **Read** Leanne is third in the line.
There are 6 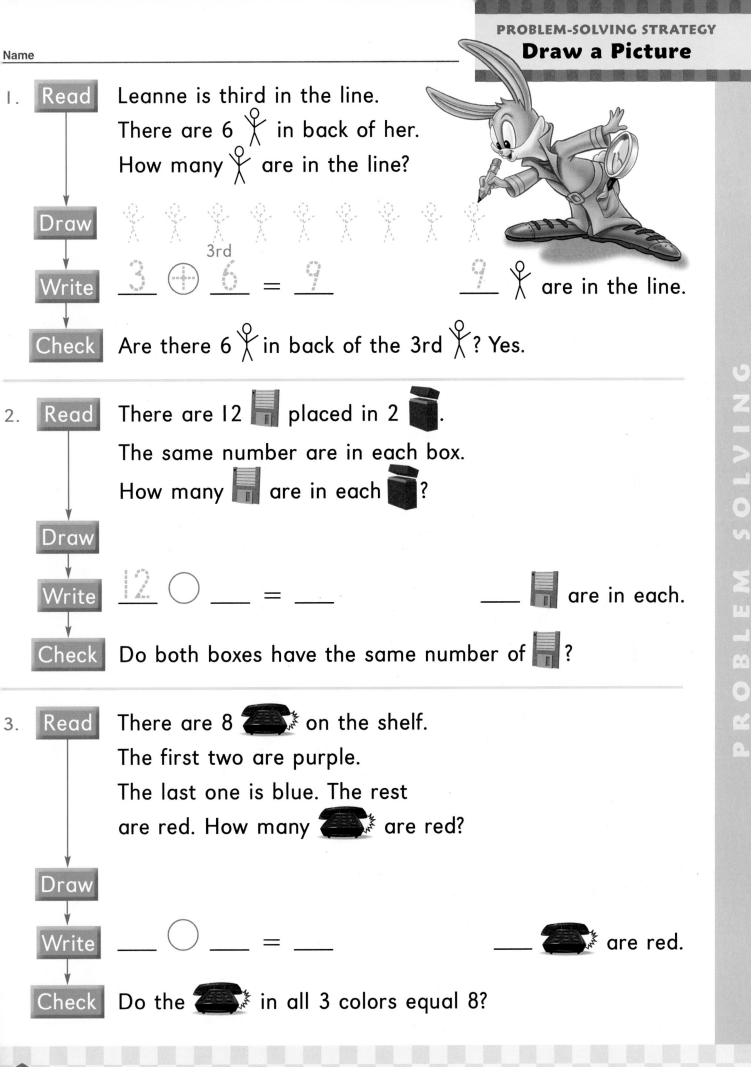 in back of her.
How many are in the line?

Draw

Write 3rd
___ ⊕ ___ = ___   ___ are in the line.
  3   6   9   9

Check Are there 6 in back of the 3rd ? Yes.

2. **Read** There are 12 placed in 2 .
The same number are in each box.
How many are in each ?

Draw

Write 12 ◯ ___ = ___   ___ are in each.

Check Do both boxes have the same number of ?

3. **Read** There are 8 on the shelf.
The first two are purple.
The last one is blue. The rest
are red. How many are red?

Draw

Write ___ ◯ ___ = ___   ___ are red.

Check Do the in all 3 colors equal 8?

PROBLEM SOLVING

4. There are 2 boxes for 8 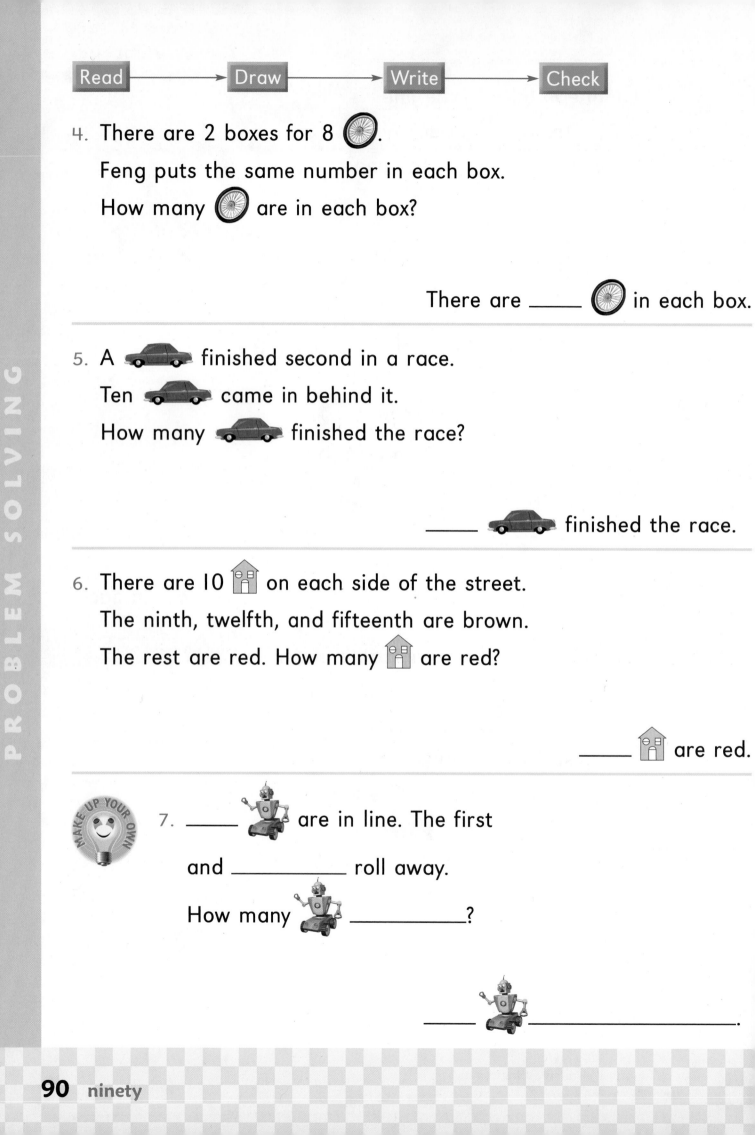.

Feng puts the same number in each box.

How many are in each box?

There are _____ in each box.

5. A finished second in a race.

Ten came in behind it.

How many finished the race?

_____ finished the race.

6. There are 10 on each side of the street.

The ninth, twelfth, and fifteenth are brown.

The rest are red. How many are red?

_____ are red.

MAKE UP YOUR OWN

7. _____ are in line. The first

and _____ roll away.

How many _____?

_____ _____.

Name

Favorite Inventions

Count by 2s.

telephone	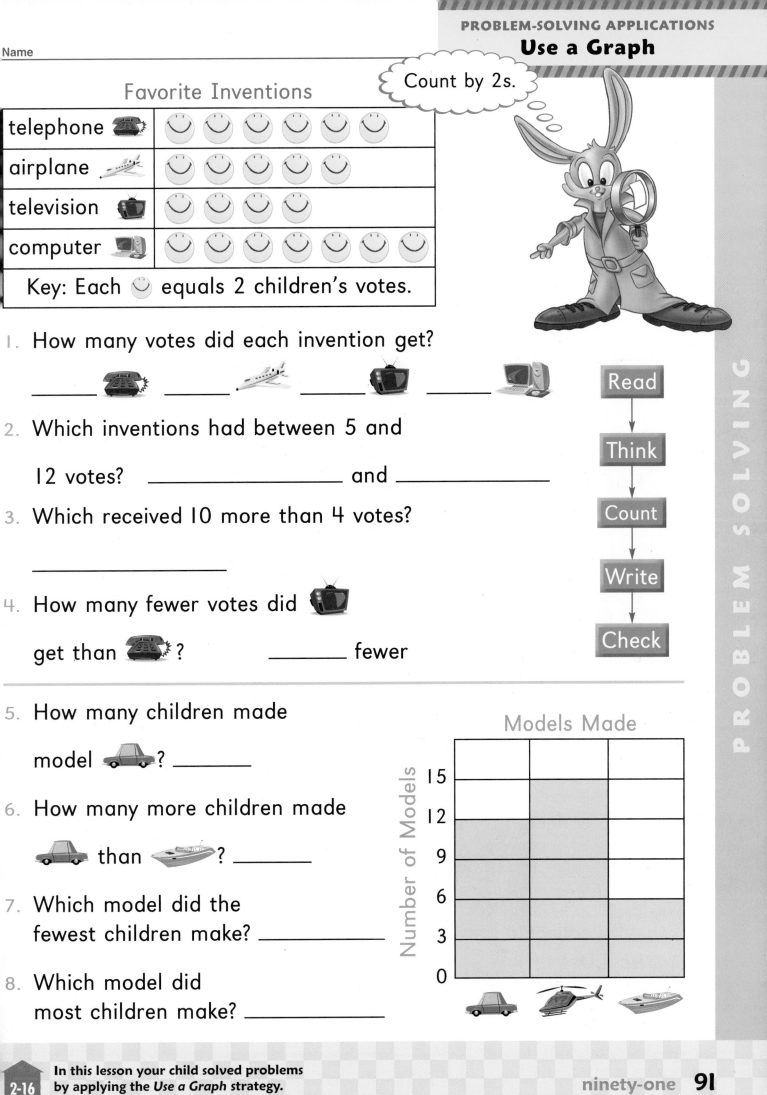	☺ ☺ ☺ ☺ ☺ ☺
airplane		☺ ☺ ☺ ☺ ☺
television		☺ ☺ ☺ ☺
computer		☺ ☺ ☺ ☺ ☺ ☺ ☺

Key: Each ☺ equals 2 children's votes.

1. How many votes did each invention get?

_____ 📞 _____ ✈️ _____ 📺 _____ 💻

Read

Think

2. Which inventions had between 5 and

12 votes? _____ and _____

Count

3. Which received 10 more than 4 votes?

_____

Write

4. How many fewer votes did 📺

get than 📞? _____ fewer

Check

5. How many children made

model 🚗? _____

Models Made

6. How many more children made

🚗 than 🚤? _____

7. Which model did the
fewest children make? _____

8. Which model did
most children make? _____

Number of Models

15
12
9
6
3
0

2-16 In this lesson your child solved problems
by applying the *Use a Graph* strategy.

ninety-one **91**

Use a strategy you have learned.

STRATEGY FILE
Draw a Picture
Use a Graph
Guess and Test
Choose the Operation

9. I have 8 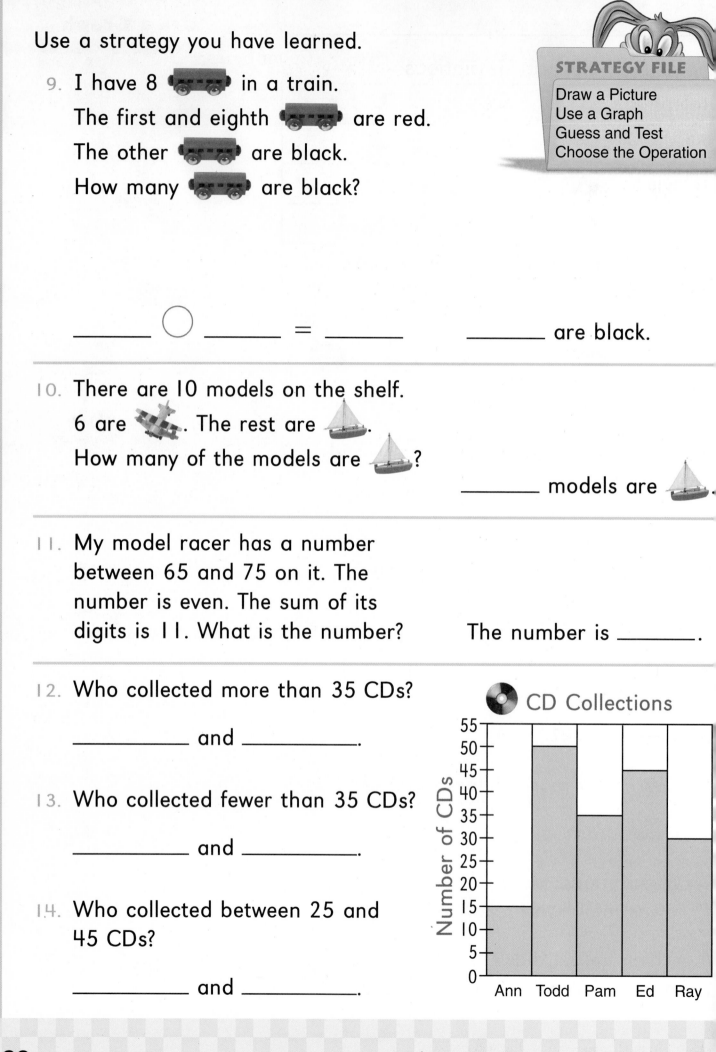 in a train.
The first and eighth are red.
The other are black.
How many are black?

_____ ◯ _____ = _____   _____ are black.

10. There are 10 models on the shelf.
6 are . The rest are .
How many of the models are ?

_____ models are .

11. My model racer has a number between 65 and 75 on it. The number is even. The sum of its digits is 11. What is the number?

The number is _____.

12. Who collected more than 35 CDs?

_____ and _____.

13. Who collected fewer than 35 CDs?

_____ and _____.

14. Who collected between 25 and 45 CDs?

_____ and _____.

CD Collections

Number of CDs

55
50
45
40
35
30
25
20
15
10
5
0

Ann Todd Pam Ed Ray

PROBLEM SOLVING

Name _____

Complete the place value, number, or word.

1. ____ tens ____ ones

____ forty-____

2. | tens | ones |
   |------|------|
   | | |

   ____

   _____

3. thirty-eight

tens	ones

____

___ + ___

4. fifteen

tens	ones

____

___ + ___

5. 70 + 6

tens	ones

____

seventy-____

6. Write the number before, after, or between.

___, 71   89, ___   98, ___, 100

7. Compare. Write < or >.

69 < 81   54 ◯ 34   38 ◯ 83   67 ◯ 69

Complete the pattern. Write the rule in the box.

8. 36, 40, 44, ___, ___ ☐

9. 95, 90, 85, ___ ☐

Color the circles.

24th ◯ ◯ 26th ◯ ◯ ◯ ◯ ◯ 31st

10. twenty-seventh (blue)

11. thirtieth (red)   12. twenty-ninth (green)   13. 25th (yellow)

14. Round 27 to the nearest ten.   27 rounds to _____.

20 21 22 23 24 25 26 **27** 28 29 30

This page reviews the mathematical content presented in Chapter 2.

REINFORCEMENT

Name _____

Connect the dots to build a machine.
Start at 0.

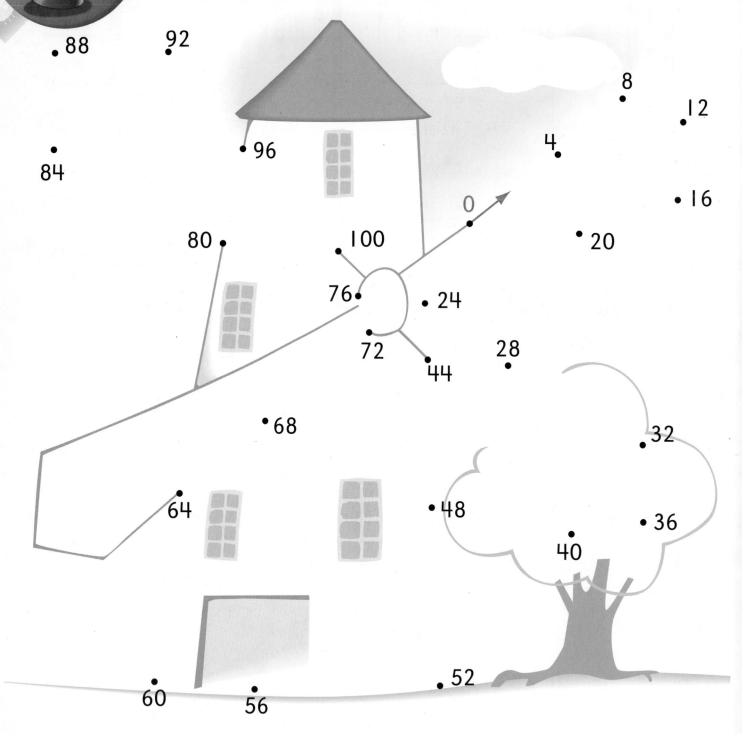

88   92

8

12

96   4

84   16

0

80   100   20

76   24

72   28

44

68   32

64   48   36

40

52

60   56

To build the machine, I counted by _____.
The sails turn to move water and more.

A _____ Indimwil

This page extends your child's
understanding of skip counting.   2

Name _____

1. Make 3 two-digit numbers using each digit once. Model each.

| 6 | 5 | 4 | 3 |

| 2 | 1 |

• Write each number in expanded form.

☐☐ = ____ + ____

☐☐ = ____ + ____   • Order your numbers:

☐☐ = ____ + ____   _____, _____, _____
                        least   greatest

2. Write the missing number. Color to match the rule.

| 7 |—| |—| 16 |—| 26 |—| |—| |

| |—| 35 |—| |—| 46 |—| |—| |

Rule: + 10, + 1

Rule: + 10, − 1

Use a separate sheet of paper.
Choose 1 of these projects.

3. Show each counting pattern on a calendar. Complete the patterns.

• Color the 4th, 8th, and 12th days (yellow).

• Color the third, sixth, and ninth days (red).

4. Make missing-number puzzles using part of a hundred chart. Give them to a classmate to solve.

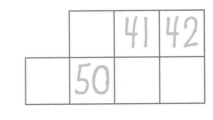

2 This page provides a variety of informal assessment opportunities in order to measure your child's understanding of Chapter 2.

ninety-five **95**

ASSESSMENT

Complete. Model each.

1. 6 tens 0 ones

_____

2.

tens	ones
7	4

_____

3. thirty-six

_____ + _____

Write the missing number.

4. _____, 51   5. 89, _____

6. 45, _____, 47

Compare. Write < or >.

7. 68 ◯ 86

8. 76 ◯ 75

Write the missing numbers. Color by 3s ▭ yellow, by 10s ▭ blue.

9.

| 76 | 66 | 56 | | |

10.

| 12 | 15 | 18 | | |

Use a ⟵┼┼┼⟶. Round to the nearest ten.

11. 89 rounds to _____.

12. 43 rounds to _____.

Match.

13. fifteenth   20th

14. twentieth   31st

15. thirty-first   15th

PROBLEM SOLVING

16. Each robot needs 4 🔋. How many 🔋 do you need for 4 robots?

_____ 🔋

Gears Used

Tyler	● ● ● ●
Sam	● ● ● ● ●
Keri	● ●

Key: Each ● equals 5 gears.

17. How many more gears did Sam use than Keri?   _____ more

18. Did they use more or less than 50 gears in all?   _____

This page is formal assessment of your child's understanding of the content presented in Chapter 2.

2

CRITICAL THINKING

Pam copied Polly Penwarden.
How many toes has Pam painted?
How many toes has she left to paint?

For more information about Chapter 3, visit the Family Information Center at **www.sadlier-oxford.com**

Internet

Dear Family,

Today your child began Chapter 3. As he/she studies addition and subtraction facts to 18, you may want to read the poem below, which was read in class. Have your child talk about some of the math ideas pictured on page 97.

Look for the 🏠 at the bottom of each skills lesson. The suggestion on the page gives you an opportunity to improve your child's understanding of math. You may want to have pennies and other countables available for your child to use throughout this chapter.

Home Reading Connection

High on the Wall

High on the wall
Where the pennywort grows
Polly Penwarden
Is painting her toes.

One is purple
And two are red
And two are the color
Of her golden head.

One is blue
And two are green
And the others are the colors
They've always been.

Charles Causley

Home Activity

Initial Design 13

Try this activity with your child. Ask him/her to use a marker or crayon to print the initial of his/her first name nine times. Then print the initial of your last name four times. Write an addition sentence to show how many letters in all. After you finish each lesson in this chapter, change the letters, the numbers, and/or the skill.

M M L M L M
M M M L M L M
9 + 4 = 13

Name _____

Take a survey to find out which kind
of art project classmates like best.

First use a tally mark to record each vote.

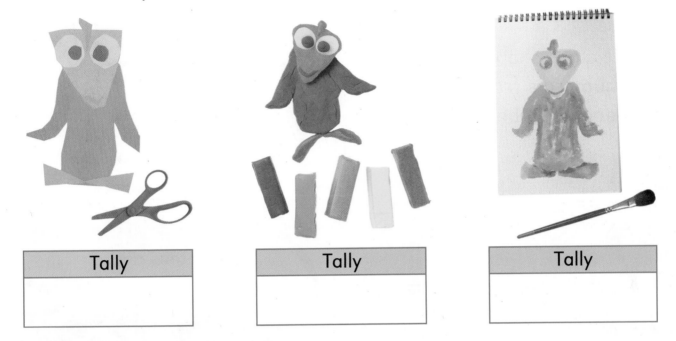

Tally

Tally

Tally

Use your data to make a bar graph.

Color one ☐ to show each vote.

Our Favorite Art Project

paper

clay

paint

0 1 2 3 4 5 6 7 8 9 10 11 12

Number of Votes

Look at the bar graph. Then write your answers.

Project favored by most: _____

Project favored by least: _____

3
PORTFOLIO You can put this in your
Math Portfolio.

ninety-nine **99**

C O N N E C T I O N S

These ancient numbers are called Roman numerals.

The Romans wrote the numbers from 11 to 20 using addition.

The art class shaped these numerals from //. Write each number.

XIII = 10 + 3 = ___

XV = ___ + ___ = ___

XIX = ___ + ___ = ___

XIV = ___ + ___ = ___

Think of tens and ones.

Draw / to make Roman numerals for each.

11 = 10 + ___
X

16 = ___ + ___

12 = ___ + ___

18 = ___ + ___

20 = ___ + ___

17 = ___ + ___

You can put this in your
Math Portfolio. PORTFOLIO

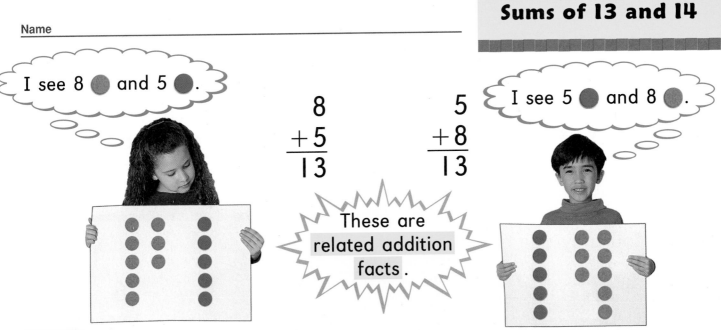

I see 8 ● and 5 ●.

$$8 \atop +5 \over 13$$

$$5 \atop +8 \over 13$$

I see 5 ● and 8 ●.

These are related addition facts.

TALK IT OVER

Does the sum change or stay the same when the order of the addends changes? Model this on a ten-frame.

Find the sum. Write the related addition fact.

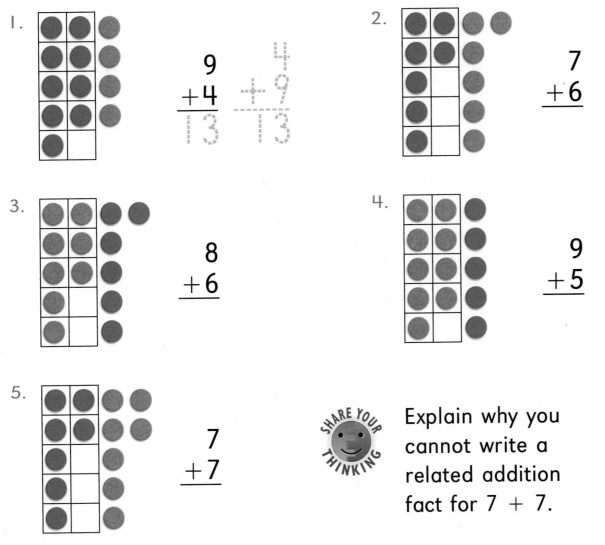

1. $9 \atop +4$ $+9 \atop 13$

2. $7 \atop +6$

3. $8 \atop +6$

4. $9 \atop +5$

5. $7 \atop +7$

SHARE YOUR THINKING

Explain why you cannot write a related addition fact for 7 + 7.

3-1   Model one of the groups above and ask your child to name the two related addition facts.

one hundred one **101**

You can model related addition facts on a +++++.

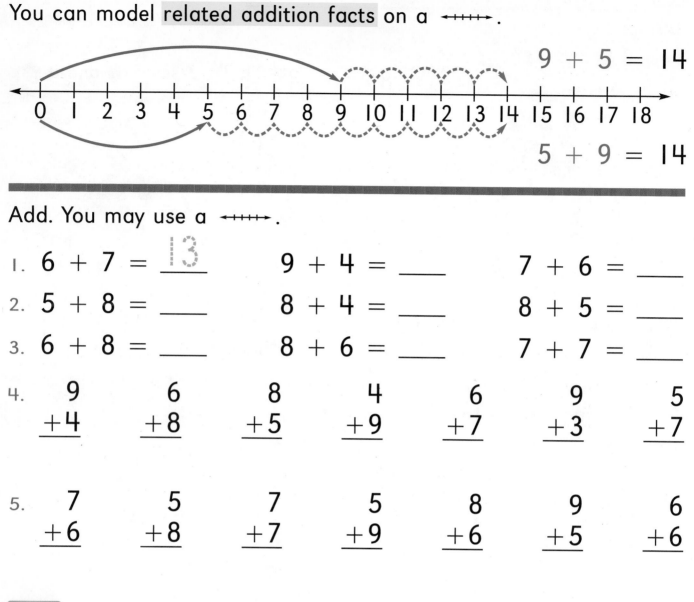

$9 + 5 = 14$

$5 + 9 = 14$

Add. You may use a +++++.

1. $6 + 7 = \underline{13}$ $9 + 4 = \underline{}$ $7 + 6 = \underline{}$

2. $5 + 8 = \underline{}$ $8 + 4 = \underline{}$ $8 + 5 = \underline{}$

3. $6 + 8 = \underline{}$ $8 + 6 = \underline{}$ $7 + 7 = \underline{}$

4.
$$\begin{array}{ccccccc} 9 & 6 & 8 & 4 & 6 & 9 & 5 \\ +4 & +8 & +5 & +9 & +7 & +3 & +7 \end{array}$$

5.
$$\begin{array}{ccccccc} 7 & 5 & 7 & 5 & 8 & 9 & 6 \\ +6 & +8 & +7 & +9 & +6 & +5 & +6 \end{array}$$

 In 1–5 ring related facts in each row.

 6. Gil painted 7 models.
Jon painted 1 less than Gil.
How many models did
they paint altogether?   _____ models altogether

7. In the museum line 4 🕴 are ahead of you.
There are 8 🕴 behind you.
How many 🕴 are there in all?   _____ 🕴 in all

Name

You can write related subtraction facts for this ten-frame.

Take away one part or the other part.

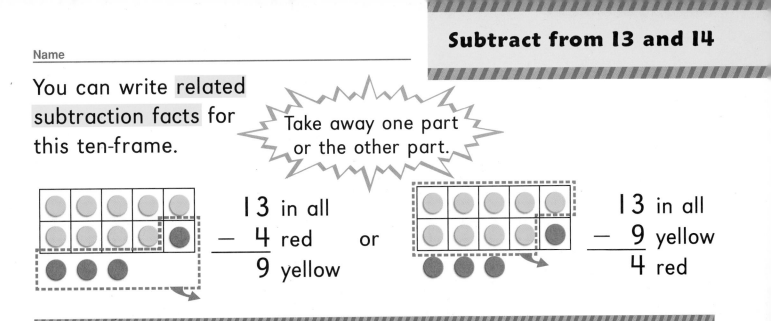

13 in all
− 4 red
9 yellow

or

13 in all
− 9 yellow
4 red

Ring the part taken away. Find the difference.
Write the related subtraction fact.

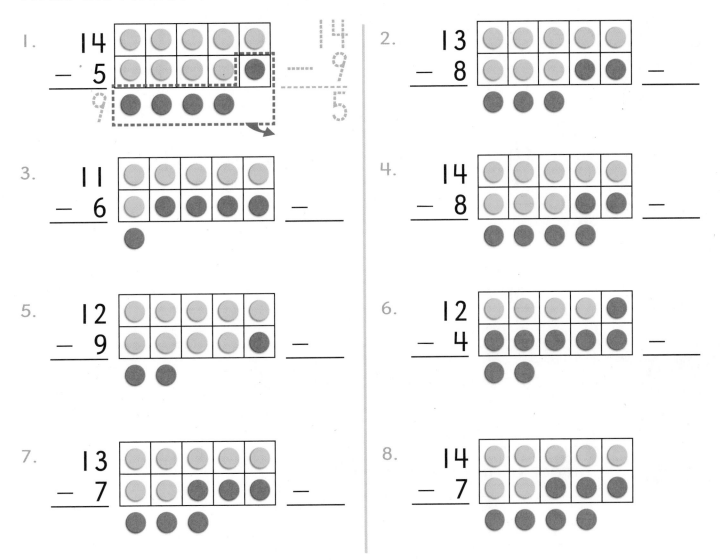

1. 14
    − 5

2. 13
    − 8

3. 11
    − 6

4. 14
    − 8

5. 12
    − 9

6. 12
    − 4

7. 13
    − 7

8. 14
    − 7

SHARE YOUR THINKING When you write related subtraction facts, do you write the whole or part first? Why?

Use a ⟷ to model related subtraction sentences.

Go to 14. Count back 8.   14 − 8 = 6

0 1 2 3 4 5 6 7 8 9 10 11 12 13 14

Go to 14. Count back 6.   14 − 6 = 8

Find the difference. You may use a ⟷.

1. 14 − 7 = _7_   14 − 9 = ___   14 − 5 = ___
2. 13 − 8 = ___   13 − 5 = ___   13 − 7 = ___
3. 13 − 4 = ___   13 − 6 = ___   13 − 9 = ___

4.
| 13 | 14 | 13 | 14 | 13 | 14 |
| −6 | −9 | −8 | −6 | −7 | −7 |

5.
| 14 | 13 | 12 | 13 | 11 | 14 |
| −8 | −4 | −9 | −9 | −8 | −5 |

In 1−5 ring the related facts in each row.

Subtract across. Subtract down.

6.
12	8	4
6	4	
6		

7.
13	5	
8	1	

8.
14	9	
7	5	

Name _____

Add by making 10.
Use a ten-frame to help.

$$\begin{array}{r} 9 \\ +7 \\ \hline 16 \end{array}$$

10 + 6

Move ● to make 10.

$$\begin{array}{r} 8 \\ +7 \\ \hline 15 \end{array}$$

10 + 5

Find the sum. Write the related addition fact.

1. $\begin{array}{r} 7 \\ +8 \\ \hline 15 \end{array}$ $\begin{array}{r} 8 \\ +7 \\ \hline 15 \end{array}$

2. $\begin{array}{r} 9 \\ +6 \\ \hline \end{array}$

3. $\begin{array}{r} 8 \\ +8 \\ \hline \end{array}$

4. $\begin{array}{r} 7 \\ +9 \\ \hline \end{array}$

5. $\begin{array}{r} 6 \\ +9 \\ \hline \end{array}$ $\begin{array}{r} 5 \\ +7 \\ \hline \end{array}$ $\begin{array}{r} 6 \\ +8 \\ \hline \end{array}$ $\begin{array}{r} 9 \\ +4 \\ \hline \end{array}$

6. $\begin{array}{r} 8 \\ +5 \\ \hline \end{array}$ $\begin{array}{r} 3 \\ +9 \\ \hline \end{array}$ $\begin{array}{r} 2 \\ +9 \\ \hline \end{array}$ $\begin{array}{r} 9 \\ +5 \\ \hline \end{array}$

---

7. Explain how you can make 10 on a ten-frame when one addend is 8.

3-3 Have your child use a ten-frame to explain how to make 10 when adding 8 + 7.

one hundred five **105**

Write the related addition facts shown on the number line.

1.

```
← | + | + | + | + | + | + | + | + | + | + | + | + | + | + | + | + | + | →
  0   1   2   3   4   5   6   7   8   9  10  11  12  13  14  15  16  17  18
```

2.

1. 9 _+ 6_ = _15_

2. 6 _+ 9_ = ___

Find the sum. Use a number line.

3. 8 + 8 = _16_ | 9 + 7 = ___ | 6 + 6 = ___

4. 6 + 9 = ___ | 7 + 9 = ___ | 7 + 7 = ___

5. 7 + 8 = ___ | 8 + 7 = ___ | 9 + 6 = ___

6.
$$\begin{array}{ccccccc} 8 & 5 & 7 & 8 & 5 & 9 & 4 \\ +7 & +8 & +9 & +8 & +9 & +7 & +9 \end{array}$$

7.
$$\begin{array}{ccccccc} 9 & 3 & 2 & 7 & 7 & 6 & 7 \\ +6 & +9 & +8 & +6 & +8 & +9 & +5 \end{array}$$

 SECOND LOOK In 3–7 ✔ even sums.

 CHALLENGE Add across. Add down.

8.

5	1	6
2	7	
7		

9.

4	3	
2	5	

10.

3	4	
5	4	

 MAKE UP YOUR OWN 11. Use grid paper to make up another addition square.

Name _____

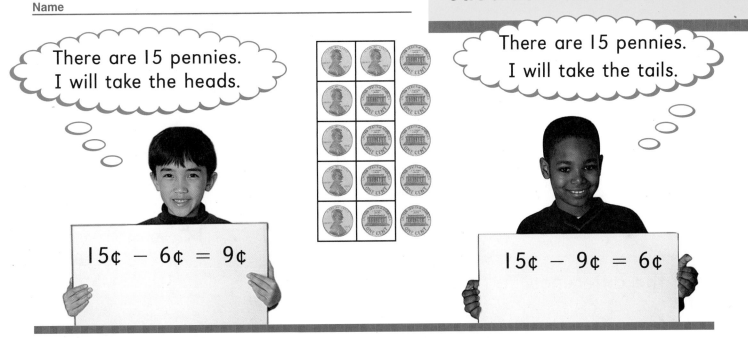

There are 15 pennies. I will take the heads.

$15¢ − 6¢ = 9¢$

There are 15 pennies. I will take the tails.

$15¢ − 9¢ = 6¢$

Subtract. Use ⬤ and a ▦.
Write the related subtraction facts.

1. $\begin{array}{r} 15¢ \\ -\ 8¢ \\ \hline 7¢ \end{array}$ $\begin{array}{r} 15¢ \\ -\ 7¢ \\ \hline 8¢ \end{array}$

2. $\begin{array}{r} 14¢ \\ -\ 9¢ \\ \hline ¢ \end{array}$

3. $\begin{array}{r} 14¢ \\ -\ 8¢ \\ \hline ¢ \end{array}$

4. $\begin{array}{r} 12¢ \\ -\ 8¢ \\ \hline ¢ \end{array}$

5. $\begin{array}{r} 16¢ \\ -\ 9¢ \\ \hline ¢ \end{array}$

6. $\begin{array}{r} 13¢ \\ -\ 4¢ \\ \hline ¢ \end{array}$

7. $\begin{array}{r} 15¢ \\ -\ 9¢ \\ \hline ¢ \end{array}$

8. $\begin{array}{r} 16¢ \\ -\ 8¢ \\ \hline ¢ \end{array}$

9. $\begin{array}{r} 13¢ \\ -\ 5¢ \\ \hline ¢ \end{array}$

Find the difference.

10. $16 − 7 = \underline{\ \ }$ $15 − 8 = \underline{\ \ }$ $14 − 7 = \underline{\ \ }$

11. $14 − 6 = \underline{\ \ }$ $15 − 7 = \underline{\ \ }$ $16 − 8 = \underline{\ \ }$

12. $15 − 6 = \underline{\ \ }$ $12 − 6 = \underline{\ \ }$ $12 − 9 = \underline{\ \ }$

 In 10–12 ✔ the doubles facts.

3-4   Using pennies, have your child model and explain a subtraction fact for 16¢ in all.

one hundred seven **107**

Write the related subtraction facts modeled on the ⟵┄┄⟶.

1. $15 - 9 = \underline{6}$

0 1 2 3 4 5 6 7 8 9 10 11 12 13 14 15 16 17 18

2. $15 - 6 = \underline{}$

Find the difference. You may use a number line.

3. $16 - 8 = \underline{8}$ $15 - 7 = \underline{}$ $16 - 9 = \underline{}$

4. $13 - 6 = \underline{}$ $15 - 8 = \underline{}$ $14 - 7 = \underline{}$

5. $14 - 8 = \underline{}$ $16 - 7 = \underline{}$ $15 - 9 = \underline{}$

6.
$\begin{array}{r} 15 \\ -\ 8 \\ \hline \end{array}$
$\begin{array}{r} 16 \\ -\ 9 \\ \hline \end{array}$
$\begin{array}{r} 15 \\ -\ 9 \\ \hline \end{array}$
$\begin{array}{r} 14 \\ -\ 6 \\ \hline \end{array}$
$\begin{array}{r} 14 \\ -\ 5 \\ \hline \end{array}$
$\begin{array}{r} 16 \\ -\ 7 \\ \hline \end{array}$

7.
$\begin{array}{r} 13 \\ -\ 9 \\ \hline \end{array}$
$\begin{array}{r} 16 \\ -\ 8 \\ \hline \end{array}$
$\begin{array}{r} 14 \\ -\ 9 \\ \hline \end{array}$
$\begin{array}{r} 15 \\ -\ 6 \\ \hline \end{array}$
$\begin{array}{r} 13 \\ -\ 5 \\ \hline \end{array}$
$\begin{array}{r} 15 \\ -\ 7 \\ \hline \end{array}$

8. Teri spent 15¢ on two stickers. One cost 8¢. How much did the other sticker cost? ___¢

9. What number doubled is 16? ____

10. What number doubled is 10? ____

11. What number doubled is 14? ____

? ▢ 16

DOUBLES MACHINE

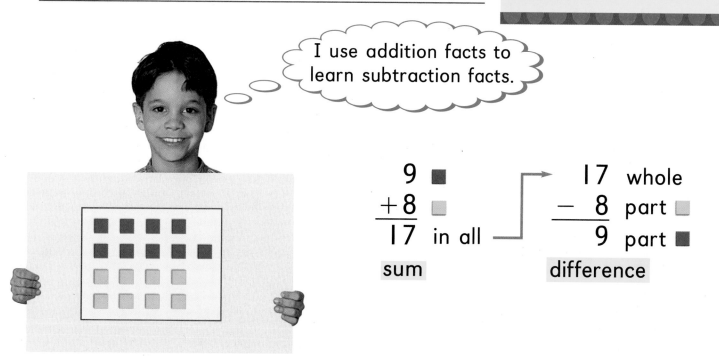

I use addition facts to learn subtraction facts.

9 ■		17	whole
+8 ▢		− 8	part ▢
17 in all		9	part ■

sum   difference

TALK IT OVER How are these facts related? Which subtraction fact does 9 + 9 = 18 help you learn?

Find the sum. Complete the related subtraction fact.

1. 8   2. 9   3. 7
   +9 −9   +9 −9   +8 −8
   17   9   7

4. 9   5. 6   6. 5
   +5 −5   +7 −7   +8 −8

7. 6   8. 9   9. 8
   +9 −9   +8 −8   +6 −6

10. 8   11. 7   12. 9
   +8 −   +4 −   +7 −

3-5 Provide your child with 4 addition facts and have him/her name the related subtraction facts.

one hundred nine **109**

Find the difference.
You may use the
number line.

$17 - 9 = \underline{8}$

$8 + 9 = 17$

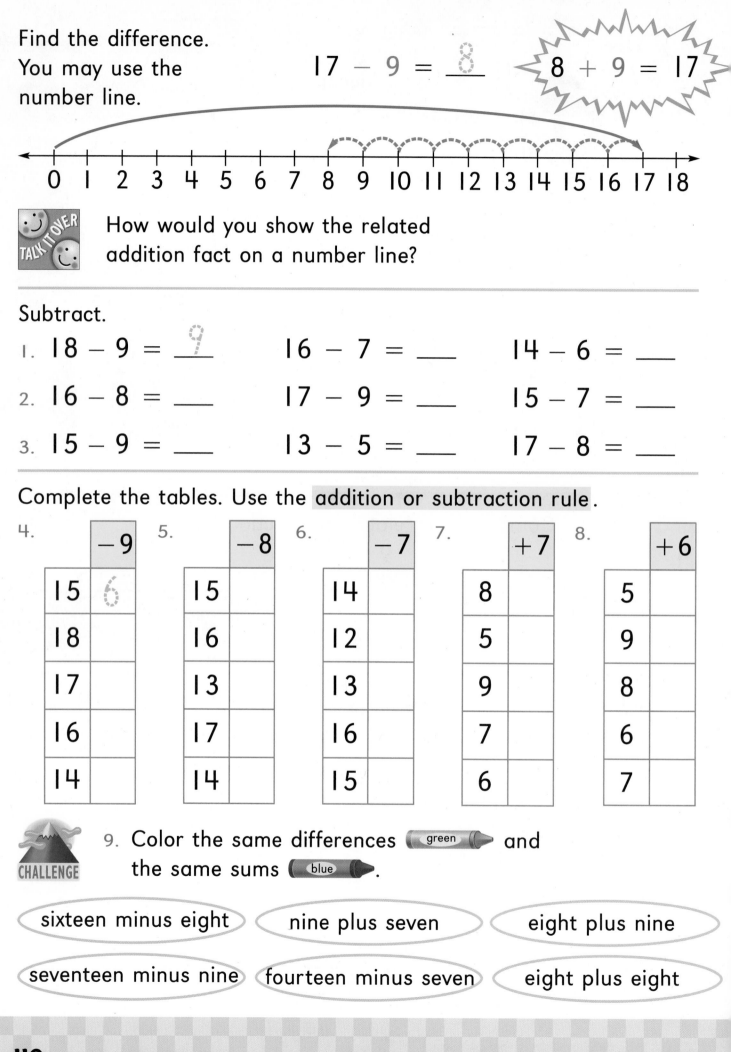

TALK IT OVER

How would you show the related
addition fact on a number line?

Subtract.

1. $18 - 9 = \underline{9}$ $16 - 7 = \underline{}$ $14 - 6 = \underline{}$

2. $16 - 8 = \underline{}$ $17 - 9 = \underline{}$ $15 - 7 = \underline{}$

3. $15 - 9 = \underline{}$ $13 - 5 = \underline{}$ $17 - 8 = \underline{}$

Complete the tables. Use the addition or subtraction rule.

4.

	−9
15	6
18	
17	
16	
14	

5.

	−8
15	
16	
13	
17	
14	

6.

	−7
14	
12	
13	
16	
15	

7.

	+7
8	
5	
9	
7	
6	

8.

	+6
5	
9	
8	
6	
7	

CHALLENGE

9. Color the same differences (green) and
the same sums (blue).

(sixteen minus eight)   (nine plus seven)   (eight plus nine)

(seventeen minus nine)   (fourteen minus seven) (eight plus eight)

Make an Addition Table

> In an addition table, the numbers in the first column and first row are addends.

1st Fill in the missing sums.

2nd Color the sum of each double red.

+	0	1	2	3	4	5	6	7	8	9
0	0	1	2	3	4	5	6	7	8	9
1	1	2	3							
2		3		5	6		8			
3	3		5		7					
4				7						13
5		6			9					
6	6		8							
7				10						
8		9								17
9										

List the other names for 13, 16, and 17.

Sum 13	9 + 4				
Sum 16	9 + 7		Sum 17		

Name _____

| 1st | Write the missing numbers in each path. |

| 2nd | Use the key to color each path. |

Key	
10 more	red
3s	blue
1 less	yellow
4s	green

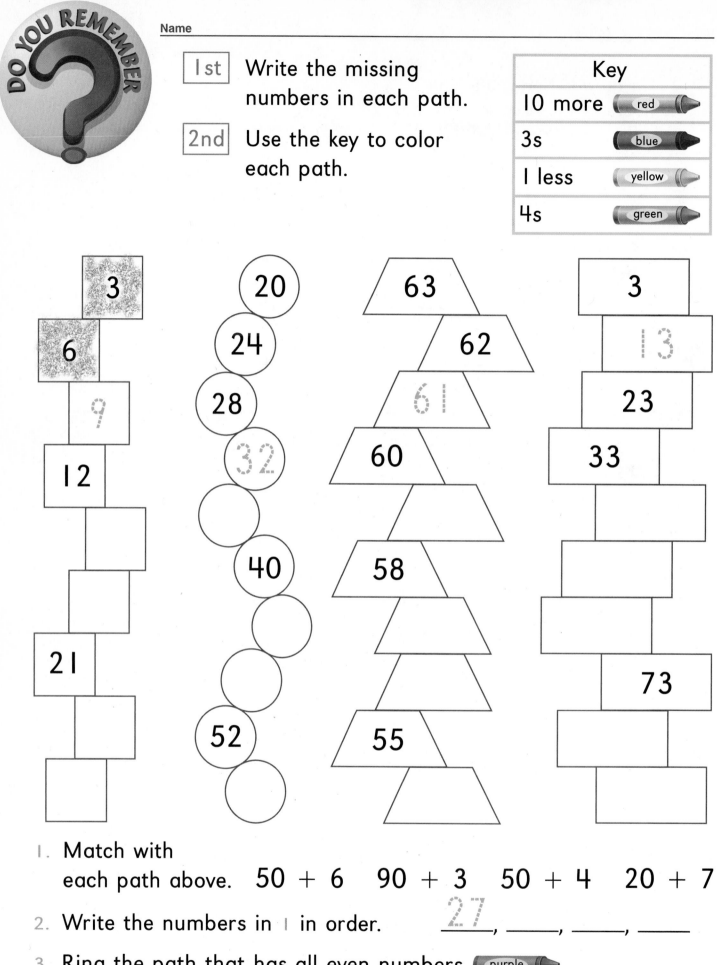

1. Match with each path above.   50 + 6   90 + 3   50 + 4   20 + 7

2. Write the numbers in 1 in order.   27 ___, ____, ____, ____

3. Ring the path that has all even numbers purple ,
   all odd numbers orange .

This page reviews the mathematical content presented in Chapter 2.

3

Name _____

To check subtraction, add.

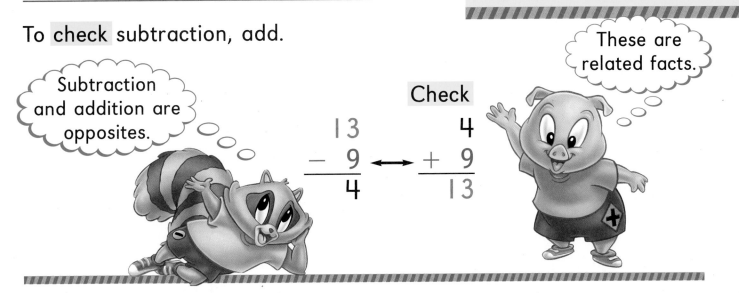

Subtraction and addition are opposites.

Check

These are related facts.

$$13 - 9 = 4 \longleftrightarrow 4 + 9 = 13$$

Subtract. Add to check.

1.
$$17 - 8 = 9$$
$$9 + 8 = 17$$

2.
$$15 - 9 = $$
$$+ 9 = 15$$

3.
$$12 - 4 = $$
$$+ 4 = 12$$

4.
$$13 - 6 = $$
$$+ 6$$

5.
$$16 - 7 = $$
$$+ 7$$

6.
$$14 - 9 = $$
$$+ 9$$

7.
$$11 - 5 = $$
$$+ $$
$$11$$

8.
$$14 - 8 = $$
$$+ $$
$$14$$

9.
$$16 - 7 = $$
$$+ $$
$$16$$

10.
$$12 - 7 = $$
$$+ $$

11.
$$13 - 5 = $$
$$+ $$

12.
$$15 - 8 = $$
$$+ $$

In 1–12 ✔ near doubles facts.

Why do addition facts end with a sum and subtraction facts begin with a sum?

3-7 Tell your child to use pennies to model 16 – 9 and to check by adding.

one hundred thirteen **113**

Subtract. Check by adding.

1. $13 - 8 = \underline{5}$ $16 - 9 = \underline{}$ $14 - 5 = \underline{}$

   $\underline{5} + 8 = 13$ $\underline{} + 9 = 16$ $\underline{} + 5 = 14$

2. $15 - 7 = \underline{}$ $17 - 9 = \underline{}$ $12 - 8 = \underline{}$

   $\underline{} + 7 = \underline{}$ $\underline{} + 9 = \underline{}$ $\underline{} + 8 = \underline{}$

3. $11 - 3 = \underline{}$ $14 - 8 = \underline{}$ $17 - 8 = \underline{}$

   $\underline{} + \underline{} = 11$ $\underline{} + \underline{} = 14$ $\underline{} + \underline{} = 17$

4. $13 - 9 = \underline{}$ $14 - 6 = \underline{}$ $12 - 9 = \underline{}$

   $\underline{} + \underline{} = \underline{}$ $\underline{} + \underline{} = \underline{}$ $\underline{} + \underline{} = \underline{}$

5. $18 - 9 = \underline{}$ $14 - 7 = \underline{}$ $16 - 8 = \underline{}$

   $\underline{} + \underline{} = \underline{}$ $\underline{} + \underline{} = \underline{}$ $\underline{} + \underline{} = \underline{}$

 CRITICAL THINKING

6. Write other names for each.

Difference 9	15 − 6	13 −	11 −	16 −	14 −	17 −

Sum 14	6 +	9 +	7 +	8 +	5 +	10 +

Difference 6	13 −	11 −	15 −	12 −	14 −	10 −

Name _____

Here are different ways I add more than 2 numbers.

Add up.
Add down.
Add doubles.
Add to make 10.

```
 5              3
 2 ---→ 7       3 ---→ 6
+6    +6       4    4        6
      13      +6   +6 ---→ +10
                             16
```

Ring the addends you add first. Find the sum.

1.
```
 6      2      6      5      1      3      4
 3      2      3      5      3      2      5
+4     +8     +7     +4     +9     +6     +3
13
```

2.
```
 3      4      3      4      5      6      6
 5      3      3      4      4      3      2
+5     +9     +8     +9     +9     +3     +6
```

3.
```
 4      3      2      1      3      3      2
 3      1      1      4      3      4      4
 2      4      3      2      3      2      2
+5     +8     +7     +8     +8     +7     +5
```

TALK IT OVER

Add up to find each sum in 1.
Is the sum still the same? Why?

3-8 Have your child write 3 + 3 + 4 + 1 vertically and find the sum by adding up or using doubles.

one hundred fifteen **115**

Find the sum. Ring doubles and tens.

1.
```
 4      4      6      2      1      3      5
 4      1      1      2      4      2      3
 1      2      2      3      1      4      5
+9     +7     +6     +9     +8     +3     +3
───    ───    ───    ───    ───    ───    ───
18
```

2.
```
 3      2      3      2      5      1      8
 2      5      5      1      0      2      2
 2      2      1      4      4      4      3
+7     +4     +9     +8     +7     +6     +3
───    ───    ───    ───    ───    ───    ───
```

3. 5 + 1 + 7 = 13   3 + 2 + 3 + 8 = ___

4. 7 + 2 + 6 = ___   1 + 1 + 5 + 6 = ___

5. 4 + 4 + 2 = ___   8 + 1 + 8 + 1 = ___

PROBLEM SOLVING

5¢   7¢   4¢

6. Tamara buys 2 🐙 and 1 🌷.

   Sommer buys 2 🌷 and 1 🐟.

   Who spends more? _____ spends more.

CHALLENGE Write the missing addends.

7. 15 = 3 + ___ + ___   8. 15 = 3 + ___ + ___

9. 15 = 3 + ___ + ___   10. 15 = 3 + ___ + ___

Subtract to answer these questions.

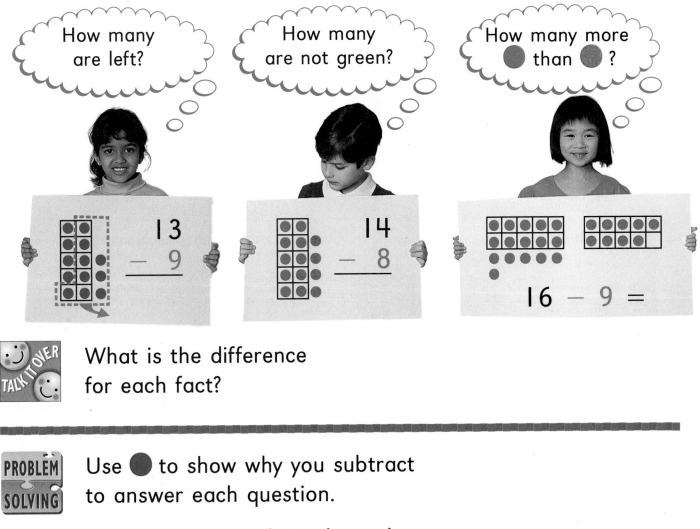

How many
are left?

How many
are not green?

How many more
● than ● ?

13
− 9

14
− 8

16 − 9 =

TALK IT OVER What is the difference
for each fact?

PROBLEM SOLVING Use ● to show why you subtract
to answer each question.

1. The 17 children in art class planned
   a party. 9 of them made signs.
   How many did not make signs?

   _____ did not make signs.

2. Thea had 14 🎈.
   She put 5 around the door.
   How many 🎈 has she left?

   She has _____ 🎈 left.

3. There are 13 girls and 8 boys in
   craft class. How many more girls
   than boys are in craft class?

   There are _____ more girls than boys.

3-9 Tell your child subtraction problems
like those above and have her/him use
counters to model them.

one hundred seventeen **117**

Here are other subtraction questions.

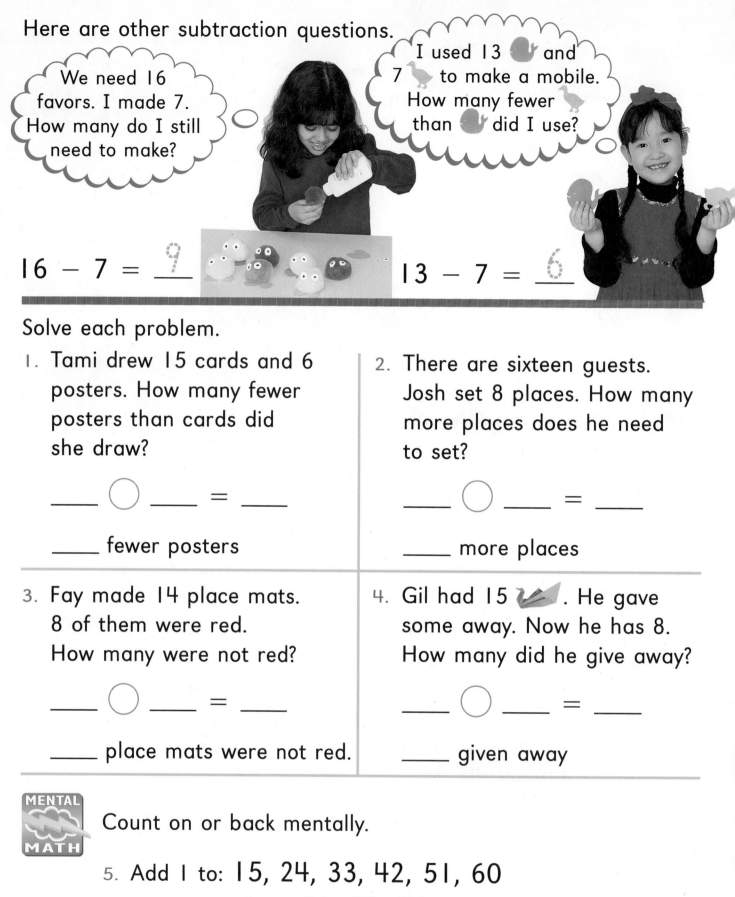

We need 16 favors. I made 7. How many do I still need to make?

I used 13 🐌 and 7 🦆 to make a mobile. How many fewer 🦆 than 🐌 did I use?

16 − 7 = _9_

13 − 7 = _6_

Solve each problem.

1. Tami drew 15 cards and 6 posters. How many fewer posters than cards did she draw?

____ ◯ ____ = ____

____ fewer posters

2. There are sixteen guests. Josh set 8 places. How many more places does he need to set?

____ ◯ ____ = ____

____ more places

3. Fay made 14 place mats. 8 of them were red. How many were not red?

____ ◯ ____ = ____

____ place mats were not red.

4. Gil had 15 🛶. He gave some away. Now he has 8. How many did he give away?

____ ◯ ____ = ____

____ given away

MENTAL MATH Count on or back mentally.

5. Add 1 to: 15, 24, 33, 42, 51, 60

6. Subtract 1 from: 91, 82, 73, 64, 55, 46

7. Add 10 to: 75, 64, 53, 42, 31, 20

8. Subtract 10 from: 91, 82, 73, 64, 55, 46

Name _____

Shana is 14 blocks from the store. She rides 9 blocks. How many more blocks does she need to ride to get there?

You can count up to find the difference.

5 blocks

0 1 2 3 4 5 6 7 8 ⑨ 10 11 12 13 ⑭ 15 16 17 18

14 − 9 = 5   **Check** Does 9 + 5 = 14? Yes.

Ring the whole and the part on the ←┼┼┼→.
Count up to find the difference. Then check.

1. 13 − 8 = __5__   **Check** 8 + 5 = 13

0 1 2 3 4 5 6 7 ⑧ 9 10 11 12 ⑬ 14 15 16 17 18

2. 16 − 7 = ___

0 1 2 3 4 5 6 7 8 9 10 11 12 13 14 15 16 17 18

3. 15 − 9 = ___

0 1 2 3 4 5 6 7 8 9 10 11 12 13 14 15 16 17 18

4. 14 − 8 = ___

0 1 2 3 4 5 6 7 8 9 10 11 12 13 14 15 16 17 18

You can count up to find
how many fewer or how many more.

Darla has 17 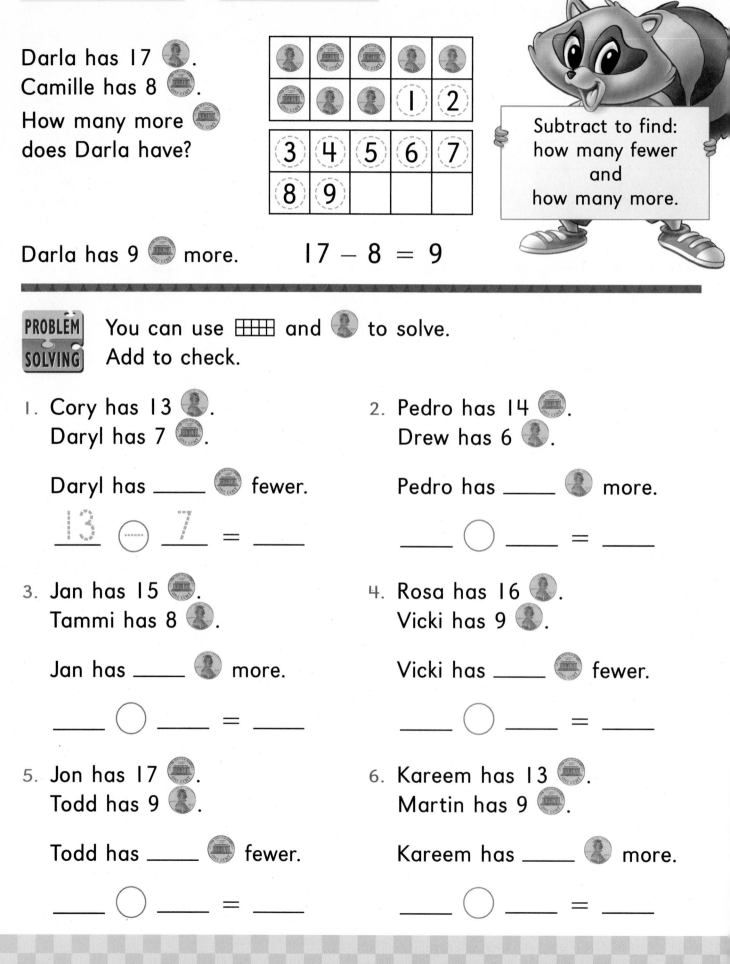.
Camille has 8.
How many more
does Darla have?

			1	2

3	4	5	6	7
8	9			

Subtract to find:
how many fewer
and
how many more.

Darla has 9 more. $17 - 8 = 9$

PROBLEM SOLVING   You can use ⊞ and 🪙 to solve.
Add to check.

1. Cory has 13.
   Daryl has 7.

   Daryl has ____ fewer.

   $13 \bigcirc 7 = \underline{\quad}$

2. Pedro has 14.
   Drew has 6.

   Pedro has ____ more.

   $\underline{\quad} \bigcirc \underline{\quad} = \underline{\quad}$

3. Jan has 15.
   Tammi has 8.

   Jan has ____ more.

   $\underline{\quad} \bigcirc \underline{\quad} = \underline{\quad}$

4. Rosa has 16.
   Vicki has 9.

   Vicki has ____ fewer.

   $\underline{\quad} \bigcirc \underline{\quad} = \underline{\quad}$

5. Jon has 17.
   Todd has 9.

   Todd has ____ fewer.

   $\underline{\quad} \bigcirc \underline{\quad} = \underline{\quad}$

6. Kareem has 13.
   Martin has 9.

   Kareem has ____ more.

   $\underline{\quad} \bigcirc \underline{\quad} = \underline{\quad}$

Name _____

There are 8 🐦 in the birdhouse.
How many more 🐦 are needed
to fill all 14 homes?

$$8 + \boxed{} = 14$$

Count on to find
the missing addend.

$$8 + \underline{6} = 14$$

8, 9, 10, 11, 12, 13, 14

6 more 🐦 are needed.

TALK IT OVER Is the missing addend a part or a whole?
What subtraction fact can you use to find
the missing addend?

Find the missing addend.

1. $7 + \boxed{} = 16$ $8 + \boxed{} = 15$ $9 + \boxed{} = 14$

2. $9 + \boxed{} = 13$ $8 + \boxed{} = 13$ $7 + \boxed{} = 13$

3. $9 + \boxed{} = 18$ $8 + \boxed{} = 16$ $7 + \boxed{} = 14$

4. $8 + \boxed{} = 17$ $7 + \boxed{} = 15$ $6 + \boxed{} = 13$

5. $4 + \boxed{} = 13$ $5 + \boxed{} = 14$ $6 + \boxed{} = 15$

6. $9 + \boxed{} = 17$ $9 + \boxed{} = 16$ $9 + \boxed{} = 15$

MATH JOURNAL 7. Each row in 1—6 makes a pattern. Choose 2 rows
and write what comes next in each.

3-11 Have your child show how she/he counts on
by tallying to find the missing addends in
exercises 1 and 2.   one hundred twenty-one **121**

You can tally as you count on to find missing addends.

1.
$$\begin{array}{r} 4 \quad |||| \\ +\ 9 \quad \cancel{||||}\ |||| \\ \hline 13 \end{array}$$

2.
$$\begin{array}{r} 8 \quad \cancel{||||}\ ||| \\ +\ \square \\ \hline 14 \end{array}$$

3.
$$\begin{array}{r} \square \\ +\ 5 \quad \cancel{||||} \\ \hline 13 \end{array}$$

4.
$$\begin{array}{r} 7 \quad \cancel{||||}\ || \\ +\ \square \\ \hline 14 \end{array}$$

5.
$$\begin{array}{r} 9 \quad \cancel{||||}\ |||| \\ +\ \square \\ \hline 16 \end{array}$$

6.
$$\begin{array}{r} \square \\ +\ 6 \quad \cancel{||||}\ | \\ \hline 13 \end{array}$$

7.
$$\begin{array}{r} \square \\ +\ 7 \quad \cancel{||||}\ || \\ \hline 15 \end{array}$$

8.
$$\begin{array}{r} 9 \quad \cancel{||||}\ |||| \\ +\ \square \\ \hline 18 \end{array}$$

9.
$$\begin{array}{r} \square \\ +\ 6 \quad \cancel{||||}\ | \\ \hline 15 \end{array}$$

SHARE YOUR THINKING — How does knowing related facts help you find a missing addend?

PROBLEM SOLVING

10. Mia has 17 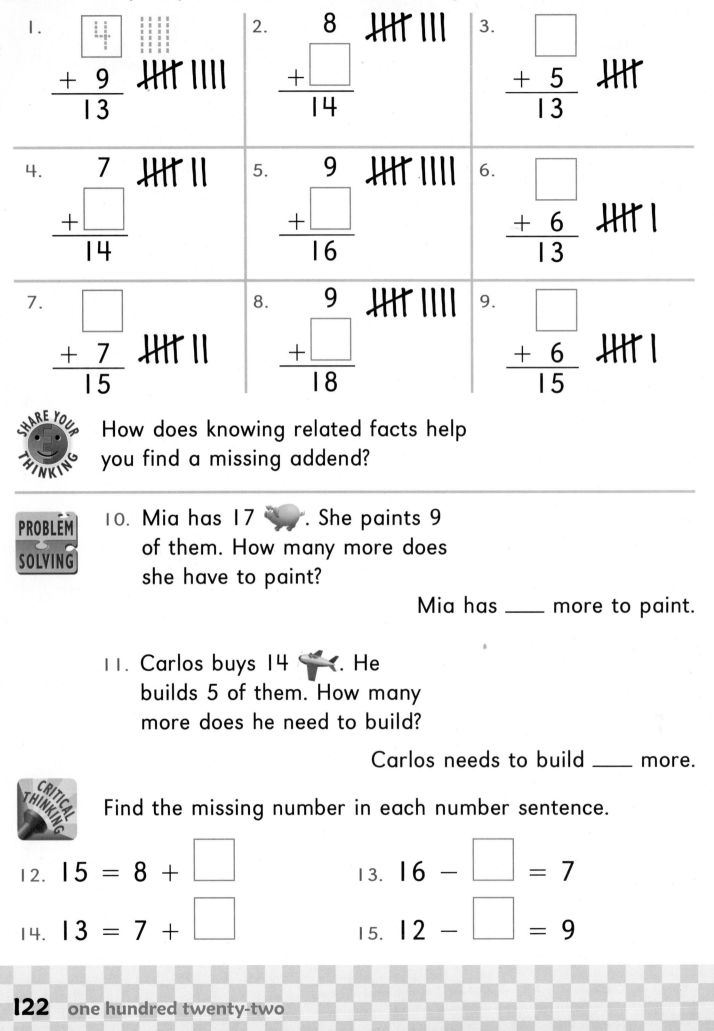. She paints 9 of them. How many more does she have to paint?

Mia has ____ more to paint.

11. Carlos buys 14 ✈. He builds 5 of them. How many more does he need to build?

Carlos needs to build ____ more.

CRITICAL THINKING — Find the missing number in each number sentence.

12. $15 = 8 + \square$

13. $16 - \square = 7$

14. $13 = 7 + \square$

15. $12 - \square = 9$

You can write addition and subtraction sentences for this pattern.

$4 + 9 = 13$ 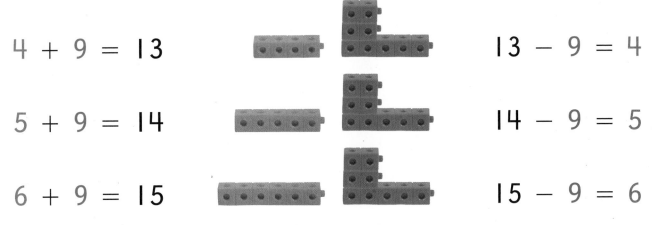 $13 - 9 = 4$

$5 + 9 = 14$ $14 - 9 = 5$

$6 + 9 = 15$ $15 - 9 = 6$

 What happens to the sum when one addend stays the same and the other increases? What happens to the difference when the whole increases by 1 and the part taken away stays the same?

Add. Write the next number sentence.

1. $6 + 6 = \underline{12}$	2. $9 + 8 = \underline{}$	3. $9 + 4 = \underline{}$
$7 + 7 = \underline{14}$	$7 + 8 = \underline{}$	$8 + 5 = \underline{}$
$8 + 8 = \underline{}$	$5 + 8 = \underline{}$	$7 + 6 = \underline{}$
$\underline{9} + \underline{} = \underline{}$	$\underline{} + \underline{} = \underline{}$	$\underline{} + \underline{} = \underline{}$
4. $12 - 6 = \underline{}$	5. $17 - 8 = \underline{}$	6. $13 - 9 = \underline{}$
$14 - 7 = \underline{}$	$15 - 8 = \underline{}$	$13 - 8 = \underline{}$
$16 - 8 = \underline{}$	$13 - 8 = \underline{}$	$13 - 7 = \underline{}$
$\underline{} - \underline{} = \underline{}$	$\underline{} - \underline{} = \underline{}$	$\underline{} - \underline{} = \underline{}$

 In 1–6 which patterns are related? How can you tell?

3-12   Tell your child to make up one addition and one subtraction pattern similar to those above.

one hundred twenty-three **123**

Write the missing numbers in each pattern.

Watch for + and −.

1.
$$\begin{array}{r} 5 \\ +5 \\ \hline 10 \end{array}$$
$$\begin{array}{r} 5 \\ +6 \\ \hline 11 \end{array}$$
$$\begin{array}{r} 5 \\ +7 \\ \hline \end{array}$$
$$\begin{array}{r} \\ +8 \\ \hline \end{array}$$
$$\begin{array}{r} \\ + \\ \hline \end{array}$$
$$\begin{array}{r} \\ + \\ \hline \end{array}$$

2.
$$\begin{array}{r} 9 \\ +7 \\ \hline \end{array}$$
$$\begin{array}{r} 8 \\ +7 \\ \hline \end{array}$$
$$\begin{array}{r} 7 \\ +7 \\ \hline \end{array}$$
$$\begin{array}{r} 6 \\ + \\ \hline \end{array}$$
$$\begin{array}{r} \\ + \\ \hline \end{array}$$
$$\begin{array}{r} \\ + \\ \hline \end{array}$$

3.
$$\begin{array}{r} 5 \\ +9 \\ \hline \end{array}$$
$$\begin{array}{r} 6 \\ +8 \\ \hline \end{array}$$
$$\begin{array}{r} 7 \\ +7 \\ \hline \end{array}$$
$$\begin{array}{r} \\ +6 \\ \hline \end{array}$$
$$\begin{array}{r} \\ + \\ \hline \end{array}$$
$$\begin{array}{r} \\ + \\ \hline \end{array}$$

4.
$$\begin{array}{r} 12 \\ -3 \\ \hline \end{array}$$
$$\begin{array}{r} 12 \\ -4 \\ \hline \end{array}$$
$$\begin{array}{r} 12 \\ -5 \\ \hline \end{array}$$
$$\begin{array}{r} \\ -6 \\ \hline \end{array}$$
$$\begin{array}{r} \\ - \\ \hline \end{array}$$
$$\begin{array}{r} \\ - \\ \hline \end{array}$$

5.
$$\begin{array}{r} 16 \\ -7 \\ \hline \end{array}$$
$$\begin{array}{r} 15 \\ -6 \\ \hline \end{array}$$
$$\begin{array}{r} 14 \\ -5 \\ \hline \end{array}$$
$$\begin{array}{r} 13 \\ - \\ \hline \end{array}$$
$$\begin{array}{r} \\ - \\ \hline \end{array}$$
$$\begin{array}{r} \\ - \\ \hline \end{array}$$

6.
$$\begin{array}{r} 8 \\ -3 \\ \hline \end{array}$$
$$\begin{array}{r} 10 \\ -4 \\ \hline \end{array}$$
$$\begin{array}{r} 12 \\ -5 \\ \hline \end{array}$$
$$\begin{array}{r} 14 \\ - \\ \hline \end{array}$$
$$\begin{array}{r} \\ - \\ \hline \end{array}$$
$$\begin{array}{r} \\ - \\ \hline \end{array}$$

CRITICAL THINKING

7. Double any number. Is the sum odd or even?

_____

8. Add 9 to an odd number. Is the sum odd or even?

_____

Name

A fact family uses the same numbers for the parts and for the whole.

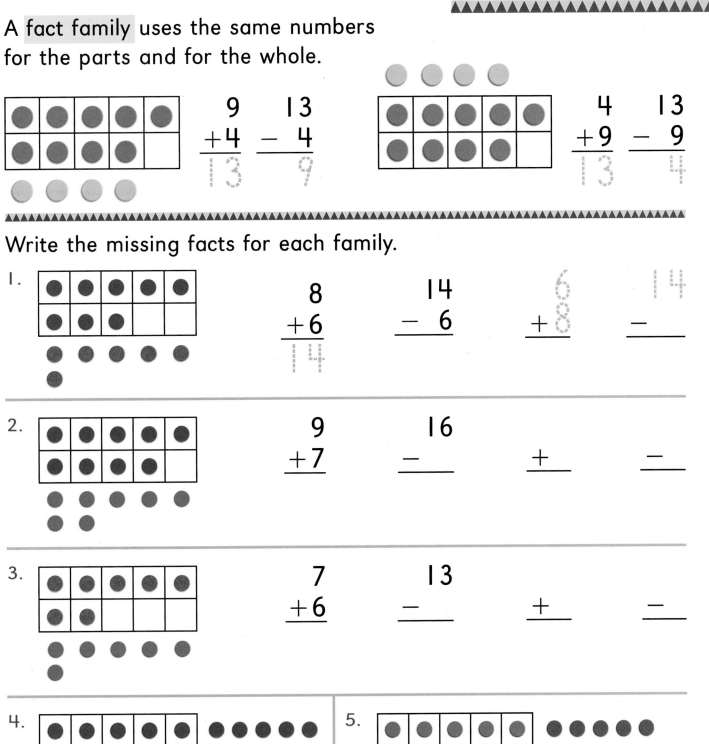

$$9 \qquad 13$$
$$+4 \qquad -\ 4$$
$$\overline{13} \qquad \overline{9}$$

$$4 \qquad 13$$
$$+9 \qquad -\ 9$$
$$\overline{13} \qquad \overline{4}$$

Write the missing facts for each family.

1.

$$8 \qquad 14 \qquad 6 \qquad 14$$
$$+6 \qquad -\ 6 \qquad +\ 8 \qquad -\ \_$$
$$\overline{14}$$

2.

$$9 \qquad 16$$
$$+7 \qquad -\ \_ \qquad +\ \_ \qquad -\ \_$$

3.

$$7 \qquad 13$$
$$+6 \qquad -\ \_ \qquad +\ \_ \qquad -\ \_$$

4.

$$9 + 9 = \underline{\ \ }$$

$$18 - \underline{\ \ } = \underline{\ \ }$$

5.

$$7 + 7 = \underline{\ \ }$$

$$\underline{\ \ } - \underline{\ \ } = \underline{\ \ }$$

All fact families do not have 4 facts. Why?

3-13   Ask your child to explain the difference between fact families with 2 facts and those with 4 facts and to give an example of each.

one hundred twenty-five **125**

✘ the number that does not belong.
Write the fact family.

1.

8	7
1̶7̶	15

$$\begin{array}{r} 8 \\ +7 \\ \hline 15 \end{array}$$ $$\begin{array}{r} 15 \\ -\ 7 \\ \hline \end{array}$$ $+$ _____ $-$ _____

2.

14	6
15	9

6

$+$ _____ $-$ _____ $+$ _____ $-$ _____

3.

5	13
8	14

$+$ _____ $-$ _____ $+$ _____ $-$ _____

4.

8	15
17	9

$+$ _____ $-$ _____ $+$ _____ $-$ _____

5.

13	9
5	14

$+$ _____ $-$ _____ $+$ _____ $-$ _____

PROBLEM SOLVING   6. My two addends are the same. Their sum is 12. What are my addends? Complete my family of facts.

Fact Family

$+$ _____ $-$ _____

Addends: _____, _____

Name

1. **Read**
There are 9 big 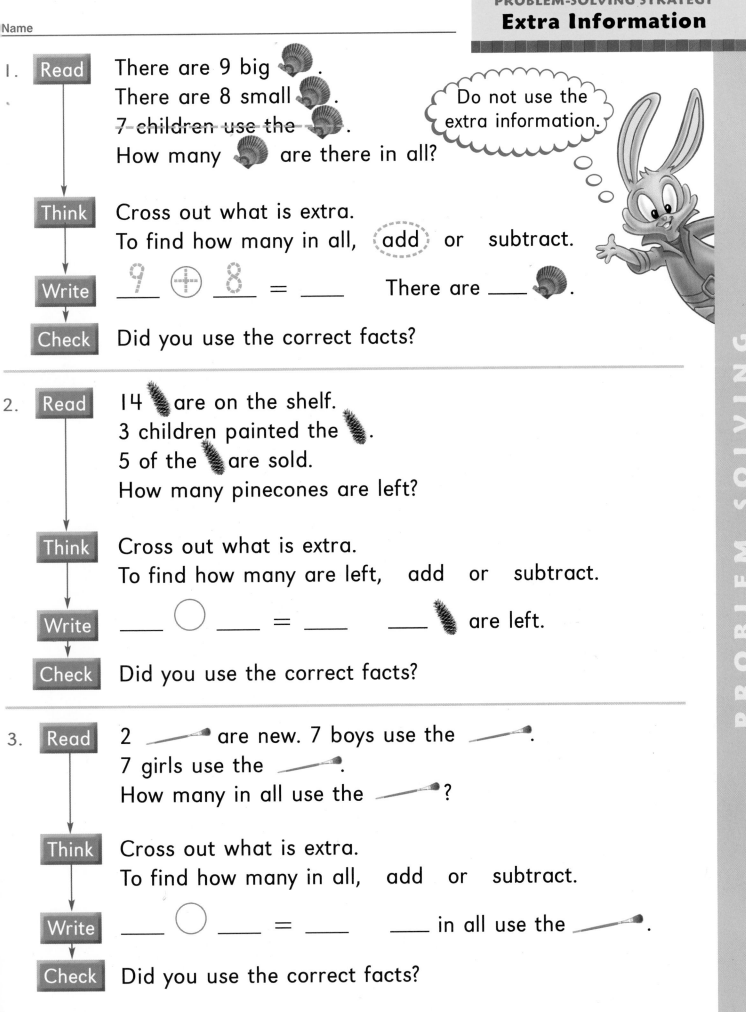.
There are 8 small .
7 children use the .
How many are there in all?

Do not use the extra information.

Think
Cross out what is extra.
To find how many in all, add or subtract.

Write
9 ⊕ 8 = ___   There are ___ .

Check
Did you use the correct facts?

2. **Read**
14 are on the shelf.
3 children painted the .
5 of the are sold.
How many pinecones are left?

Think
Cross out what is extra.
To find how many are left, add or subtract.

Write
___ ◯ ___ = ___   ___ are left.

Check
Did you use the correct facts?

3. **Read**
2 are new. 7 boys use the .
7 girls use the .
How many in all use the ?

Think
Cross out what is extra.
To find how many in all, add or subtract.

Write
___ ◯ ___ = ___   ___ in all use the .

Check
Did you use the correct facts?

PROBLEM SOLVING

3-14 Have your child explain how he/she determined the extra information in each problem above.

one hundred twenty-seven **127**

Cross out what is extra.

4. Tonya made 9 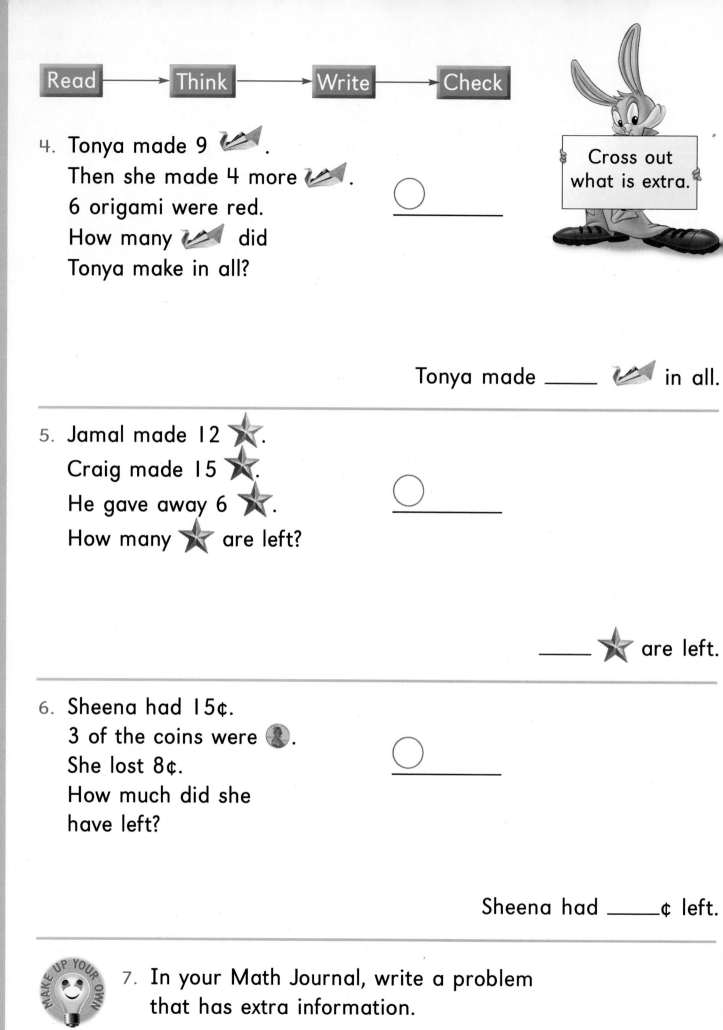.
   Then she made 4 more.
   6 origami were red.
   How many did
   Tonya make in all?

   ◯ _____

   Tonya made _____ in all.

5. Jamal made 12 ★.
   Craig made 15 ★.
   He gave away 6 ★.
   How many ★ are left?

   ◯ _____

   _____ ★ are left.

6. Sheena had 15¢.
   3 of the coins were 🪙.
   She lost 8¢.
   How much did she
   have left?

   ◯ _____

   Sheena had _____¢ left.

7. In your Math Journal, write a problem
   that has extra information.

PROBLEM SOLVING

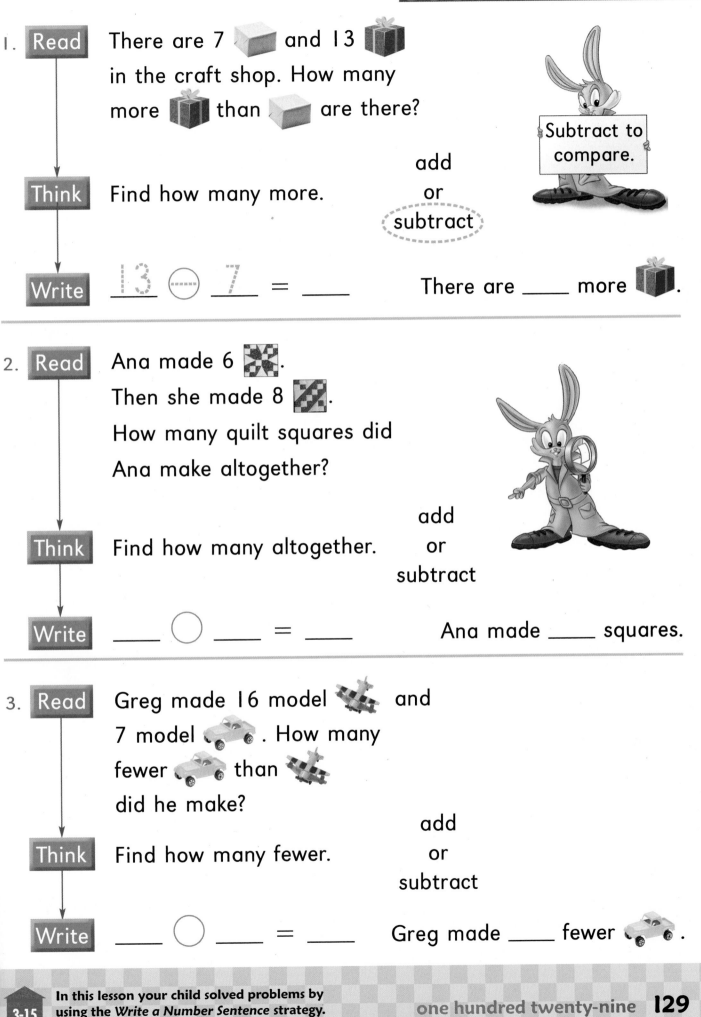

1. **Read** There are 7 ▢ and 13 🎁 in the craft shop. How many more 🎁 than ▢ are there?

Think Find how many more.

add
or
(subtract)

Subtract to compare.

Write 13 ⊖ 7 = ____   There are ____ more 🎁.

2. **Read** Ana made 6 ▨.
Then she made 8 ▨.
How many quilt squares did
Ana make altogether?

Think Find how many altogether.

add
or
subtract

Write ____ ◯ ____ = ____   Ana made ____ squares.

3. **Read** Greg made 16 model ✈ and
7 model 🚗. How many
fewer 🚗 than ✈
did he make?

Think Find how many fewer.

add
or
subtract

Write ____ ◯ ____ = ____   Greg made ____ fewer 🚗.

PROBLEM SOLVING

Use a strategy you have learned.

STRATEGY FILE

Extra Information
Guess and Test
Use a Graph
Write a Number Sentence

4. My number is between 76 and 81. The sum of its digits is 15. What is my number?

77, 78, 79, 80

_____ ⃝ _____ = 15

My number is _____.

5. Daria bought 13 silk flowers. 6 were 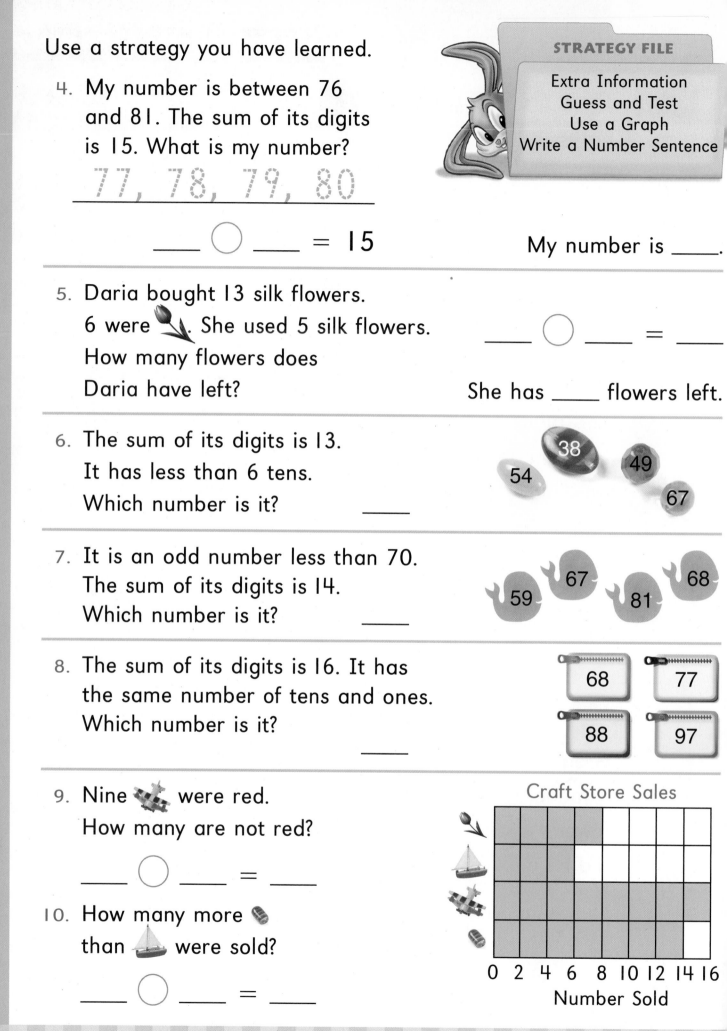. She used 5 silk flowers. How many flowers does Daria have left?

_____ ⃝ _____ = _____

She has _____ flowers left.

6. The sum of its digits is 13. It has less than 6 tens. Which number is it? _____

54   38   49   67

7. It is an odd number less than 70. The sum of its digits is 14. Which number is it? _____

59   67   81   68

8. The sum of its digits is 16. It has the same number of tens and ones. Which number is it? _____

68   77   88   97

9. Nine ✈ were red. How many are not red?

_____ ⃝ _____ = _____

10. How many more 🚗 than ⛵ were sold?

_____ ⃝ _____ = _____

Craft Store Sales

0 2 4 6 8 10 12 14 16
Number Sold

Find the sum.

1.
$$\begin{array}{r} 8 \\ +7 \\ \hline 15 \end{array}$$
$$\begin{array}{r} 4 \\ +9 \\ \hline \end{array}$$
$$\begin{array}{r} 6 \\ +8 \\ \hline \end{array}$$
$$\begin{array}{r} 9 \\ +7 \\ \hline \end{array}$$
$$\begin{array}{r} 8 \\ +5 \\ \hline \end{array}$$
$$\begin{array}{r} 9 \\ +9 \\ \hline \end{array}$$

2.
$$\begin{array}{r} 1 \\ 5 \\ +7 \\ \hline \end{array}$$
$$\begin{array}{r} 2 \\ 6 \\ +9 \\ \hline \end{array}$$
$$\begin{array}{r} 6 \\ 1 \\ +9 \\ \hline \end{array}$$
$$\begin{array}{r} 2 \\ 2 \\ 3 \\ +7 \\ \hline \end{array}$$
$$\begin{array}{r} 3 \\ 3 \\ 3 \\ +6 \\ \hline \end{array}$$
$$\begin{array}{r} 4 \\ 2 \\ 2 \\ +6 \\ \hline \end{array}$$

Find the difference.

3.
$$\begin{array}{r} 14 \\ -5 \\ \hline \end{array}$$
$$\begin{array}{r} 15 \\ -6 \\ \hline \end{array}$$
$$\begin{array}{r} 14 \\ -7 \\ \hline \end{array}$$
$$\begin{array}{r} 15 \\ -8 \\ \hline \end{array}$$
$$\begin{array}{r} 13 \\ -8 \\ \hline \end{array}$$
$$\begin{array}{r} 14 \\ -9 \\ \hline \end{array}$$

4.
$$\begin{array}{r} 17 \\ -9 \\ \hline \end{array}$$
$$\begin{array}{r} 13 \\ -4 \\ \hline \end{array}$$
$$\begin{array}{r} 14 \\ -6 \\ \hline \end{array}$$
$$\begin{array}{r} 16 \\ -9 \\ \hline \end{array}$$
$$\begin{array}{r} 13 \\ -6 \\ \hline \end{array}$$
$$\begin{array}{r} 18 \\ -9 \\ \hline \end{array}$$

Find the the missing addends.

5. $7 + \boxed{} = 13$

6. $\boxed{} + 5 = 14$

Write the fact family.

7. $8 + 8 =$ ___

_____

PROBLEM SOLVING

8. There are 15 🫙.
6 jars are white.
Nine jars break. How many 🫙 are left?

___ 🫙

9. Ian made 9 posters.
Sara made 8 posters.
How many posters did they make in all?

___ posters

REINFORCEMENT

Add or subtract from
left to right.

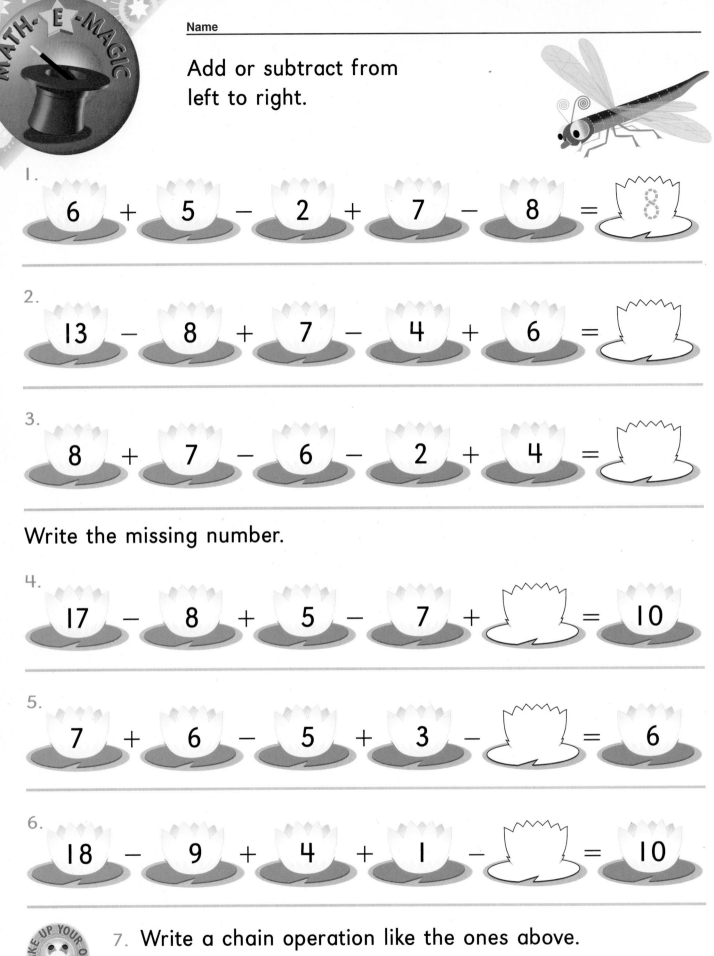

1. 6 + 5 − 2 + 7 − 8 = 8

2. 13 − 8 + 7 − 4 + 6 =

3. 8 + 7 − 6 − 2 + 4 =

Write the missing number.

4. 17 − 8 + 5 − 7 + ▢ = 10

5. 7 + 6 − 5 + 3 − ▢ = 6

6. 18 − 9 + 4 + 1 − ▢ = 10

7. Write a chain operation like the ones above.
Then cover 1 number.

This page extends your child's
understanding of addition and subtraction
facts using chain operations.

3

Name _____

1. Model to show how you count on or make 10 to add.

$$8 + 5$$ $$6 + 9$$ $$9 + 4$$ $$8 + 8$$

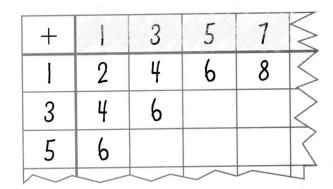

6 7 8 9 10 13

2. Find the missing numbers.
   Explain your thinking.

   $$7 + \boxed{} = 15 \qquad 8 + \boxed{} = 14$$

3. Subtract. Model the pattern. Write the next 2 facts.

   $$18 - 9$$ $$16 - 9$$ $$14 - 9$$ $$-\!\!-\!\!-$$ $$-\!\!-\!\!-$$

PORTFOLIO   Choose one of these projects.
Use a separate sheet of paper.

4. Make 4 number lines.
   Use one for each number
   sentence in the fact family
   for 17.

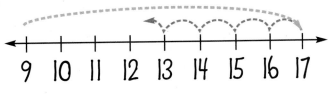

9 10 11 12 13 14 15 16 17

5. Make up an Addition Table
   like this one to show the sum
   of odd numbers from 1 to 9.

+	1	3	5	7
1	2	4	6	8
3	4	6		
5	6			

Name _____

Add.

1.
$$\begin{array}{r} 8 \\ +9 \\ \hline \end{array} \qquad \begin{array}{r} 8 \\ +6 \\ \hline \end{array} \qquad \begin{array}{r} 7 \\ +7 \\ \hline \end{array} \qquad \begin{array}{r} 6 \\ +8 \\ \hline \end{array} \qquad \begin{array}{r} 4 \\ 4 \\ +7 \\ \hline \end{array} \qquad \begin{array}{r} 5 \\ 3 \\ +5 \\ \hline \end{array}$$

2.
$$\begin{array}{r} 9 \\ +5 \\ \hline \end{array} \qquad \begin{array}{r} 7 \\ +9 \\ \hline \end{array} \qquad \begin{array}{r} 5 \\ +9 \\ \hline \end{array} \qquad \begin{array}{r} 8 \\ +7 \\ \hline \end{array} \qquad \begin{array}{r} 8 \\ +8 \\ \hline \end{array} \qquad \begin{array}{r} 4 \\ +9 \\ \hline \end{array}$$

3. $4 + 3 + 2 + 9 = $ _____ $\qquad 2 + 1 + 9 + 1 = $ _____

Subtract.

4.
$$\begin{array}{r} 13 \\ -9 \\ \hline \end{array} \qquad \begin{array}{r} 17 \\ -8 \\ \hline \end{array} \qquad \begin{array}{r} 13 \\ -4 \\ \hline \end{array} \qquad \begin{array}{r} 18 \\ -9 \\ \hline \end{array} \qquad \begin{array}{r} 13 \\ -7 \\ \hline \end{array} \qquad \begin{array}{r} 14 \\ -5 \\ \hline \end{array}$$

5.
$$\begin{array}{r} 15 \\ -9 \\ \hline \end{array} \qquad \begin{array}{r} 15 \\ -7 \\ \hline \end{array} \qquad \begin{array}{r} 14 \\ -9 \\ \hline \end{array} \qquad \begin{array}{r} 16 \\ -8 \\ \hline \end{array} \qquad \begin{array}{r} 14 \\ -7 \\ \hline \end{array} \qquad \begin{array}{r} 13 \\ -5 \\ \hline \end{array}$$

SECOND LOOK In 1–3 ring related facts in each row.
In 4 and 5 ring doubles facts.

Subtract. Add to check.

6. $15 - 8 = $ _____ $\qquad 13 - 8 = $ _____ $\qquad 16 - 9 = $ _____

_____ $\qquad$ _____ $\qquad$ _____

PROBLEM SOLVING

7. Lou made 14 ⭐.
and 8 🛩. How
many more ⭐ than
🛩 did Lou make?

_____ ⭐ more

8. Selma bought 7 ◻.
Robert bought 8 ✏.
Téa bought 6 ◻.
How many ◻ did
they buy in all?

_____ ◻ in all

This page is a formal assessment of your child's understanding of the content presented in Chapter 3.

3

Name _____

Mark the ○ for your answer.

Listening Section

Favorite Animals

tiger							
elephant							
monkey							

0   2   4   6   8   10   12   14
Number of Children

Ⓐ ○ tiger
   ○ monkey
   ○ elephant

Ⓑ ○ 6 children
   ○ 4 children
   ○ 3 children

1. How many ▪?

   ○ twelve   ○ fifteen
   ○ twenty   ○ thirteen

2. Which place is red?

   ○ twelfth   ○ eleventh
   ○ tenth   ○ eighth

3. Which shows a count by 4s pattern?

   ○ 10, 20, 30,...   ○ 12, 15, 18,...
   ○ 12, 16, 20,...   ○ 12, 14, 16,...

4. What is the number word?

   ○ ninety
   ○ eighty
   ○ seventy
   ○ sixty

5. Find the missing addend.

   $7 + \boxed{} = 11$

   18   16   3   4
   ○   ○   ○   ○

6. What is between?

   29, ___, 31

   30   32   28   40
   ○   ○   ○   ○

7. Find the missing number.

   $9 - \boxed{} = 9$ $9 + \boxed{} = 9$

   18   0   1   9
   ○   ○   ○   ○

8. Which is a sum of 10?

   $\begin{array}{r}5\\+3\\\hline\end{array}$ $\begin{array}{r}2\\+7\\\hline\end{array}$ $\begin{array}{r}1\\+9\\\hline\end{array}$ $\begin{array}{r}4\\+7\\\hline\end{array}$
   ○   ○   ○   ○

REINFORCEMENT

Mark the ◯ for your answer.

9. Find the sum.

$$5 \\ +7$$

- ◯ 12
- ◯ 14
- ◯ 15
- ◯ 17

10. Add.

$$4 \\ 5 \\ +4$$

- ◯ 8
- ◯ 9
- ◯ 11
- ◯ 13

11. Check by adding.

$$15 - 6 = \underline{\quad}$$

$$\begin{array}{cccc} 6 & 9 & 15 & 11 \\ +8 & +6 & +\ 6 & +\ 4 \end{array}$$

◯ ◯ ◯ ◯

12. Find the difference.

$$18 - 9 = \underline{\quad}$$

11	10	9	8
◯	◯	◯	◯

13. Subtract.

$$15¢ - 8¢ = \underline{\quad}$$

6¢	7¢	8¢	9¢
◯	◯	◯	◯

14. 5 tens = _____

4	14	40	50
◯	◯	◯	◯

15. Compare.

14 ◯ 17

>	<	=	cannot tell
◯	◯	◯	◯

16. How many tens and ones?

63 = ___

- ◯ 6 tens 3 ones
- ◯ 9 tens 0 ones
- ◯ 3 tens 6 ones
- ◯ 6 tens 0 ones

17.

```
<----|++++++++++|++++++++++|---->
    10          20          30
```

23 is closer to ____.

10	20	30	cannot tell
◯	◯	◯	◯

18. Nine 🐯 are in the field.
4 of them are running.
The rest are eating.
How many 🐯 are eating?

6	5	4	3
◯	◯	◯	◯

19. The zoo has 14 🐵, 9 🦒
and 6 🐯. How many more
🐵 than 🐯 are there?

12	9	8	7
◯	◯	◯	◯

Money and Time

TIME TRAVEL
MUSEUM

Hours: 11:00 A.M.
to 9:00 P.M.

MONEY OF THE
20TH CENTURY

CRITICAL
THINKING

On November 1, you save 5 pennies.
On November 2, you save 10 pennies.
On November 3, you save 15 pennies.
If this pattern continues, when
will you have more than 100 pennies?

137

For more information about Chapter 4, visit the Family Information Center at **www.sadlier-oxford.com**

Internet

Dear Family,

Today your child began Chapter 4. As your child studies about money and time, you may want to read the poem below, which was read in class. Have your child talk about some of the math ideas pictured on page 137.

Look for the 🏠 at the bottom of each skills lesson. The suggestion on the page gives you an opportunity to improve your child's understanding of math. You may want to have pennies, nickels, dimes, and quarters available for your child to use throughout the chapter.

Home Activity

Money and Time

As your child studies money, cut out supermarket ads for various items that cost under three dollars. Have your child tell what bills and coins he/she would use to buy the items.

As your child studies time, show a time to the hour on a paper clock. Then tell a story about what happens at that time, including how long the activity takes. Pass the clock to your child and ask him/her to adjust it to show the new time and to continue the story, including how much time passes. Start with activities on the hour and half hour and adjust the clock as your child learns more about how to read time.

Home Reading Connection

How Many Seconds in a Minute?

How many seconds in a minute?
Sixty, and no more in it.

How many minutes in an hour?
Sixty for sun and shower.

How many hours in a day?
Twenty-four for work and play.

How many days in a week?
Seven both to hear and speak.

How many weeks in a month?
Four, as the swift moon runn'th.

How many months in a year?
Twelve the almanack makes clear.

How many years in an age?
One hundred says the sage.

How many ages in time?
No one knows the rhyme.

Christina G. Rossetti

Name _____

Learn the Time Trippers' Dance.
Estimate how many
times in 1 minute
you can do each
step. Then check
your estimate.

> We can
> do step 1 for
> one minute.

How long is a minute?

skip count by 5

1 minute = _____ seconds

To **estimate**,
guess about
how long.

Dance Step	Estimate	Actual
1. With a partner, swing 2 times 'round.		
2. With your right hand, touch the ground.		
3. Take 2 slides forward without a sound.		
4. From 10 to 0, all count down.		

FINDING TOGETHER

5. With a partner, make up a dance step.
   Estimate how many times in one minute
   you can do the step, then dance to
   check your estimate.

You can put this in your Math Portfolio.

These coins have a picture of a president of the United States on the side we call heads.

Solve and fill in the letter code to name each president.

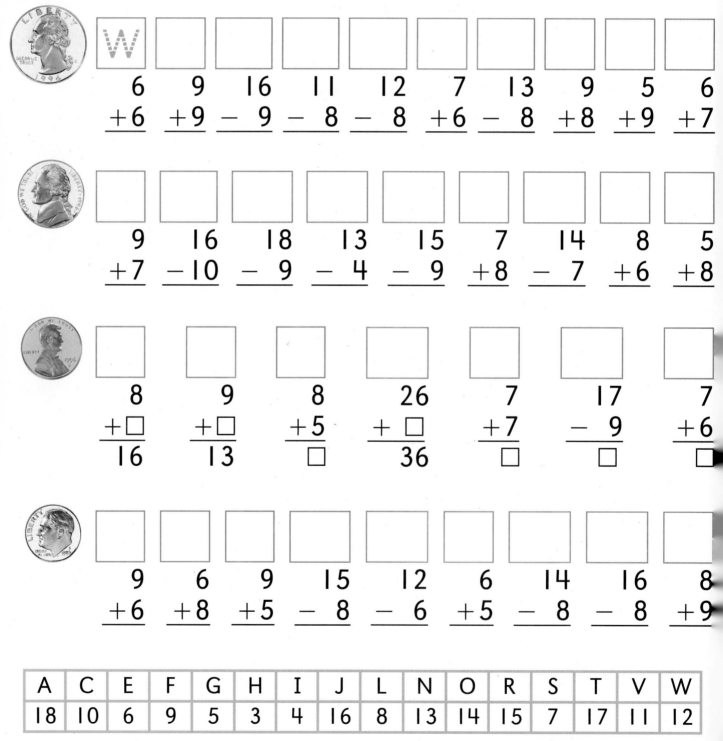

W									
6 +6	9 +9	16 − 9	11 − 8	12 − 8	7 +6	13 − 8	9 +8	5 +9	6 +7

9 +7	16 −10	18 − 9	13 − 4	15 − 9	7 +8	14 − 7	8 +6	5 +8

8 +□ 16	9 +□ 13	8 +5 □	26 + □ 36	7 +7 □	17 − 9 □	7 +6 □

9 +6	6 +8	9 +5	15 − 8	12 − 6	6 +5	14 − 8	16 − 8	8 +9

A	C	E	F	G	H	I	J	L	N	O	R	S	T	V	W
18	10	6	9	5	3	4	16	8	13	14	15	7	17	11	12

You can put this in your Math Portfolio.

Name _____

I penny   I nickel   I dime

I sort coins before I count.

I cent   5 cents   10 cents ← value
I¢   5¢   10¢

10¢   20¢   25¢   30¢   35¢   36¢ ← amount

Count on. Write the amounts.

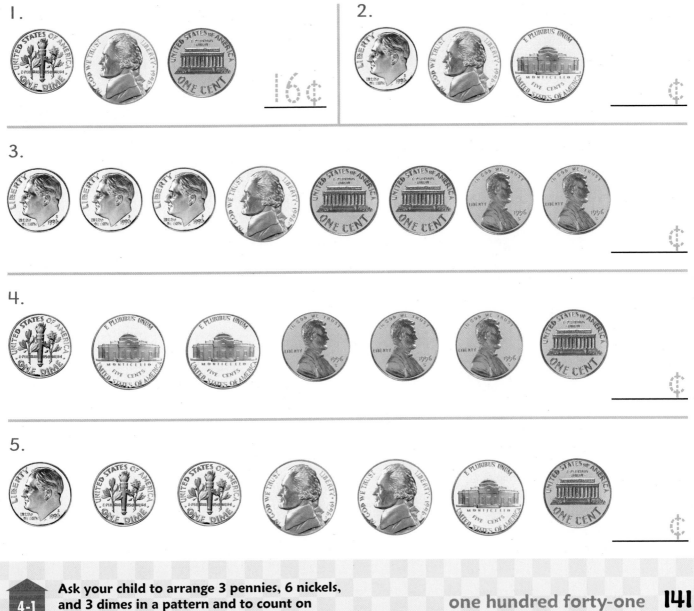

1. _16¢_

2. _____ ¢

3. _____ ¢

4. _____ ¢

5. _____ ¢

Write the amount.

1.  _____ ¢

2.  _____ ¢

3.  _____ ¢

SECOND LOOK In 1–3 ✔ the greatest amount.

Ring the amount.

4. 45¢

5. 51¢

6. 24¢

FINDING TOGETHER 7. List 3 ways to show 17¢ and 3 ways to show 35¢.

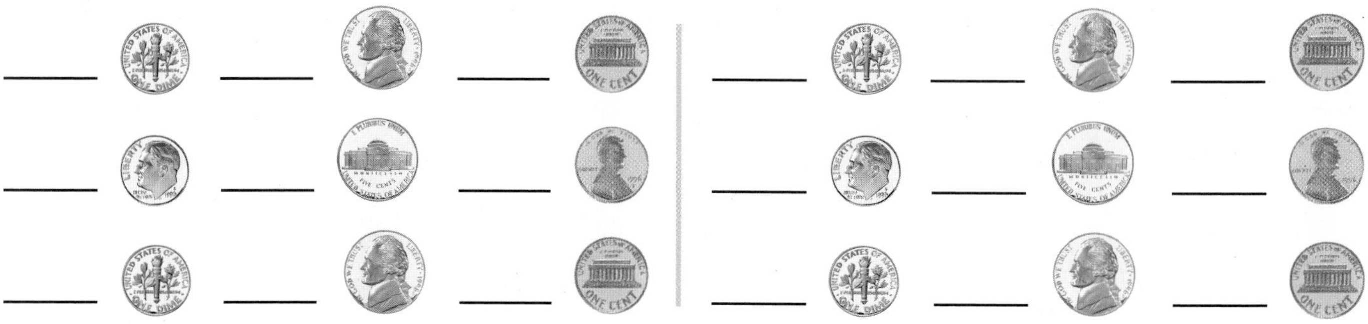

_____   _____   _____ | _____   _____   _____

_____   _____   _____ | _____   _____   _____

_____   _____   _____ | _____   _____   _____

Name _____

I quarter

25 cents

25¢

2 quarters equal 50¢.

3 quarters equal 75¢.

TALK IT OVER

Name 3 different ways to show 25¢.

Count on. Write the amount.

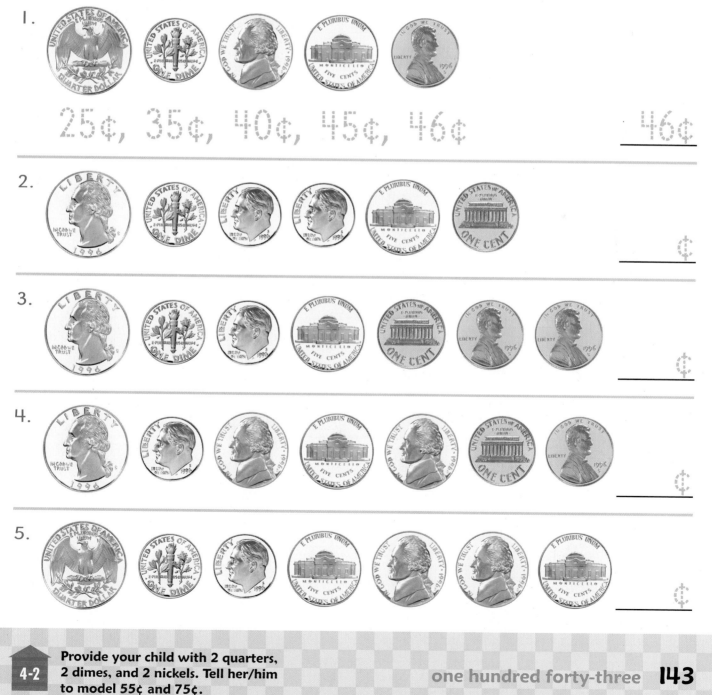

1. 25¢, 35¢, 40¢, 45¢, 46¢   46¢

2. _____ ¢

3. _____ ¢

4. _____ ¢

5. _____ ¢

4-2 Provide your child with 2 quarters, 2 dimes, and 2 nickels. Tell her/him to model 55¢ and 75¢.

one hundred forty-three **143**

Write the amount.

1.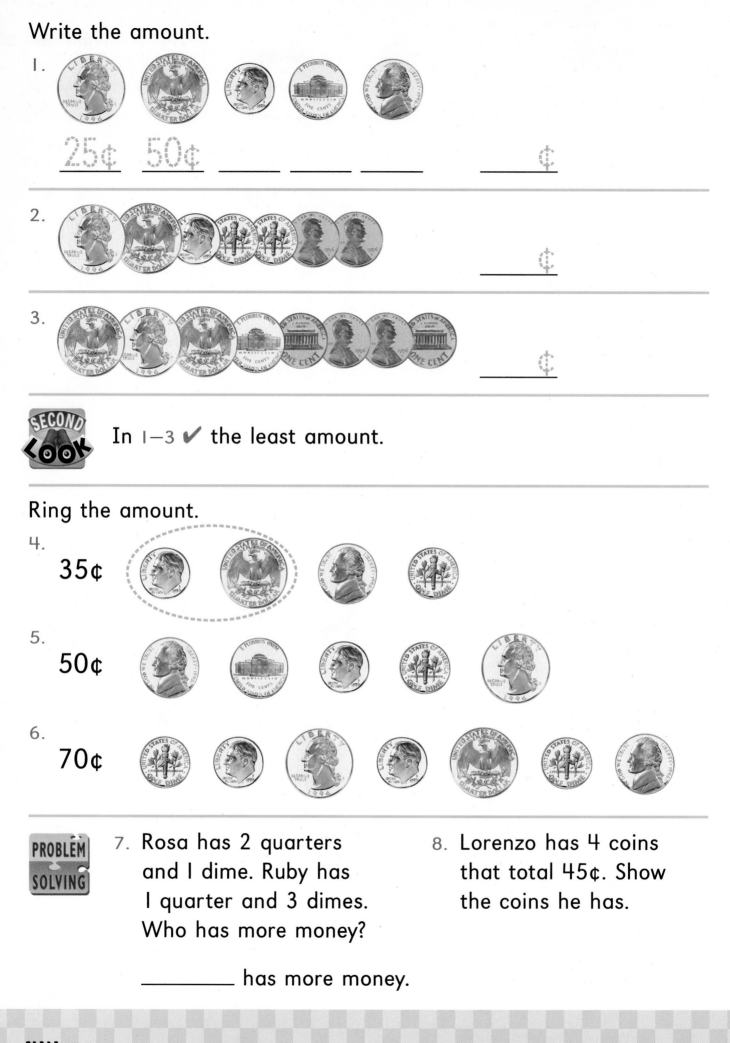

25¢   50¢   _____   _____   _____ ¢

2. _____ ¢

3. _____ ¢

SECOND LOOK In 1–3 ✔ the least amount.

Ring the amount.

4. 35¢

5. 50¢

6. 70¢

PROBLEM SOLVING

7. Rosa has 2 quarters and 1 dime. Ruby has 1 quarter and 3 dimes. Who has more money?

_____ has more money.

8. Lorenzo has 4 coins that total 45¢. Show the coins he has.

Name _____

I half dollar   2 quarters

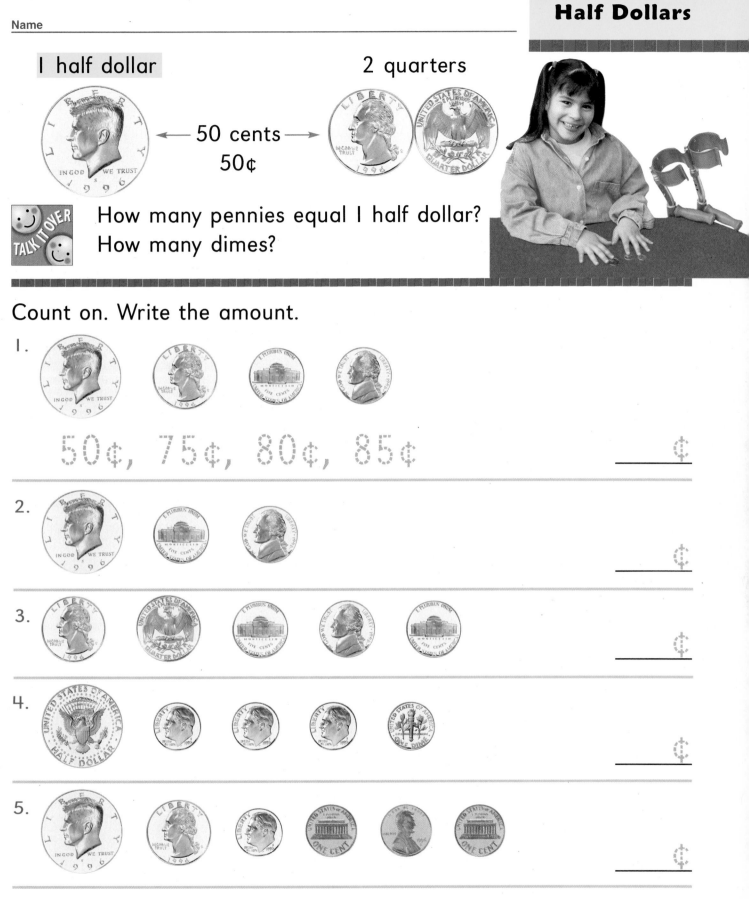

←— 50 cents —→
50¢

TALK IT OVER

How many pennies equal 1 half dollar?
How many dimes?

Count on. Write the amount.

1.

$$50¢, \quad 75¢, \quad 80¢, \quad 85¢ \qquad \underline{}¢$$

2.

____¢

3.

____¢

4.

____¢

5.

____¢

SHARE YOUR THINKING

How does counting on by 10s, 5s, and 1s
help you find the total amount? In 1–5
where did you count on by 10s? 5s? 1s?

4-3 Prompt your child to show amounts equal
to a half dollar using dimes, and then a
combination of dimes and nickels.

one hundred forty-five **145**

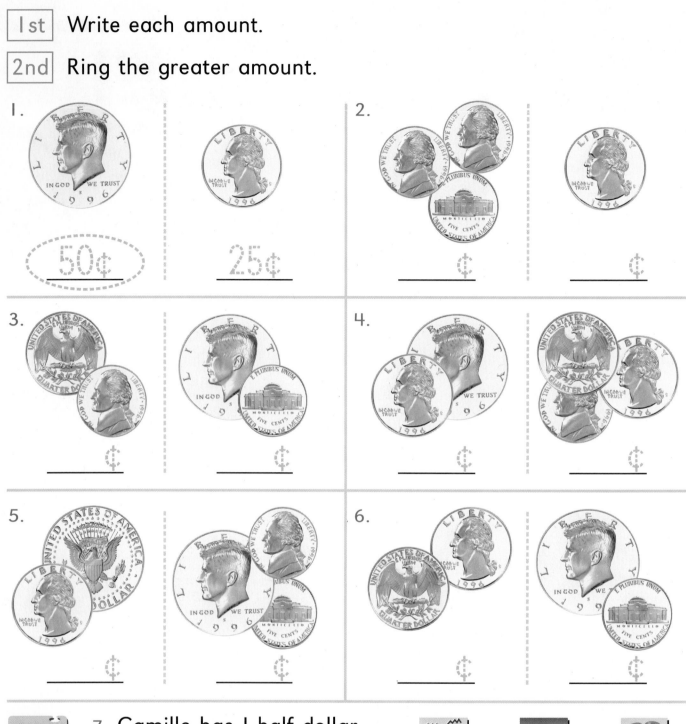

| 1st | Write each amount. |
| 2nd | Ring the greater amount. |

1. (50¢)   25¢

2. _____ ¢   _____ ¢

3. _____ ¢   _____ ¢

4. _____ ¢   _____ ¢

5. _____ ¢   _____ ¢

6. _____ ¢   _____ ¢

 PROBLEM SOLVING

7. Camille has 1 half dollar, 3 dimes, and 2 nickels. Ring two cards she can buy.

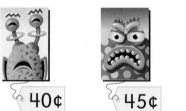   40¢   45¢   50¢

 FINDING TOGETHER

8. What is the value of this coin pattern?

 _____ ¢

Make a pattern in your Math Journal worth between 50¢ and 99¢.

I have 🪙 more.

36¢

I have 🪙 less.

Draw or ✗ coins to make each group equal the first amount.

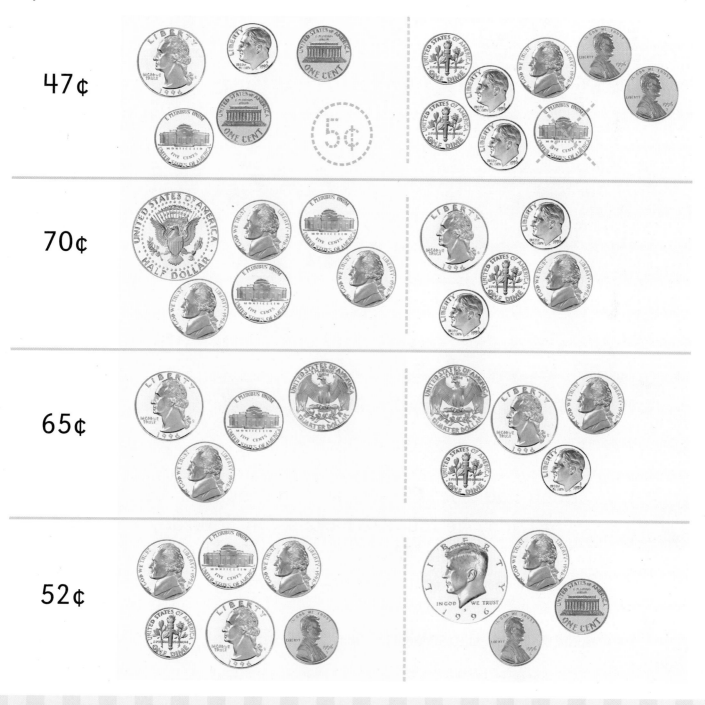

47¢

70¢

65¢

52¢

4-4 Discuss with your child how he/she used trading and comparing in the exercises above.

one hundred forty-seven **147**

1. 57¢

   Apple — 50¢   Yes   No

2. ___¢

   Juice — 35¢   Yes   No

3. ___¢

   Pineapple — 65¢   Yes   No

4. ___¢

   Sandwich — 99¢   Yes   No

 CRITICAL THINKING

5. John had 25¢. On Thursday he found a nickel. On Friday he found 2 nickels. On Saturday he found 3 nickels. If this pattern continues, how much money will he have on Sunday?   ___¢

 MAKE UP YOUR OWN

6. Make up a problem like 1–4.

Name _____

I can count up from the price to make change.

You pay with a 🪙.
Count up from 21¢.

22¢, 23¢, 24¢, 25¢

Your change is 4¢.

Count up from the price. Draw the change. Write the amount.

Price	Amount Paid	Change
1. 17¢	dime, nickel	⊙1¢ ⊙1¢ ⊙1¢ Your change is __3¢__.
2. 33¢	nickel, dime, dime, dime	Your change is _____ ¢.
3. 26¢	quarter, nickel	Your change is _____ ¢.

4-5 Invite your child to show how to find the change for a 71¢ item paid for with 3 quarters.

one hundred forty-nine **149**

Count up from the price. Draw
the change. Write the amount.

Price	Amount Paid	Change
1.		
2.		
3.		

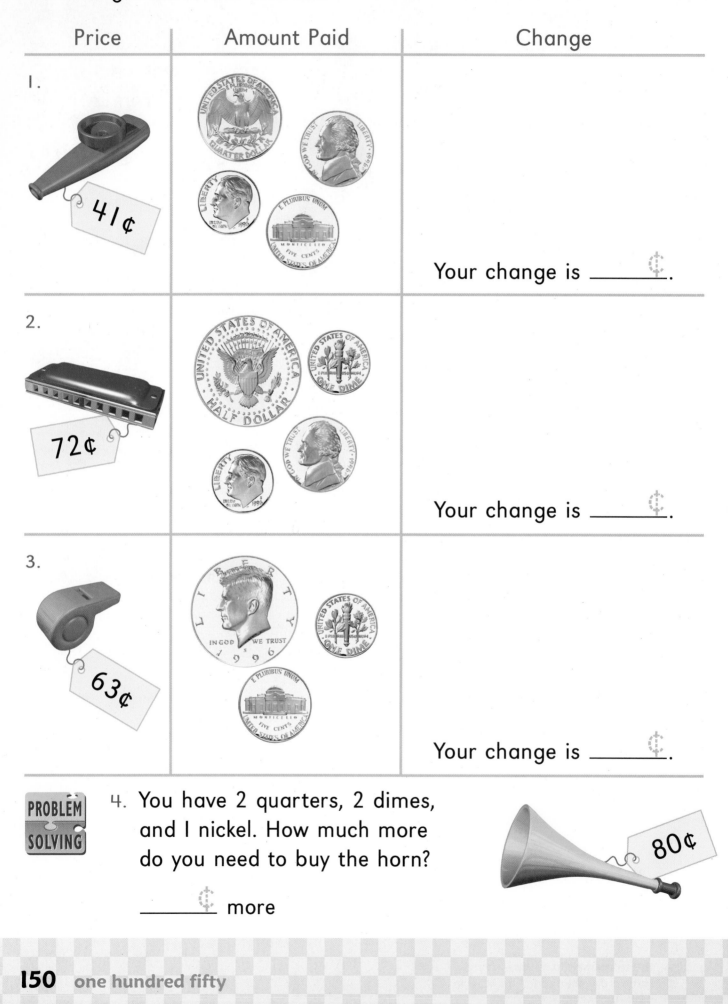

1. 41¢

Your change is _____ ¢.

2. 72¢

Your change is _____ ¢.

3. 63¢

Your change is _____ ¢.

PROBLEM SOLVING

4. You have 2 quarters, 2 dimes, and I nickel. How much more do you need to buy the horn?

_____ ¢ more

80¢

Name _____

1 dollar

100 cents
100¢ = $1.00
↑
dollar sign

Write the **decimal point** between the dollars and the cents.

$1.00

Ring the amounts equal to $1.00.

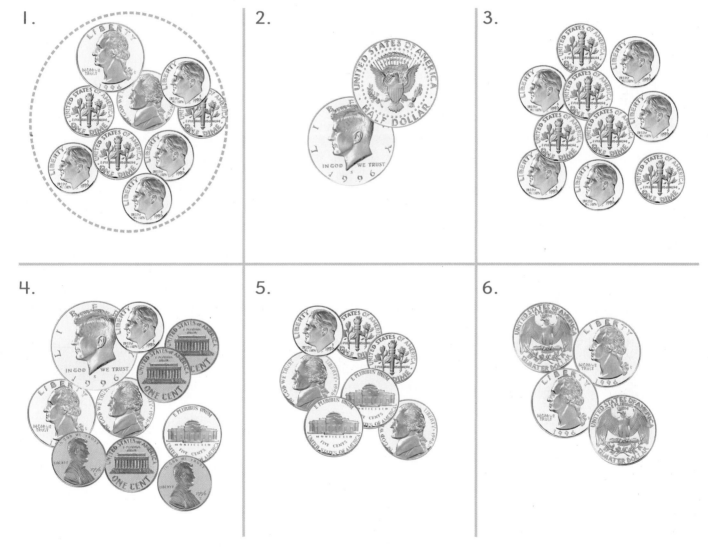

1.

2.

3.

4.

5.

6.

 TALK IT OVER Which amount shown above is less than $1.00? Name the coins you need to make it equal to 1 dollar.

✔ the correct amount.

1. one dollar and twenty cents

$1.20 _____

2. one dollar and fifty cents

$1.50 _____

Write the amount.

3.  $ _____ . _____

4.  $ _____ . _____

5. Write: one dollar and eighty-five cents. $ _____ . _____

    one dollar and sixty cents. $ _____ . _____

    one dollar and forty-two cents. $ _____ . _____

CHALLENGE

I have 2 dollar bills, 1 quarter, and 1 penny.

I have 2 quarters and 4 pennies.

$2.26
two dollars and
twenty-six cents

$0.54

fifty-four cents

Count the dollars and cents. Write the amount.

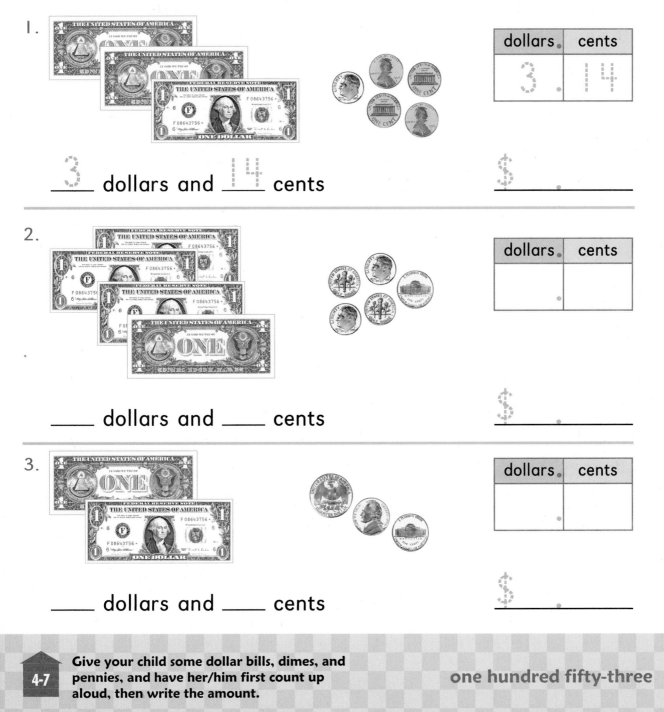

1.

dollars.	cents
3	14

___3___ dollars and __14__ cents $_____

2.

dollars.	cents

_____ dollars and _____ cents $_____

3.

dollars.	cents

_____ dollars and _____ cents $_____

4-7 Give your child some dollar bills, dimes, and pennies, and have her/him first count up aloud, then write the amount.

one hundred fifty-three **153**

Write the number of bills and coins.
Then write the amount.

1. Beth has 2 dollars and 3 pennies.

dollars.	dimes	pennies	amount
2.	0	3	$2.03

2. You save 3 dollars and 3 dimes.

dollars.	dimes	pennies	amount
			$.

3. Willis has 4 dollars and 9 pennies.

dollars.	dimes	pennies	amount
			$.

4. Claude spent 2 dollars and 5 dimes.

dollars.	dimes	pennies	amount
			$.

5. Vetta saved 3 dollars.

dollars.	dimes	pennies	amount
			$.

PROBLEM SOLVING   Write the amount.

6. 5 dollars and 8 cents $5.08

7. 67 cents   _____

8. 7 dollars and 10 cents   _____

9. 5 cents   _____

SECOND LOOK   In 6–9 ✔ the greatest amount. Ring the least amount.

Write the sums or differences.
Use the code to color.

less than 9 [color]   greater than 15 [color]

9 [color]   14 [color]   13 [color]   15 [color]

$$9 + 9 = \square$$

$$8 + 7 = \square$$

$$8 + 9 = \square$$

$$9 + 5 = \square$$

$$7 + 7 = \square$$

$$9 + 6 = 15$$

$$14 - 6 = \square$$

$$14 - 9 = \square$$

$$13 - 9 = \square$$

$$17 - 8 = \square$$

$$18 - 9 = \square$$

$$14 - 7 = \square$$

$$13 - 7 = \square$$

$$8 + 8 = \square$$

$$13 - 8 = \square$$

$$15 - 6 = \square$$

$$16 - 7 = \square$$

$$14 - 8 = \square$$

$$9 + 4 = \square$$

$$9 + 7 = \square$$

$$8 + 5 = \square$$

$$13 - 5 = \square$$

Name _____

About 1 minute

About 1 hour

About 1 day

Estimate. Ring about how long.

1.
(2 minutes)   2 hours

2.
5 minutes   5 hours

3.
2 minutes   2 hours

4.
10 minutes   10 hours

5.
2 hours   2 days

6.
5 hours   5 days

MATH JOURNAL

7. Draw 2 activities that take about the same time.

After describing an activity that takes 2 hours, ask your child to say an activity of comparable length.

4-8

minute hand

hour hand

3:00

3 o'clock

3:30

three thirty
half past 3
30 minutes after 3

TALK IT OVER

What is the difference in the position of the hour and minute hands at 3:00 and 3:30?

Write the time.

1. _____ **11** o'clock

11:00

2. half past _____

____:____

3. _____ o'clock

____:____

4. half past _____

____:____

5. half past _____

____:____

6. _____ o'clock

____:____

7. half past _____

____:____

8. half past _____

____:____

9. **4:30**

_____ minutes after _____

____:____

10. **11:30**

_____ thirty

____:____

11. **9:30**

_____ minutes after _____

____:____

4-9 Engage your child in a discussion about how 8:30 can also be called half past 8 or 30 minutes after 8.

one hundred fifty-seven **157**

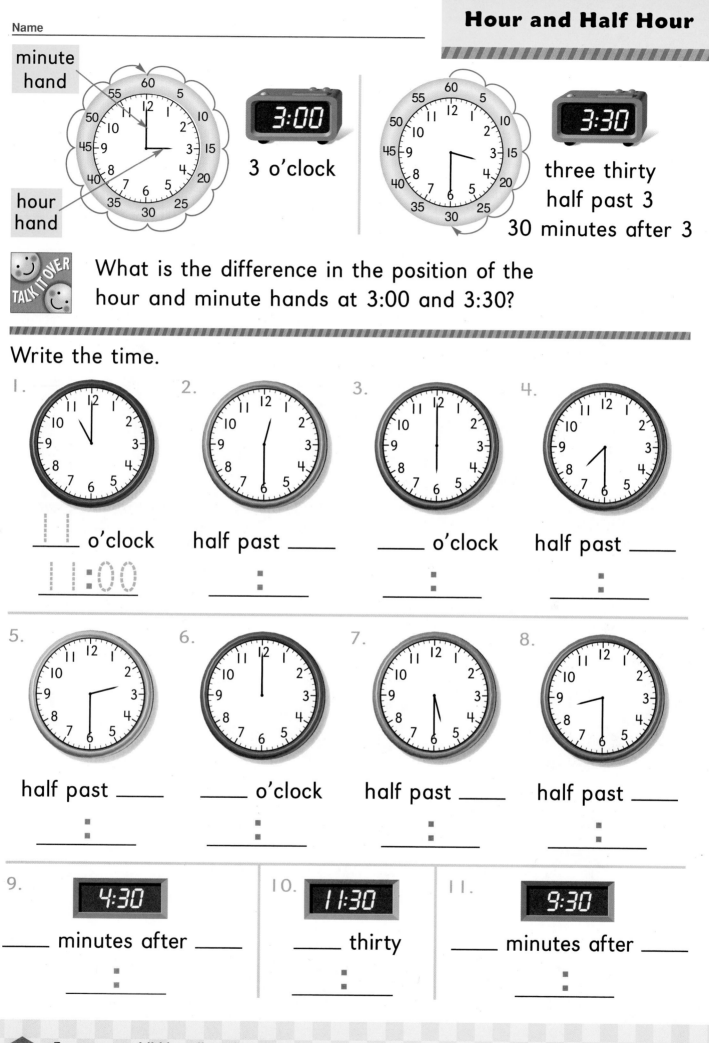

Write the time.

1.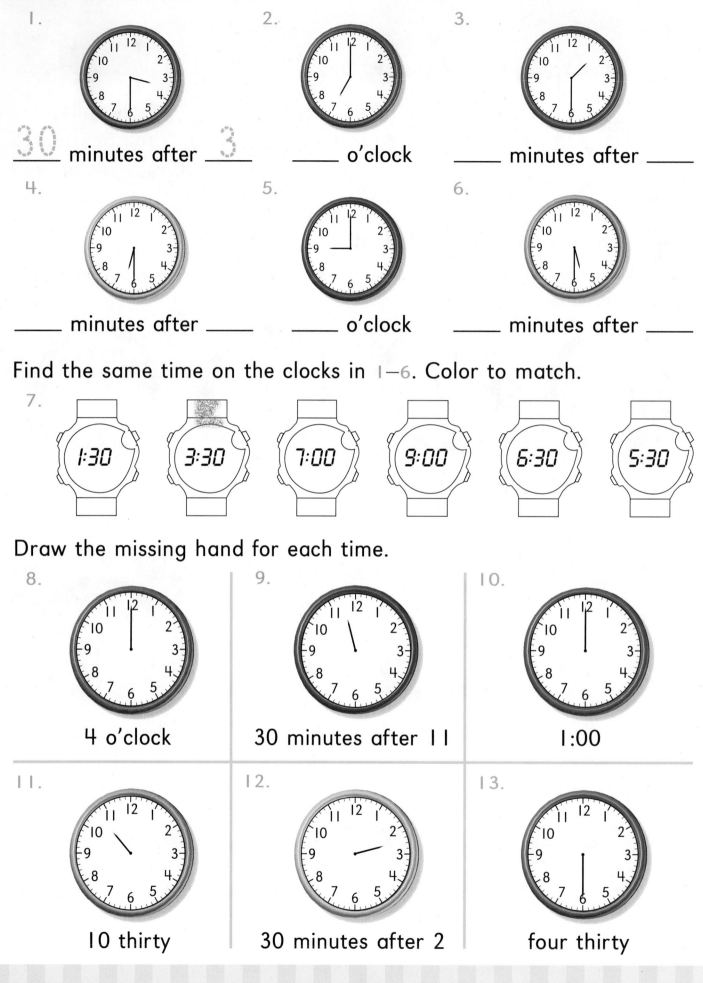

__30__ minutes after __3__

2.

_____ o'clock

3.

_____ minutes after _____

4.

_____ minutes after _____

5.

_____ o'clock

6.

_____ minutes after _____

Find the same time on the clocks in 1–6. Color to match.

7.

1:30   3:30   7:00   9:00   6:30   5:30

Draw the missing hand for each time.

8.

4 o'clock

9.

30 minutes after 11

10.

1:00

11.

10 thirty

12.

30 minutes after 2

13.

four thirty

1 hour has 60 minutes.

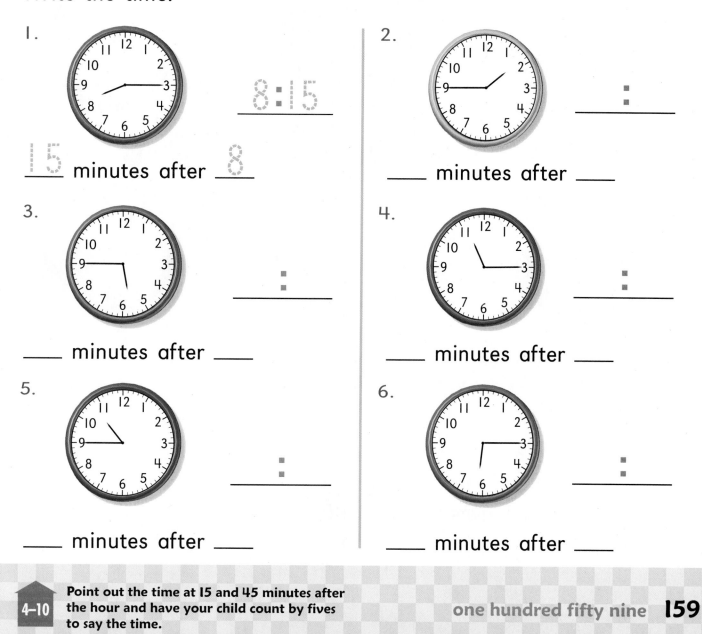

15 minutes after 4
4:15
four fifteen

half past 4
4:30
four thirty

45 minutes after 4
4:45
four forty-five

TALK IT OVER   In this pattern, what time comes next?

Write the time.

1.

8:15
_____
15 minutes after 8

2.

_____
____ minutes after ____

3.

_____
____ minutes after ____

4.

_____
____ minutes after ____

5.

_____
____ minutes after ____

6.

_____
____ minutes after ____

4–10   Point out the time at 15 and 45 minutes after the hour and have your child count by fives to say the time.

one hundred fifty nine   **159**

Write the time shown on the clock.
Compare the times. Ring early or late.

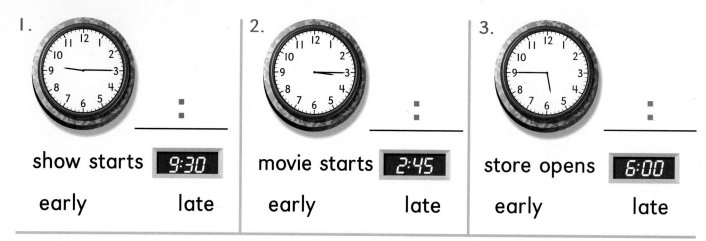

1.

show starts   `9:30`

early   late

2.

movie starts   `2:45`

early   late

3.

store opens   `6:00`

early   late

Draw the minute hand to show each time.

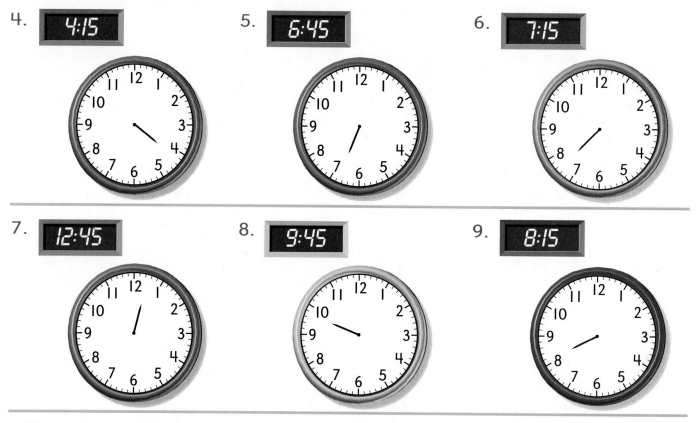

4. `4:15`

5. `6:45`

6. `7:15`

7. `12:45`

8. `9:45`

9. `8:15`

PROBLEM SOLVING

10. The movie started at 8:15.
Johnny arrived at 9:15.
How late is Johnny?

_____ late

11. Tina has a lesson at 5 o'clock.
It takes a half hour to get there.
If she leaves home at 4:00, how
early or late will she be?

_____ minutes _____

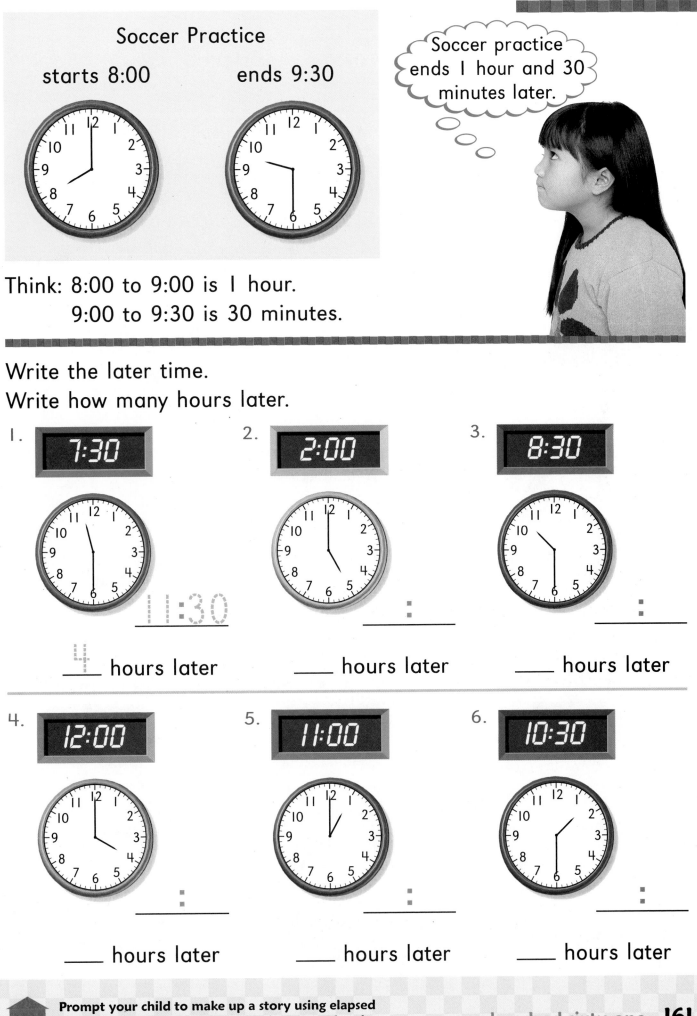

Soccer Practice

starts 8:00   ends 9:30

Soccer practice ends 1 hour and 30 minutes later.

Think: 8:00 to 9:00 is 1 hour.
9:00 to 9:30 is 30 minutes.

Write the later time.
Write how many hours later.

1. **7:30**

11:30

4 hours later

2. **2:00**

___:___

___ hours later

3. **8:30**

___:___

___ hours later

4. **12:00**

___:___

___ hours later

5. **11:00**

___:___

___ hours later

6. **10:30**

___:___

___ hours later

Write and draw the later time.

I hour later than **6:30** is __7:30__

2 hours later than **3:30** is __:__

I hour later than **12:00** is __:__

2 hours later than **4:00** is __:__

3 hours later than **12:30** is __:__

I hour and 30 minutes later than **3:00** is __:__

CHALLENGE Write and draw the later time.

2 hours later than **9:15** is __:__

2 hours later than **10:45** is __:__

45 minutes later than **4:30** is __:__

Name _____

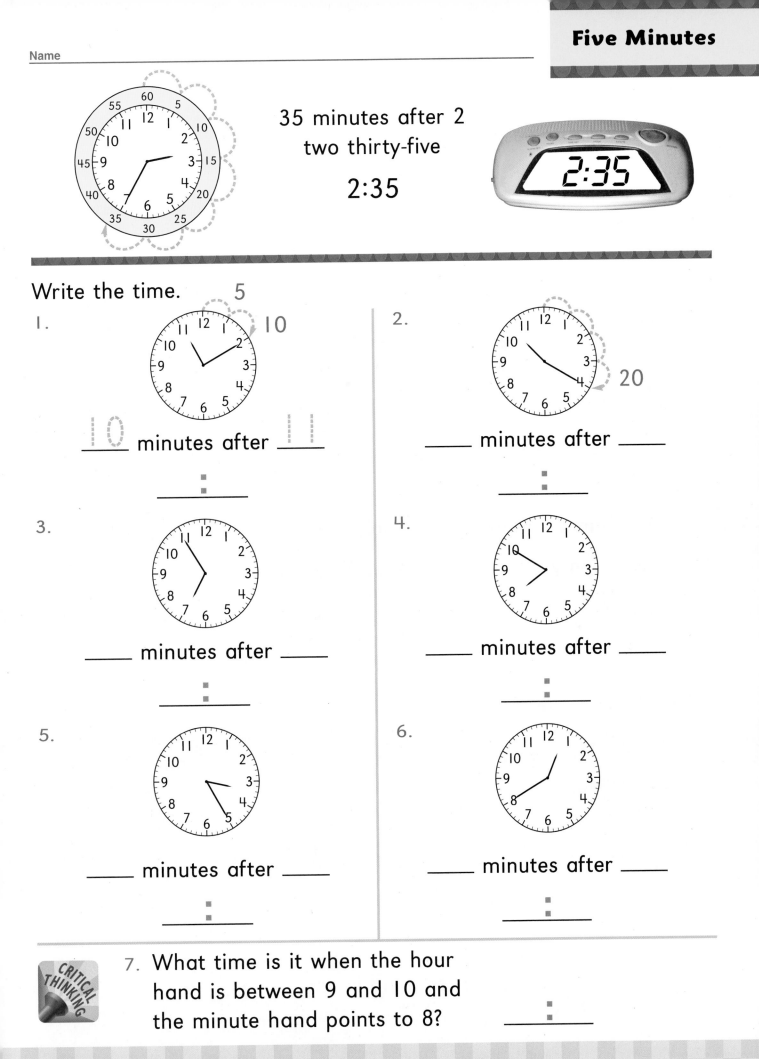

35 minutes after 2
two thirty-five

2:35

Write the time.

1.

_____ minutes after _____
(10 ... 11)

____ : ____

2.

____ minutes after ____

____ : ____

3.

____ minutes after ____

____ : ____

4.

____ minutes after ____

____ : ____

5.

____ minutes after ____

____ : ____

6.

____ minutes after ____

____ : ____

CRITICAL THINKING

7. What time is it when the hour hand is between 9 and 10 and the minute hand points to 8?

____ : ____

4-12 Tell your child to skip count by five to show how she/he read the times above.

one hundred sixty-three **163**

Write the time.

1.

7:05

2.

_____

3.

_____

4.

_____

5.

_____

6.

_____

7.

_____

8.

_____

Draw the minute hand on the clock.

9. nine twenty

10. three fifty

11. six forty

---

12. 8:10

13. seven fifteen

14. 30 minutes after 8

MENTAL MATH

15. How many minutes after the hour is it
when the minute hand is on:
3, 6, 9, 2, 4, 8, 10?

Name _____

A calendar shows the 12 months in a year.

1st **January**
S M T W T F S
          1 2 3
4 5 6 7 8 9 10
11 12 13 14 15 16 17
18 19 20 21 22 23 24
25 26 27 28 29 30 31

2nd **February**
S M T W T F S
1 2 3 4 5 6 7
8 9 10 11 12 13 14
15 16 17 18 19 20 21
22 23 24 25 26 27 28

3rd **March**
S M T W T F S
1 2 3 4 5 6 7
8 9 10 11 12 13 14
15 16 17 18 19 20 21
22 23 24 25 26 27 28
29 30 31

4th **April**
S M T W T F S
          1 2 3 4
5 6 7 8 9 10 11
12 13 14 15 16 17 18
19 20 21 22 23 24 25
26 27 28 29 30

May
S M T W T F S
                1 2
3 4 5 6 7 8 9
10 11 12 13 14 15 16
17 18 19 20 21 22 23
24/31 25 26 27 28 29 30

June
S M T W T F S
   1 2 3 4 5 6
7 8 9 10 11 12 13
14 15 16 17 18 19 20
21 22 23 24 25 26 27
28 29 30

July
S M T W T F S
            1 2 3 4
5 6 7 8 9 10 11
12 13 14 15 16 17 18
19 20 21 22 23 24 25
26 27 28 29 30 31

August
S M T W T F S
                     1
2 3 4 5 6 7 8
9 10 11 12 13 14 15
16 17 18 19 20 21 22
23/30 24/31 25 26 27 28 29

September
S M T W T F S
      1 2 3 4 5
6 7 8 9 10 11 12
13 14 15 16 17 18 19
20 21 22 23 24 25 26
27 28 29 30

October
S M T W T F S
               1 2 3
4 5 6 7 8 9 10
11 12 13 14 15 16 17
18 19 20 21 22 23 24
25 26 27 28 29 30 31

November
S M T W T F S
1 2 3 4 5 6 7
8 9 10 11 12 13 14
15 16 17 18 19 20 21
22 23 24 25 26 27 28
29 30

December
S M T W T F S
         1 2 3 4 5
6 7 8 9 10 11 12
13 14 15 16 17 18 19
20 21 22 23 24 25 26
27 28 29 30 31

TALK IT OVER In the fifth month, what does 24/31 mean?

1. The shortest month is _February_.

2. The month just before July is _____.

3. The month between March and May is _____.

4. The month that follows June is _____.

5. The next year begins on _____.

6. Do the first and last months of the year have the same number of days? _____

7. Write the date of your birthday. _____

8. Color the months with 31 days (yellow), 30 days (green).

4-13 Invite your child to use a calendar to find the day and date of family members' birthdays.

December

Sunday	Monday	Tuesday	Wednesday	Thursday	Friday	Saturday
		1	2	3	4	5
6	7	8	9		11	12
13		15	16		18	19

PROBLEM SOLVING Complete the calendar.
Then solve each problem.

1. On which day is the thirty-first? _____

2. How many Mondays are in this month? _____

3. What is the date of the fourth Friday? _____

4. What is the day after December 6? _____

5. On which day and date
   did the last month end? _____

6. What are the day and date
   2 weeks after December 10? _____

7. What is the day and date
   1 week before December 28? _____

8. How many days are in 2 weeks? _____ days

FINDING TOGETHER 9. Make a calendar for the first month of
   the next year. Ring New Year's Day.
   Make up problems like 1—8 for your classmates.

The Computer and the Internet

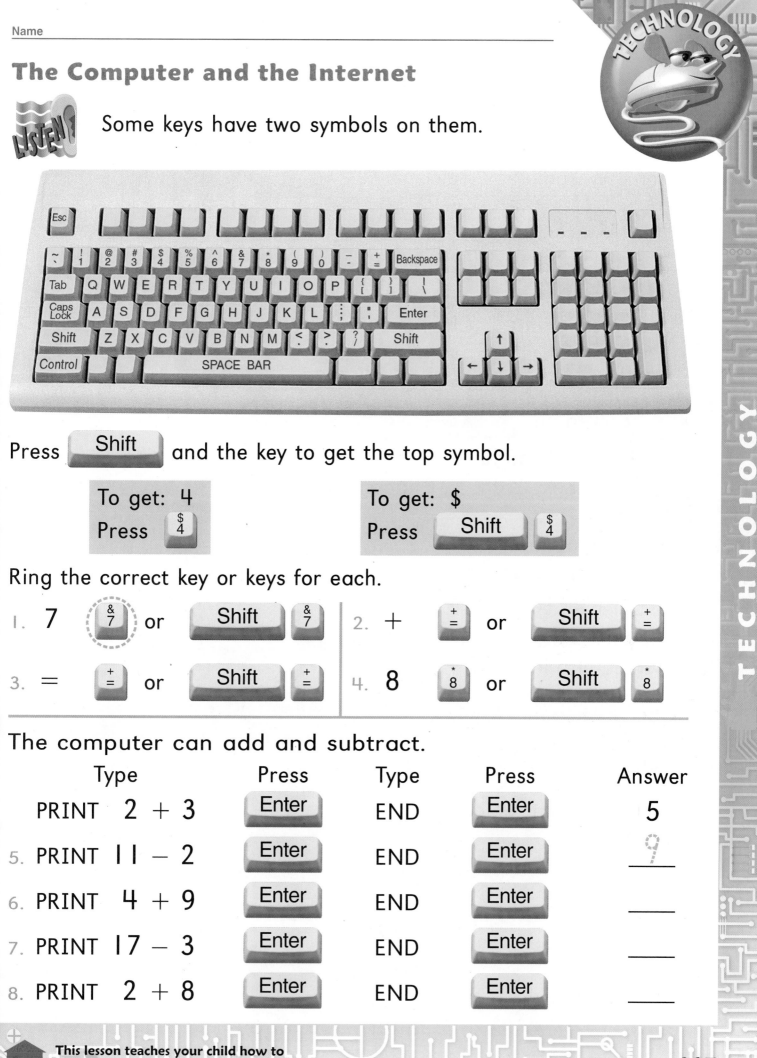

TECHNOLOGY

Some keys have two symbols on them.

Press **Shift** and the key to get the top symbol.

To get: 4
Press $\begin{smallmatrix}\$\\4\end{smallmatrix}$

To get: $
Press **Shift** $\begin{smallmatrix}\$\\4\end{smallmatrix}$

Ring the correct key or keys for each.

1. 7 $\begin{smallmatrix}\&\\7\end{smallmatrix}$ or **Shift** $\begin{smallmatrix}\&\\7\end{smallmatrix}$

2. + $\begin{smallmatrix}+\\=\end{smallmatrix}$ or **Shift** $\begin{smallmatrix}+\\=\end{smallmatrix}$

3. = $\begin{smallmatrix}+\\=\end{smallmatrix}$ or **Shift** $\begin{smallmatrix}+\\=\end{smallmatrix}$

4. 8 $\begin{smallmatrix}*\\8\end{smallmatrix}$ or **Shift** $\begin{smallmatrix}*\\8\end{smallmatrix}$

The computer can add and subtract.

Type	Press	Type	Press	Answer
PRINT 2 + 3	Enter	END	Enter	5
5. PRINT 11 − 2	Enter	END	Enter	9
6. PRINT 4 + 9	Enter	END	Enter	___
7. PRINT 17 − 3	Enter	END	Enter	___
8. PRINT 2 + 8	Enter	END	Enter	___

4-14 This lesson teaches your child how to use symbols on a computer keyboard, as well as how to use the Internet.

one hundred sixty-seven **167**

You can find information using a computer.

Millions of documents and information can be found on the World Wide Web.

Enter www.sadlier-oxford.com

World Wide Web   your textbook   company

Click on Student Activity Center.

Find Activities, Grade 2, Chapter 4.

TALK IT OVER What happened when you clicked on Student Activity Center? How did you find Activities, Grade 2, Chapter 4?

Click on Home.

Send Sadlier an e-mail. Tell us what you liked most about Chapter 4 or anything else in math.

Click on Mathematics K-8.

Click on Math@Sadlier.com

Type **Math** for the subject.

Enter your message.

Click **Send** to send us your e-mail.

We hope to hear from you.

Name _____

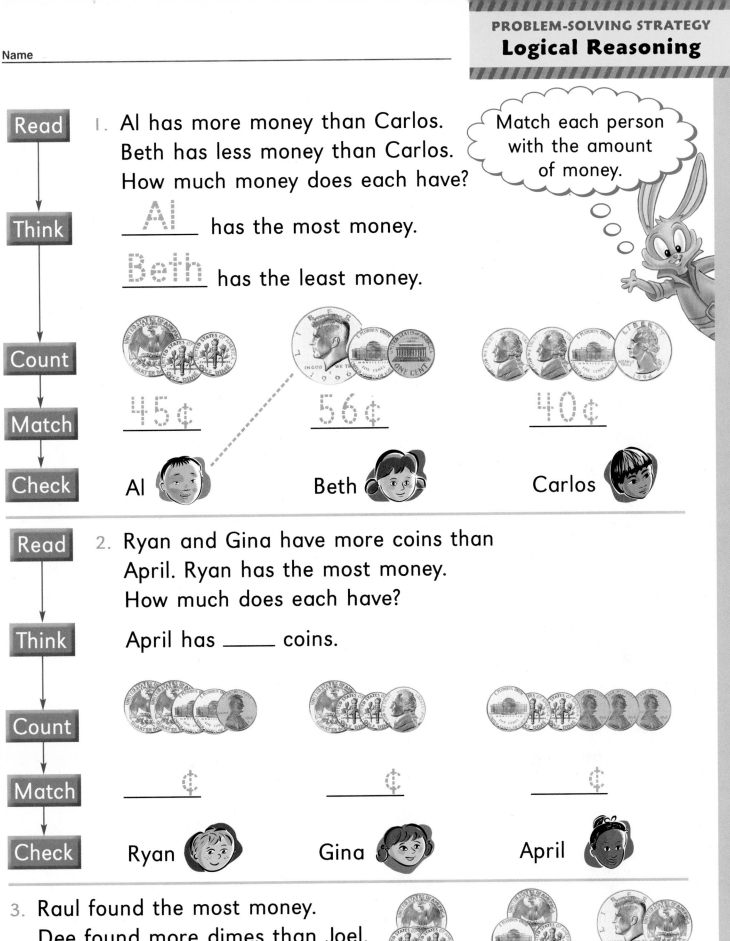

Read | 1. Al has more money than Carlos.
Beth has less money than Carlos.
How much money does each have?

> Match each person with the amount of money.

Think | _____Al_____ has the most money.

_____Beth_____ has the least money.

Count

Match | 45¢   56¢   40¢

Check | Al   Beth   Carlos

Read | 2. Ryan and Gina have more coins than April. Ryan has the most money. How much does each have?

Think | April has _____ coins.

Count

Match | _____ ¢   _____ ¢   _____ ¢

Check | Ryan   Gina   April

---

3. Raul found the most money.
Dee found more dimes than Joel.
How much did Joel find?

Joel found _____ ¢.   _____ ¢   _____ ¢   _____ ¢

PROBLEM SOLVING

4-15
Have your child explain how he/she solved the problems using *Logical Reasoning*.

one hundred sixty-nine **169**

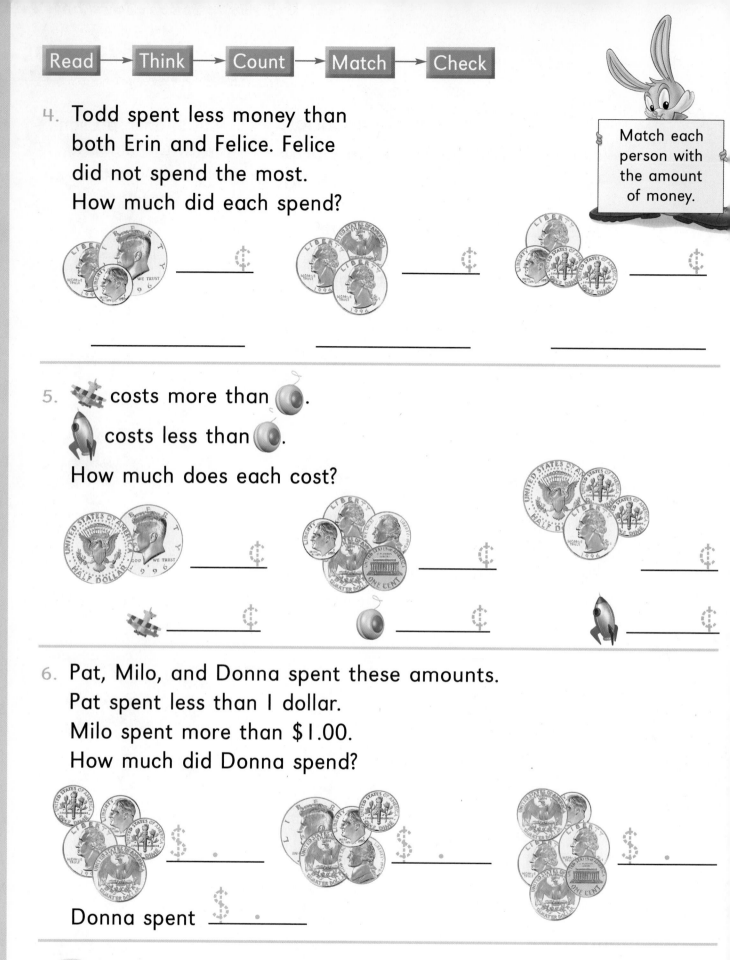

4. Todd spent less money than both Erin and Felice. Felice did not spend the most. How much did each spend?

_____ ¢   _____ ¢   _____ ¢

_____   _____   _____

Match each person with the amount of money.

5. ✈ costs more than ⊙.
   🚀 costs less than ⊙.
   How much does each cost?

_____ ¢   _____ ¢   _____ ¢

✈ _____ ¢   ⊙ _____ ¢   🚀 _____ ¢

6. Pat, Milo, and Donna spent these amounts.
   Pat spent less than 1 dollar.
   Milo spent more than $1.00.
   How much did Donna spend?

$ _____ . $ _____ . $ _____ .

Donna spent $ _____ .

Write a problem like 4 or 5 in your Math Journal.

Name _____

Read → Think → Draw → Write → Check

Model the first time. Then count on by hours and minutes.

1. Lisa began shopping at half past 11. She finished 1 hour 30 minutes later. What time was it then?

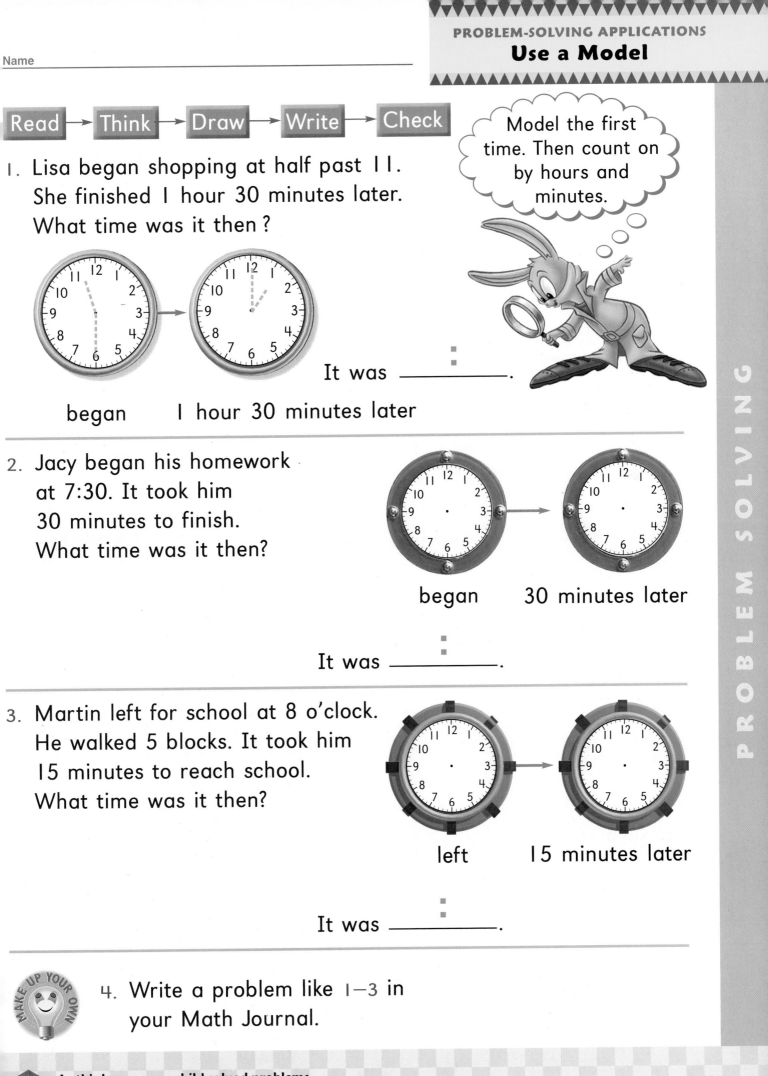

It was _____ : _____ .

began   1 hour 30 minutes later

2. Jacy began his homework at 7:30. It took him 30 minutes to finish. What time was it then?

began   30 minutes later

It was _____ : _____ .

3. Martin left for school at 8 o'clock. He walked 5 blocks. It took him 15 minutes to reach school. What time was it then?

left   15 minutes later

It was _____ : _____ .

MAKE UP YOUR OWN

4. Write a problem like 1–3 in your Math Journal.

4-16   In this lesson your child solved problems by using the *Use a Model* strategy.

one hundred seventy-one   **171**

PROBLEM SOLVING

Use a strategy you have learned.

STRATEGY FILE

Use a Table
Extra Information
Logical Reasoning
Guess and Test

1. The value of my coin is greater than 6¢ doubled. It is less than a half dollar. What is my coin? _____

2. A museum pencil costs 77¢. Pat paid with 3 quarters and 1 nickel. Tina paid 85¢. What was Pat's change? _____¢ change

3. Ed arrived at the museum first. Luna arrived later than Sue. Sue arrived 45 minutes after Ed. What time did each person arrive?

_____

_____

_____

`10:15`

`9:00`

`9:45`

Schedule for Field Day	
10:00	parade
10:30	50-yard dash
11:00	three-legged race
11:30	wheelbarrow race
12:00	lunch
1:00	balloon toss
1:15	soccer game

4. Joy took 30 minutes for lunch. What time was it then? _____

5. The soccer game took 1 hour. What time did the game end? _____

6. How many games are there between 10 o'clock and 12 o'clock? _____

7. Russ came to the field at 9:30. How long must he wait for the parade?

_____ minutes

8. John left 3 hours after lunch. It took him 1 hour to get home. What time did he get home?

_____

Write the amount.

1. 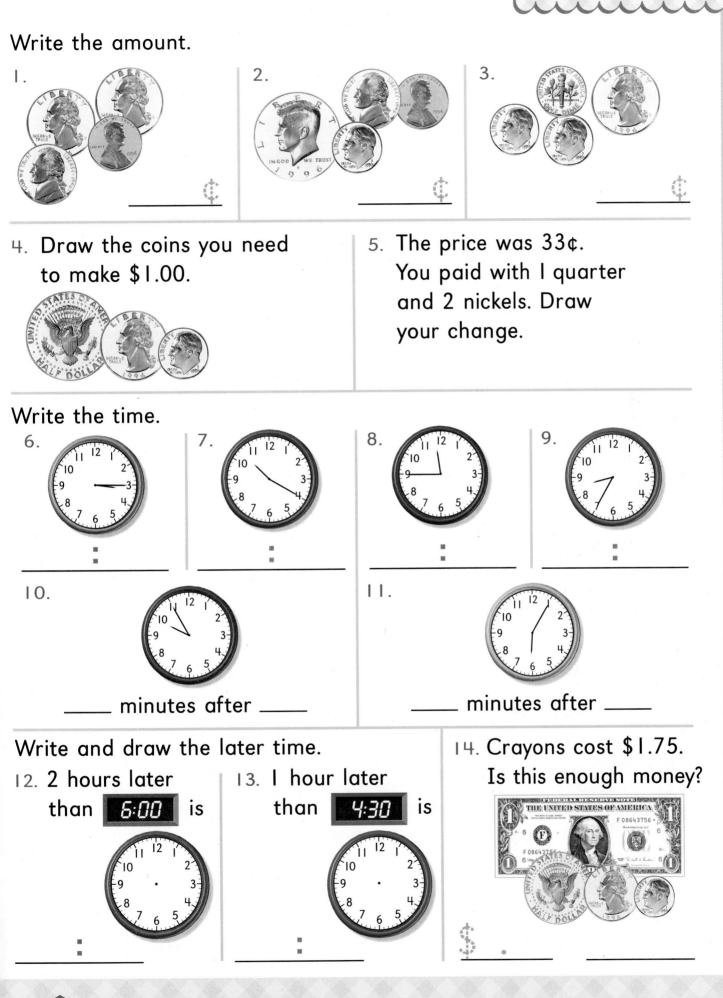 _____ ¢

2. _____ ¢

3. _____ ¢

4. Draw the coins you need to make $1.00.

5. The price was 33¢. You paid with 1 quarter and 2 nickels. Draw your change.

Write the time.

6. __ __ : __ __

7. __ __ : __ __

8. __ __ : __ __

9. __ __ : __ __

10. _____ minutes after _____

11. _____ minutes after _____

Write and draw the later time.

12. 2 hours later than **6:00** is

__ __ : __ __

13. 1 hour later than **4:30** is

__ __ : __ __

14. Crayons cost $1.75. Is this enough money?

_____

This page reviews the mathematical content presented in Chapter 4.

Write the time you do each activity.
Draw the hands on the clock.

Schedule for _____ _____
day   date

1. wake-up time

_____ minutes after _____

2. breakfast time

_____ minutes after _____

3. school time

_____ minutes after _____

4. lunch time

_____ minutes after _____

5. going-home time

_____ minutes after _____

6. dinner time

_____ minutes after _____

This page extends your child's
understanding of time schedules.

4

1. Amy has I half dollar and I quarter.
Toni has 2 quarters and I nickel.
They traded to share the money equally.
Draw the coins each has now.

Amy now has

Toni now has

2. Complete the time pattern.

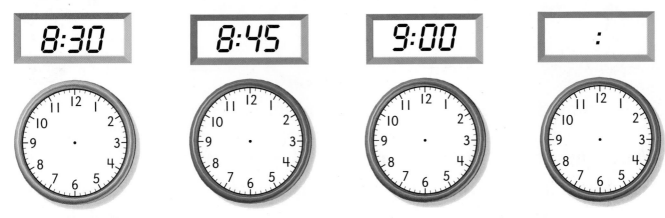

| 8:30 | 8:45 | 9:00 | : |

PORTFOLIO

Choose I of these projects.
Use a separate sheet of paper.

3. Outline a ⊞ on
a calendar.
Find each sum.
For example:

2	3
9	10

Write <, =, or >.

2 + 10 __?__ 3 + 9

4. Make each coin pattern.

- Use pennies and quarters.
6 coins to total 54¢.

- Use nickels and dimes.
8 coins to total 50¢.

4

This page provides a variety of informal
assessment opportunities in order to measure
your child's understanding of Chapter 4.

one hundred seventy-five **175**

ASSESSMENT

Name _____

Write the amount.

1. 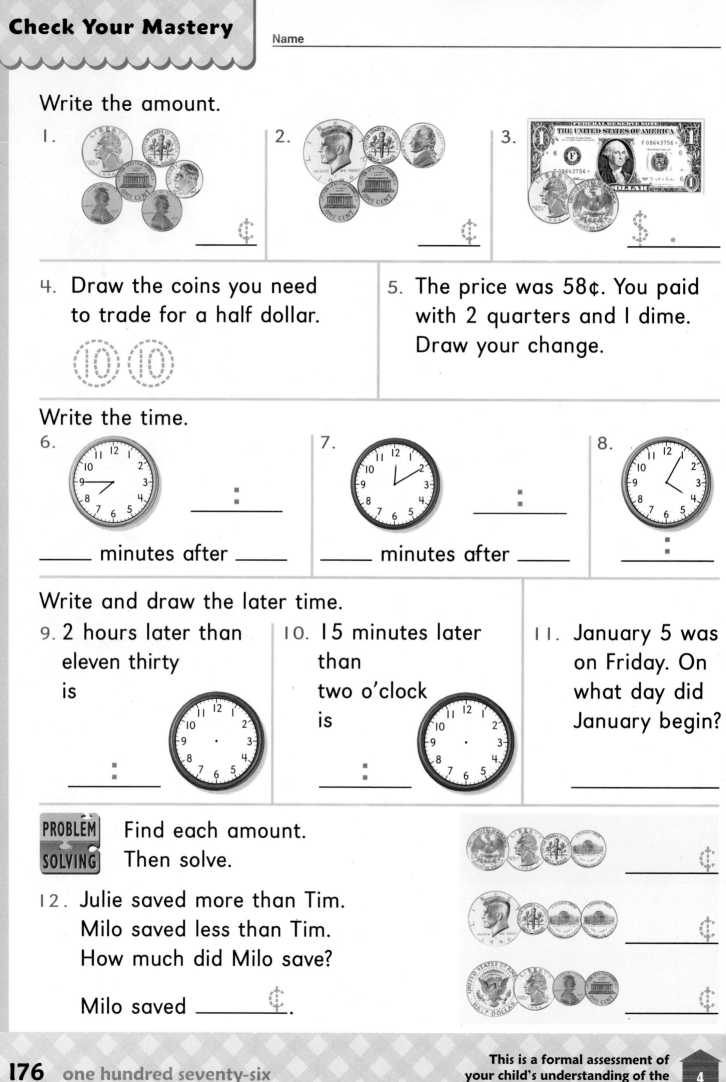 _____ ¢

2. _____ ¢

3. $ _____ . _____

4. Draw the coins you need to trade for a half dollar.

5. The price was 58¢. You paid with 2 quarters and 1 dime. Draw your change.

Write the time.

6. _____ minutes after _____

7. _____ minutes after _____

8.

Write and draw the later time.

9. 2 hours later than eleven thirty is

10. 15 minutes later than two o'clock is

11. January 5 was on Friday. On what day did January begin? _____

PROBLEM SOLVING Find each amount. Then solve.

12. Julie saved more than Tim. Milo saved less than Tim. How much did Milo save?

Milo saved _____ ¢.

_____ ¢

_____ ¢

_____ ¢

This is a formal assessment of your child's understanding of the content presented in Chapter 4.

4

Addition of Two-Digit Numbers

WELCOME TO DIGIT COUNTY

You are _____ miles from _____.

15 Stadium →

8 Center ←

20 World ←

5 County Seat →

18 Refuge ←

30 Center →

CRITICAL THINKING

Lena and her dad plan to visit two places that are 10 miles apart. What two places might they visit?

For more information about Chapter 5, visit the Family Information Center at **www.sadlier-oxford.com**

Dear Family,

Today your child began Chapter 5. As she/he studies addition of 2-digit numbers, you may want to read the poem below, which was read in class. Have your child talk about some of the math ideas pictured on page 177.

Look for the 🏠 at the bottom of each skills lesson. The suggestion on the page gives you an opportunity to improve your child's understanding of math and to reinforce her/his math language. You may want to have dimes, pennies, and other countables available for your child to use throughout this chapter.

Home Activity

How Far Is It?

Try this activity with your child. To make four license plates, divide a sheet of paper into four sections. In each section write two 2-digit addends horizontally; they will be the first two numbers of the license plate. Ask your child to add the addends and then to write the sum as the third number on the plate. When your child has finished writing the license plate numbers, have her/him use stickers or drawings to decorate it.

Ohio
20 ☆ 30 ☆ 50

Home Reading Connection

Maps

A map is a picture
Of where we are going.
The wiggly lines show us
Where rivers are flowing;
The red lines are highways
On which we will travel;
The black lines are byways
Topped sometimes with gravel.
The dots are the cities
Where gas stations are
And each capital city
Is marked with a star.

Goldie Capers Smith

Name _____

A map shows distances.
You can add to find
the distance you travel.

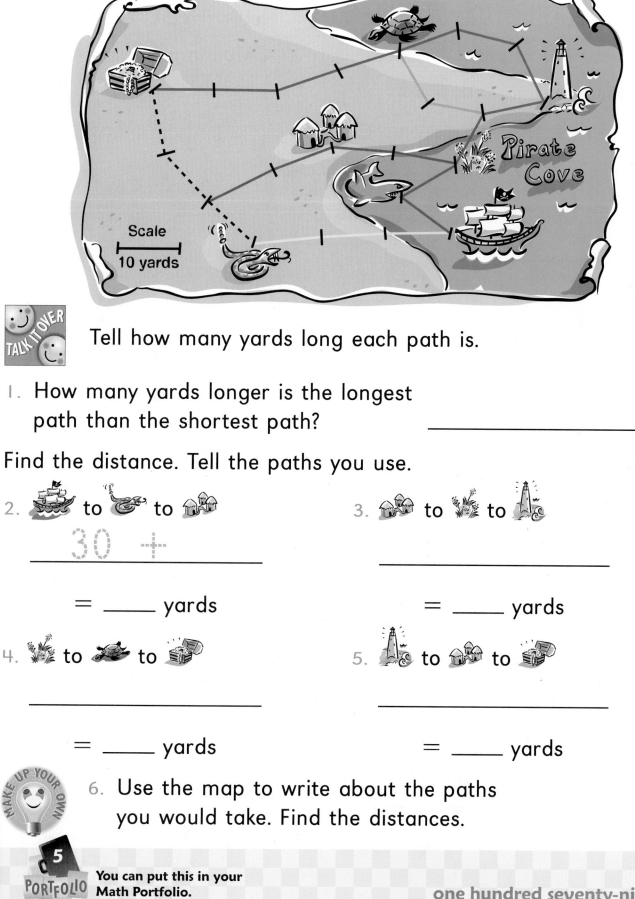

Scale
10 yards

TALK IT OVER Tell how many yards long each path is.

1. How many yards longer is the longest
   path than the shortest path? _____

Find the distance. Tell the paths you use.

2. [ship] to [snake] to [huts]

   30 + _____

   = _____ yards

3. [huts] to [grass] to [lighthouse]

   _____

   = _____ yards

4. [grass] to [turtle] to [treasure]

   _____

   = _____ yards

5. [lighthouse] to [huts] to [treasure]

   _____

   = _____ yards

6. Use the map to write about the paths
   you would take. Find the distances.

MAKE UP YOUR OWN

CONNECTIONS

Use the following code to color the states.

Alaska $50 - 0$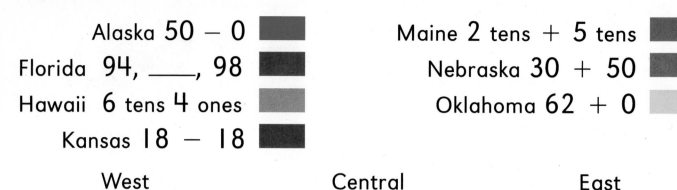

Florida $94, \underline{\quad}, 98$

Hawaii 6 tens 4 ones

Kansas $18 - 18$

Maine 2 tens $+ 5$ tens

Nebraska $30 + 50$

Oklahoma $62 + 0$

West   **Central**   **East**

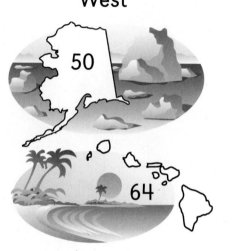

50

64

80

0

62

70

96

Find these states on a map of the United States.
Write the state names to finish the sentences below.

1. The states in the West are _____ and

   _____.

2. The states in the East are _____ and

   _____.

3. The Central states are _____,

   _____, and _____.

  Make up a number puzzle for your state
and two states on its border.

You can put this in your
Math Portfolio. PORTFOLIO

Name _____

$43 + 22 = $?

tens	ones

First add ones.

tens	ones
4	3
+ 2	2
	5

Then add tens.

tens	ones
4	3
+ 2	2
6	5

$$\begin{array}{r} 43 \\ +22 \\ \hline 65 \end{array}$$

Record the addends. Add. You may use models to check.

1. $31 + 18$

tens	ones
3	1
+ 1	8
4	9

2. $12 + 65$

tens	ones
1	2
+ 6	5

3. $15 + 23$

tens	ones
1	5
+	

Find the sum.

4.
$$\begin{array}{r} 24 \\ +63 \\ \hline 87 \end{array} \qquad \begin{array}{r} 50 \\ +27 \\ \hline \end{array} \qquad \begin{array}{r} 68 \\ +30 \\ \hline \end{array} \qquad \begin{array}{r} 36 \\ +22 \\ \hline \end{array} \qquad \begin{array}{r} 22 \\ +57 \\ \hline \end{array} \qquad \begin{array}{r} 21 \\ +34 \\ \hline \end{array}$$

5.
$$\begin{array}{r} 35 \\ +31 \\ \hline \end{array} \qquad \begin{array}{r} 26 \\ +13 \\ \hline \end{array} \qquad \begin{array}{r} 17 \\ +52 \\ \hline \end{array} \qquad \begin{array}{r} 45 \\ +44 \\ \hline \end{array} \qquad \begin{array}{r} 60 \\ +29 \\ \hline \end{array} \qquad \begin{array}{r} 16 \\ +20 \\ \hline \end{array}$$

6.
$$\begin{array}{r} 80 \\ +14 \\ \hline \end{array} \qquad \begin{array}{r} 38 \\ +61 \\ \hline \end{array} \qquad \begin{array}{r} 64 \\ +14 \\ \hline \end{array} \qquad \begin{array}{r} 70 \\ +25 \\ \hline \end{array} \qquad \begin{array}{r} 53 \\ +45 \\ \hline \end{array} \qquad \begin{array}{r} 42 \\ +12 \\ \hline \end{array}$$

5-1 Ask your child to use the digits 3, 2, 5, 4 to make three different addition problems.

one hundred eighty-one **181**

Add.

1. 33 +12 45	15 +40	12 +75	80 +13	31 +58	22 +41
2. 41 +36	23 +24	52 +16	77 +10	20 +71	65 +34
3. 20 +38	62 +24	54 +35	33 +46	49 +50	16 +51
4. 45 +13	76 +23	13 +33	22 +13	45 +42	40 +26

In 1–4 ring sums less than 60.

5. Fourteen stamps were blue.
Thirty-four stamps were green.
Each stamp cost 29¢.
How many stamps in all? _____ stamps

Compare. Write <, =, or >.

6. 30 + 10 ___ 20 + 20   7. 30 + 40 ___ 50 + 40

8. 20 + 40 ___ 40 + 10   9. 60 + 10 ___ 50 + 20

10. 87 + 11 ___ 57 + 11   11. 22 + 31 ___ 22 + 41

Count on by **ones** to add **23 + 2** mentally.

23, 24, 25

Count on by **tens** to add **23 + 20** mentally.

23, 33, 43

Add mentally.

1. Add 2 to:

6	16	26	36
8	18	___	___

2. Add 3 to:

15	25	35	45
___	___	___	___

3. Add 4 to:

4	14	24	34
___	___	___	___

4. Add 5 to:

13	23	33	43
___	___	___	___

5. Add 20 to:

6	16	26	36
___	___	___	___

6. Add 30 to:

15	25	35	45
___	___	___	___

7. Add 40 to:

4	14	24	34
___	___	___	___

8. Add 50 to:

13	23	33	43
___	___	___	___

MATH JOURNAL

9. Which is greater, 35¢ + 4¢ or 35¢ + 40¢? How do you know?

5-2 — Ask your child to tell how counting on by ones or tens makes it easy to add mentally.

one hundred eighty-three **183**

Add.

1.
```
  13        24        35        46        23        56
+ 50      + 40      + 30      + 20      + 50      + 20
  63
```

2.
```
   3        64         8        94        53        93
+ 45      +  4      + 71      +  4      +  3      +  5
  48
```

3.
```
  46        20        40         5        70        33
+  2      + 36      + 34      + 84      + 24      + 50
```

 In 1–3 ✔ the sum if you counted on by 1s.
Ring the sum if you counted on by 10s.

Write the missing sum.

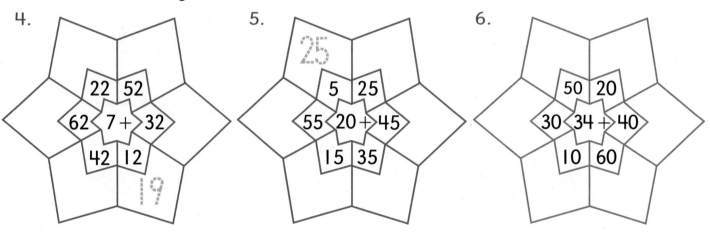

4.

```
 22  52
62  7 +  32
 42  12
     19
```

5.

```
     25
  5   25
55  20 + 45
 15   35
```

6.

```
 50  20
30  34 + 40
 10  60
```

 Use coins or solve mentally.

7. Kara has 4 dimes
1 nickel. She finds 3
more dimes. Match
the postcard
she can buy.

79¢

70¢

75¢

8. Ty has 3 quarters.
He gets 4 pennies.
Match the postcard
he can buy.

Name _____

Regroup when there are more than 9 ones.

1 ten 6 ones + 8 ones = ?

10 ones = 1 ten

1 ten 6 ones + 8 ones = 1 ten 14 ones
= 2 tens 4 ones

Add.

1. 2 tens 7 ones + 5 ones

= 2 tens 12 ones

= 3 tens 2 ones

2. 1 ten 9 ones + 7 ones

= 1 ten ___ones

= ___tens ___ones

3. 2 tens 6 ones + 5 ones

= 2 tens ___ones

= ___tens ___one

4. 1 ten 4 ones + 6 ones

= 1 ten ___ones

= ___tens ___ones

5. 3 tens 7 ones + 7 ones

= 3 tens ___ones

= ___tens ___ones

6. 2 tens 9 ones + 8 ones

= 2 tens ___ones

= ___tens ___ones

5-3 Have your child show how to regroup 1 ten 5 ones + 8 ones as 2 tens 3 ones using cereal or pasta.

one hundred eighty-five **185**

Use models. Regroup 10 ones as 1 ten.

1. 2 tens 7 ones
  +1 ten 4 ones
  ___3 tens ___11 ones
  ___4 tens ___1 one
	2. 4 tens 8 ones
	  +2 tens 2 ones
	  ___tens ___ones
	  ___tens ___ones
		3. 2 tens 6 ones
		  +2 tens 7 ones
		  ___tens ___ones
		  ___tens ___ones

4. 3 tens 5 ones
  +3 tens 5 ones
  ___tens ___ones
  ___tens ___ones
	5. 2 tens 7 ones
	  +6 tens 8 ones
	  ___tens ___ones
	  ___tens ___ones
		6. 1 ten 9 ones
		  +3 tens 9 ones
		  ___tens ___ones
		  ___tens ___ones

7. 1 ten 9 ones
  +1 ten 5 ones
  ___tens ___ones
  ___tens ___ones
	8. 3 tens 8 ones
	  +2 tens 4 ones
	  ___tens ___ones
	  ___tens ___ones
		9. 4 tens 3 ones
		  +1 ten 9 ones
		  ___tens ___ones
		  ___tens ___ones

10. 4 tens 8 ones
  +3 tens 8 ones
  ___tens ___ones
  ___tens ___ones
	11. 2 tens 3 ones
	  + 4 tens 8 ones
	  ___tens ___ones
	  ___tens ___one
		12. 6 tens 9 ones
		  +1 ten 1 one
		  ___tens ___ones
		  ___tens ___ones

PROBLEM SOLVING Use the map.

13. Lyle can use 2 different paths
from A to D. Name each path.
How long is each? ✔ the longer.

___A__ to _____ is ___ miles.

___A__ to _____ is ___ miles.

10 miles
B ●————————● A
3
miles ● C
7   20
miles   miles
D ●
5
miles   ● E

Name _____

Add 37 to 15.

Add ones. Regroup.

Then add tens.

Regroup 12 ones as 1 ten 2 ones.

TALK IT OVER Would you regroup to add 33 to 15? Why or why not?

Add. Use models to regroup.

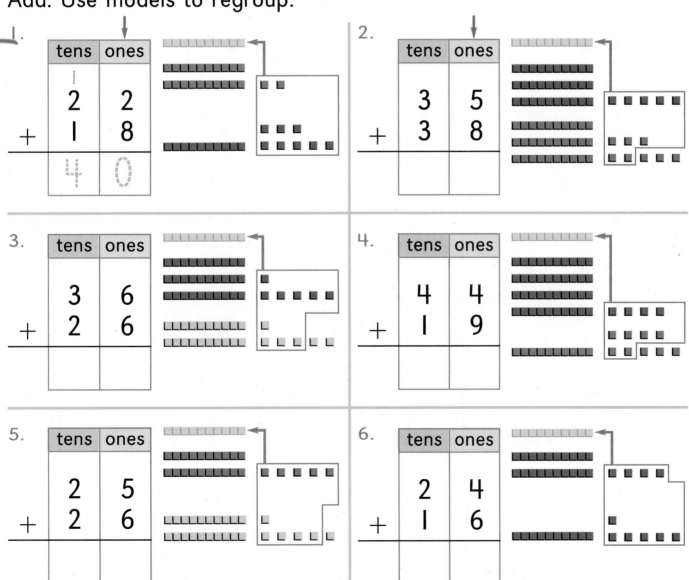

1.

tens	ones
2	2
+ 1	8
4	0

2.

tens	ones
3	5
+ 3	8

3.

tens	ones
3	6
+ 2	6

4.

tens	ones
4	4
+ 1	9

5.

tens	ones
2	5
+ 2	6

6.

tens	ones
2	4
+ 1	6

5-4 Ask your child to explain how to model regrouping to add 16 + 14.

Record the addends. Ring to show regrouping. Find the sum.

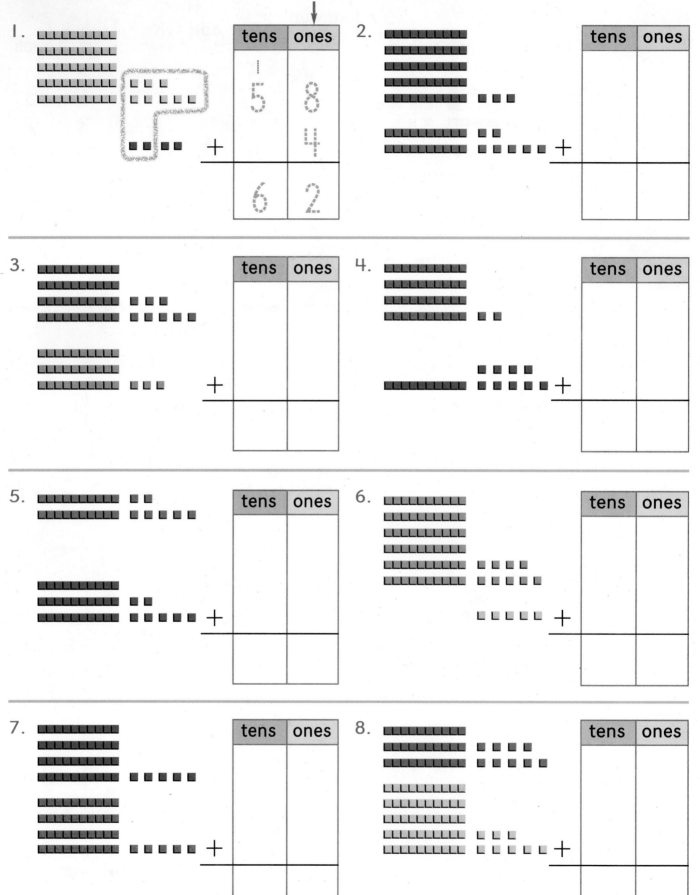

	tens	ones
	5	8
+		4
	6	2

2.

tens	ones

3.

tens	ones

4.

tens	ones

5.

tens	ones

6.

tens	ones

7.

tens	ones

8.

tens	ones

Find the sum of 58 and 17.

Add ones.

tens	ones
5	8
+ 1	7
	5

Regroup 10 ones
as 1 ten.

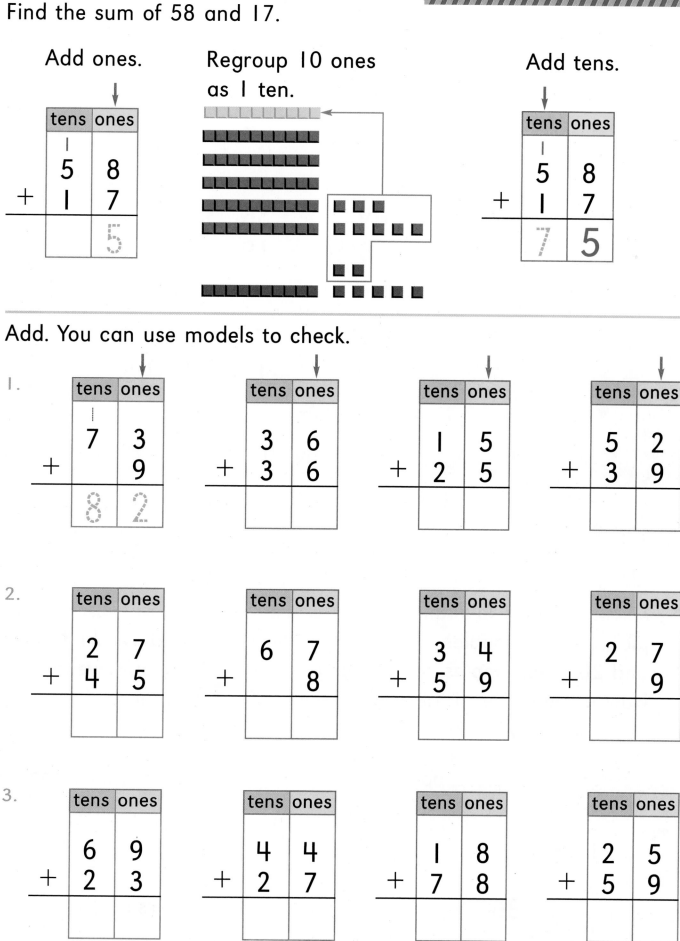

Add tens.

tens	ones
5	8
+ 1	7
7	5

Add. You can use models to check.

1.

tens	ones
7	3
+	9
8	2

tens	ones
3	6
+ 3	6

tens	ones
1	5
+ 2	5

tens	ones
5	2
+ 3	9

2.

tens	ones
2	7
+ 4	5

tens	ones
6	7
+	8

tens	ones
3	4
+ 5	9

tens	ones
2	7
+	9

3.

tens	ones
6	9
+ 2	3

tens	ones
4	4
+ 2	7

tens	ones
1	8
+ 7	8

tens	ones
2	5
+ 5	9

5-5 Have your child use the digits 2, 7, and 9
to show how to add with regrouping.

Find the sum. Regroup where needed.

1.

tens	ones
1	5
+ 1	6
3	1

tens	ones
3	8
+	3

tens	ones
5	3
+ 2	6

tens	ones
5	7
+	7

2.
$$\begin{array}{r} 44 \\ +\ 6 \\ \hline 50 \end{array}$$

$$\begin{array}{r} 47 \\ +38 \\ \hline \end{array}$$

$$\begin{array}{r} 69 \\ +\ 9 \\ \hline \end{array}$$

$$\begin{array}{r} 36 \\ +19 \\ \hline \end{array}$$

$$\begin{array}{r} 89 \\ +\ 4 \\ \hline \end{array}$$

$$\begin{array}{r} 31 \\ +49 \\ \hline \end{array}$$

3.
$$\begin{array}{r} 17 \\ +\ 4 \\ \hline \end{array}$$

$$\begin{array}{r} 29 \\ +38 \\ \hline \end{array}$$

$$\begin{array}{r} 77 \\ +\ 3 \\ \hline \end{array}$$

$$\begin{array}{r} 56 \\ +17 \\ \hline \end{array}$$

$$\begin{array}{r} 42 \\ +48 \\ \hline \end{array}$$

$$\begin{array}{r} 18 \\ +45 \\ \hline \end{array}$$

 SECOND LOOK In 2 and 3 ring sums that have 0 ones.

 PROBLEM SOLVING Use models to solve.

4. At the orchard TJ's family picked 34 red apples and 28 green apples. How many apples in all did TJ's family pick?

_____ apples

5. At the 🍴 there were 56 adults and 16 children. How many people were at the 🍴 altogether?

_____ people

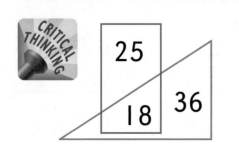

6. Find the sum of the numbers inside the rectangle. _____

7. Find the sum of the numbers inside the triangle. _____

Name _____

Change the <mark>order</mark> of addends
to check the sum.

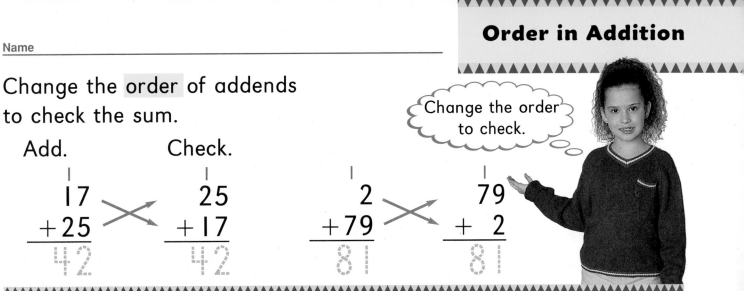

Change the order
to check.

Add.	Check.		
17	25	2	79
+25	+17	+79	+ 2
42	42	81	81

Add. Change the order to check.

1.
```
  46      16
 +16    + 46
  62      62
```

2.
```
  56
 + 5    +
```

3.
```
  63
 +19    +
```

4.
```
  52
 +28    +
```

5.
```
  37
 +18    +
```

6.
```
  84
 +12    +
```

7.
```
  76
 +18    +
```

8.
```
  17
 +69    +
```

9.
```
  29
 +48    +
```

10.
```
   5
 +93    +
```

11.
```
  49
 +29    +
```

12.
```
  66
 +17    +
```

MATH JOURNAL

13. Tell how you use the order of addends to compare.
Write < or >.

15 + 48 _____ 48 + 5   16 + 25 _____ 25 + 61

5-6   Give your child an addition example that is
incorrect and have him/her change the order
to check and correct the sum.
one hundred ninety-one   **191**

Name _____

Write the time in the ☐. Then draw to show the time 2 hours later.

1. four twenty
2. seven fifteen
3. 45 minutes after 9

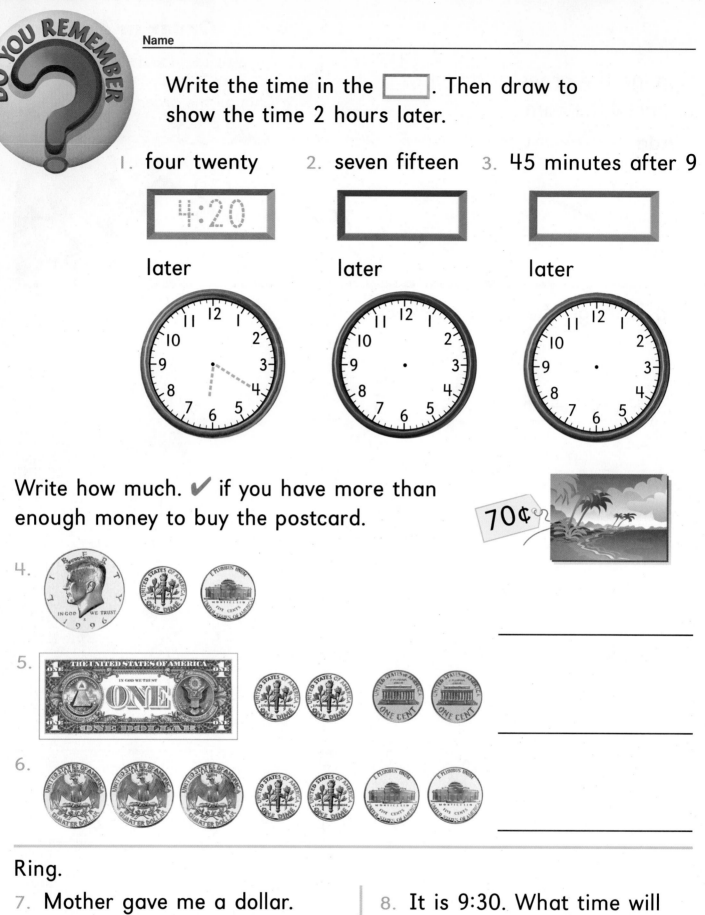

later   later   later

Write how much. ✔ if you have more than enough money to buy the postcard.

70¢

4. _____

5. _____

6. _____

Ring.

7. Mother gave me a dollar.
   I had thirty-three cents.
   How much do I have now?

   $1.13 $1.33

8. It is 9:30. What time will it be in one half hour?

   9:15   10:00

This page reviews the mathematical content presented in Chapter 4.

5

Name _____

Add 46¢ + 35¢.

Add the pennies. Regroup 10 pennies as 1 dime. Add the dimes.

dimes	pennies

	dimes	pennies
	1	
	4	6
+	3	5
	8	1

```
    1
   46¢
 + 35¢
   81¢
```

TALK IT OVER

How is adding dimes and pennies like adding tens and ones?

Find the sum. Regroup where needed.

1.
```
    1
   26¢      58¢      52¢      26¢      29¢
 + 44¢    + 29¢    + 20¢    + 62¢    +  7¢
   70¢
```

2.
```
   60¢      19¢      49¢      36¢      57¢
 + 14¢    + 64¢    + 25¢    +  7¢    + 39¢
```

3.
```
   38¢      55¢      37¢      22¢      76¢
 +  2¢    + 27¢    + 18¢    + 27¢    + 20¢
```

4.
```
   49¢       6¢      72¢       9¢      40¢
 + 19¢    + 65¢    + 18¢    + 81¢    + 38¢
```

SHARE YOUR THINKING

Which sum in each row above can you find by adding mentally? Why?

5-7   Have your child use dimes and pennies to redo exercise 4 and to explain when and how to regroup pennies as dimes.

one hundred ninety-three   **193**

Toy Museum Sale

39¢   28¢   54¢   35¢

1. Earl bought a ⛵ and a 🚙. How much did he spend in all?

_____ ¢

2. Jean bought a 🚚 and a 🚀. How much did she spend altogether?

_____ ¢

3. Calvin bought a ⛵ and a 🚀. How much did he spend in all?

_____ ¢

4. Dulce bought two 🚙. How much did she spend altogether?

_____ ¢

5. Jennifer wants to buy a 🚙 and a 🚀. She has 3 quarters. Does she have enough money?

_____

6. Rosa wants to buy 2 🚀. She has 6 dimes. Does she have enough money?

_____

CHALLENGE

7. Jose spent 82¢. Which two toys did he buy?

_____ and _____

Name _____

| 1st | Add ones. | 2nd | Add tens. |

tens	ones
2	2
2	3
+ 1	1
	6

tens	ones
2	2
2	3
+ 1	1
5	6

tens	ones
1	9
2	0
+ 1	5
	4

tens	ones
1	9
2	0
+ 1	5
5	4

Not enough ones to regroup.

Regroup 14 ones as 1 ten 4 ones.

Find the sum. Regroup where needed.

1.
```
  34      58      36      60      24      18
  21      10      42      27      31      40
+ 38    + 29    +  5    +  2    + 36    + 29
  93
```

2.
```
  74      63      14      23      37      10
  15      16      33      26      10       4
+  3    +  2    + 40    + 23    + 29    + 38
```

3.
```
  38      12      15      25      40      31
  11      35      31      32       7      16
+  7    +  4    + 23    + 34    + 36    + 38
```

 Which one in each row did not require regrouping? Why?

5-8 Put papers with the numbers 0–9 in a bag. Have your child pick 4 numbers, make 2 two-digit numbers, and add them to 11.

one hundred ninety-five **195**

Add. Regroup where needed. Look for doubles and tens.

1.
24	11	44	60	51	55
14	26	22	19	17	22
+ 8	+34	+33	+ 2	+12	+20
46					

2.
71	8	57	37	40	43
6	61	13	22	3	5
+17	+28	+20	+31	+39	+12

3.
23	32	21	81	23	34
14	35	16	8	16	41
+25	+ 5	+19	+ 6	+13	+23

4.
51	22	9	66	32	52
18	26	20	12	44	12
+21	+33	+51	+11	+15	+26

PROBLEM SOLVING

5. There were 13 people on the bus.
At Pine Street 6 more people got on.
At the next stop 21 people got on.
How many are now on the bus?   _____ people

MENTAL MATH

6. Color the greatest sum  yellow and
the sum that is least blue.

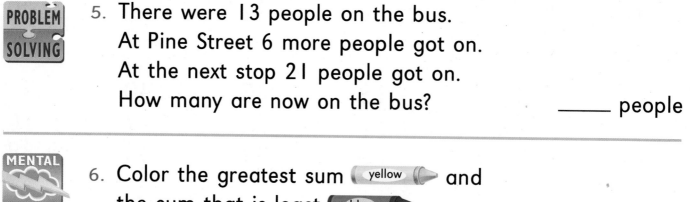

| 60 + 20 + 5 | 80 + 2 + 8 | 40 + 7 + 50 |

Name _____

I can add mentally.
I do not need to regroup.

I use paper and pencil.
I need to regroup.

21¢ + 2¢ + 10¢ = 33¢

Think: 22, 23, 33

$$\begin{array}{r} 1\\ 45\\ 13\\ +29\\ \hline 87 \end{array} \; > \; \begin{array}{r} 8 \text{ ones}\\ +9 \text{ ones}\\ \hline 17 \text{ ones} \end{array}$$

TALK IT OVER Explain how Nina counted on.
Explain how Jorgé regrouped 17 ones.

Add. Ring the sums you can do mentally.

1.
$$\begin{array}{r} 18\\ 1\\ +40\\ \hline 59 \end{array} \qquad \begin{array}{r} 24\\ 34\\ +4\\ \hline \end{array} \qquad \begin{array}{r} 14\\ 5\\ +56\\ \hline \end{array} \qquad \begin{array}{r} 12¢\\ 42¢\\ +20¢\\ \hline \end{array} \qquad \begin{array}{r} 53¢\\ 17¢\\ +21¢\\ \hline \end{array} \qquad \begin{array}{r} 10¢\\ 70¢\\ +5¢\\ \hline \end{array}$$

2.
$$\begin{array}{r} 24\\ +61\\ \hline \end{array} \qquad \begin{array}{r} 73\\ +18\\ \hline \end{array} \qquad \begin{array}{r} 33\\ +32\\ \hline \end{array} \qquad \begin{array}{r} 64¢\\ +18¢\\ \hline \end{array} \qquad \begin{array}{r} 26¢\\ +25¢\\ \hline \end{array} \qquad \begin{array}{r} 35¢\\ +18¢\\ \hline \end{array}$$

3.
$$\begin{array}{r} 48\\ +2\\ \hline \end{array} \qquad \begin{array}{r} 35\\ +47\\ \hline \end{array} \qquad \begin{array}{r} 5\\ +93\\ \hline \end{array} \qquad \begin{array}{r} 61¢\\ +2¢\\ \hline \end{array} \qquad \begin{array}{r} 78¢\\ +15¢\\ \hline \end{array} \qquad \begin{array}{r} 76¢\\ +8¢\\ \hline \end{array}$$

4.
$$\begin{array}{r} 29\\ +37\\ \hline \end{array} \qquad \begin{array}{r} 82\\ +16\\ \hline \end{array} \qquad \begin{array}{r} 48\\ +26\\ \hline \end{array} \qquad \begin{array}{r} 50¢\\ +8¢\\ \hline \end{array} \qquad \begin{array}{r} 85¢\\ +9¢\\ \hline \end{array} \qquad \begin{array}{r} 37¢\\ +6¢\\ \hline \end{array}$$

5-9 **Have your child explain why she/he can find some sums mentally.**

Find the sum. Regroup where needed.

Describe each pattern.

1.
```
  29        38        47        56        65
+  9      +  8      +  7      +  6      +  5
  38
```

2.
```
  65        44        23         2        11        20
  25        34        43        52        31        10
+  9      + 19      + 29      + 39      + 49
```

3.
```
  23        25        27        29        31        33
+ 58      + 58      + 58      + 58      + 58      + 58
```

PROBLEM SOLVING

4. Bus 1 has 22 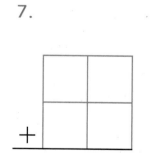, bus 2 has 43, and bus 3 has 34. How many in all are on the three buses?

_____

5. At the station 1 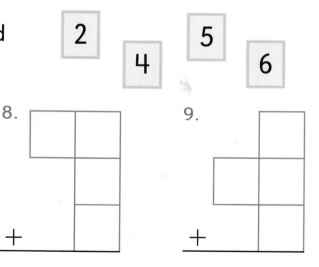 costs 39¢. How much would it cost in all to buy 2 ?

_____

CHALLENGE Use the number cards to find different sums less than 100.

2   4   5   6

6.
```
+ ___ ___
```

7.
```
+ ___ ___
```

8.
```
+ ___ ___
```

9.
```
+ ___ ___
```

Name _____

Read → Name → Think → Write → Check

The number Mia has is hidden information.

1. Reba has 25 postcards.
Mia has a dozen more than Reba.
How many postcards
do they have altogether?

$$25 + 37 = 62$$

25 + 12

They have __62__ postcards altogether.

2. There is 79¢ in the 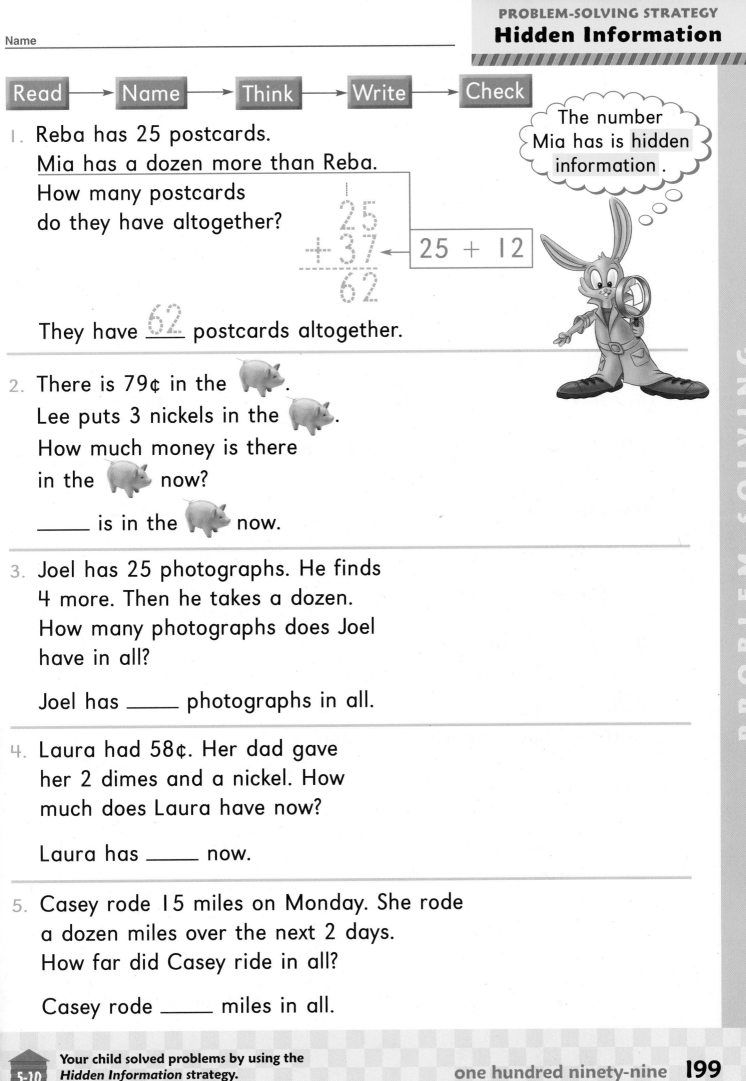.
Lee puts 3 nickels in the .
How much money is there
in the now?

_____ is in the now.

3. Joel has 25 photographs. He finds
4 more. Then he takes a dozen.
How many photographs does Joel
have in all?

Joel has _____ photographs in all.

4. Laura had 58¢. Her dad gave
her 2 dimes and a nickel. How
much does Laura have now?

Laura has _____ now.

5. Casey rode 15 miles on Monday. She rode
a dozen miles over the next 2 days.
How far did Casey ride in all?

Casey rode _____ miles in all.

PROBLEM SOLVING

5-10 Your child solved problems by using the
Hidden Information strategy.

one hundred ninety-nine **199**

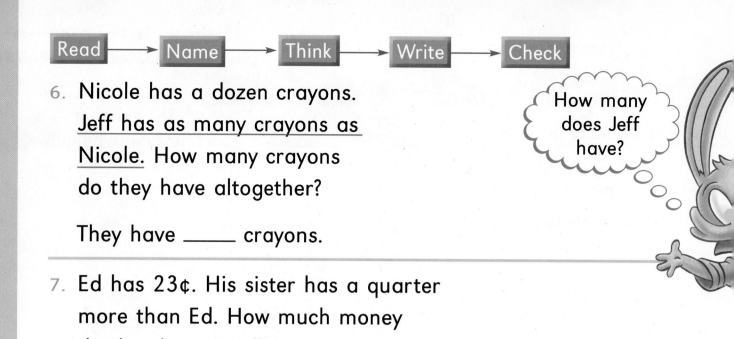

How many does Jeff have?

6. Nicole has a dozen crayons. <u>Jeff has as many crayons as Nicole.</u> How many crayons do they have altogether?

They have _____ crayons.

7. Ed has 23¢. His sister has a quarter more than Ed. How much money do they have in all?

They have _____ in all.

8. Beth scored 38 points. Carrie scored a dozen points more than Beth. How many points did they score in all?

They scored _____ points in all.

9. Tara has 26 books. Mark has 3 fewer books than Tara. Yolanda has as many books as Tara. How many books do the three children have altogether?

They have _____ books altogether.

10. I have a dozen pencils. You have ____ more. How many pencils do we have in all?

We have _____ pencils.

11. My map costs 29¢. Yours costs ____¢ more. How much do both maps cost?

Both maps cost _____¢.

PROBLEM SOLVING

Name _____

Read

1. Phil jogged 8 miles.
   Kyle jogged 5 miles.

Ask

   How many miles in all?
   or
   How many more miles
   did Phil jog than Kyle?

Check! Did you answer the question? Does your answer make sense?

Think

   Add to find how many in all. 8 ⊕ 5 = 13

   They jogged 13 miles in all.
   or
   Subtract to find how many more. 8 ⊖ 5 = 3

Write

   Phil jogged 3 more miles.

Read

2. Leah drew 13 maps.
   Katie drew 8 maps.

Ask

   _____

   _____

Think

   To answer your question,
   do you add or subtract?

Write

                        _____ maps.

Read

3. Feng spent a dime and a nickel.
   Rita spent 12¢. José spent 8¢.

Ask

   _____

   _____

Think

   To answer your question,
   do you add or subtract?

Write

                        _____.

PROBLEM SOLVING

5-11 Create problem situations like those above.
Ask your child to make up the question.

two hundred one **201**

Use a strategy you have learned.

4. On Monday 26 people rode on the ferry. The next day 2 dozen people rode on it. How many people in all rode on the ferry?

_____ people rode in all.

5. A 🔑 costs more than a ⭐.
A ⚽ costs less than a ⭐.
Match each key ring with its cost.

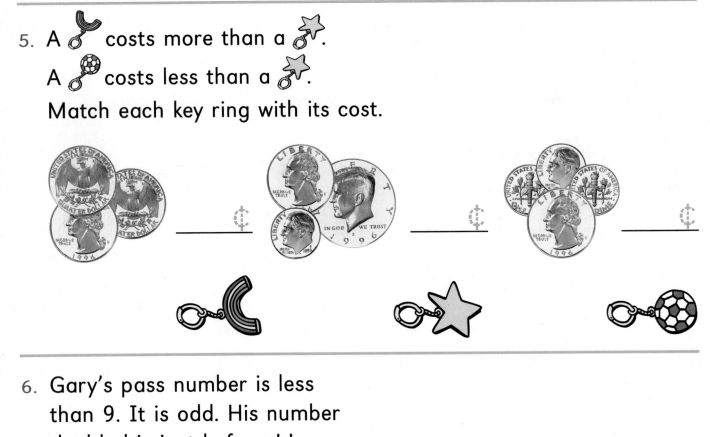

6. Gary's pass number is less than 9. It is odd. His number doubled is just before 11. What is Gary's pass number?

His number is _____.

7. Kerry saw 18 🎈 at the festival.
She saw 6 more on the way home.
8 🎈 had stripes.
How many balloons did Kerry see?

Kerry saw _____ 🎈.

Name _____

Add. Regroup where needed.

1.
78	57	24	37	23	26
+13	+23	+60	+ 5	+54	+27

91

2.
12	85	69	53	68	12
+28	+ 5	+14	+18	+28	+46

3.
65¢	24¢	46¢	24¢	84¢
+30¢	+ 7¢	+46¢	+39¢	+15¢

4.
61	12	24	76¢	23¢
14	43	24	3¢	34¢
+13	+29	+24	+10¢	+17¢

5. $14 + 20 + 1 =$ _____

6. $30 + 45 + 2 =$ _____

Add and check.

7.
52	
+37	+

8.
21	
+35	+

9.
27¢	
+63¢	+

PROBLEM SOLVING

10. Macy has 66¢. Her mom gave her a quarter. How much does Macy have now?

_____

11. Raj took 36 photos. Don took a dozen more than Raj. How many photos did they take altogether?

_____ photos

5 This page reviews the mathematical content presented in Chapter 5.

two hundred three **203**

MATH-E-MAGIC

Find the names of Christopher Columbus's 3 ships.

(2,6) means
across 2, up 6.

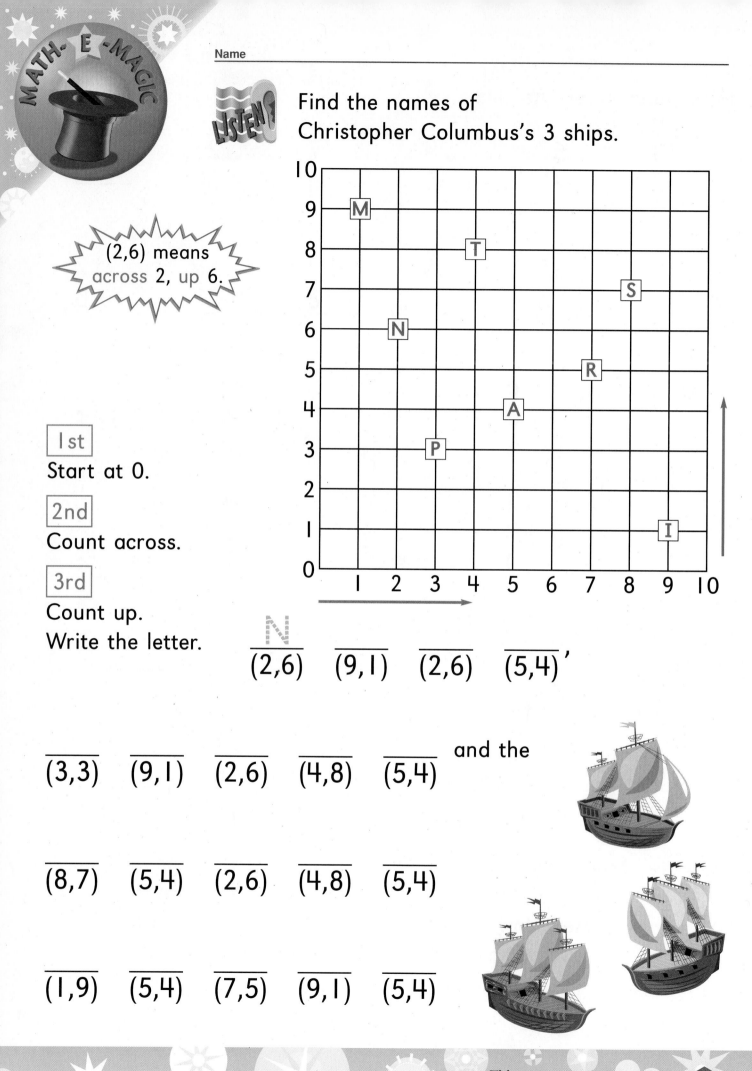

1st
Start at 0.

2nd
Count across.

3rd
Count up.
Write the letter.

$\overline{\underset{(2,6)}{N}}$ $\overline{\underset{(9,1)}{}}$ $\overline{\underset{(2,6)}{}}$ $\overline{\underset{(5,4)}{}}$'

$\overline{\underset{(3,3)}{}}$ $\overline{\underset{(9,1)}{}}$ $\overline{\underset{(2,6)}{}}$ $\overline{\underset{(4,8)}{}}$ $\overline{\underset{(5,4)}{}}$ and the

$\overline{\underset{(8,7)}{}}$ $\overline{\underset{(5,4)}{}}$ $\overline{\underset{(2,6)}{}}$ $\overline{\underset{(4,8)}{}}$ $\overline{\underset{(5,4)}{}}$

$\overline{\underset{(1,9)}{}}$ $\overline{\underset{(5,4)}{}}$ $\overline{\underset{(7,5)}{}}$ $\overline{\underset{(9,1)}{}}$ $\overline{\underset{(5,4)}{}}$

This page extends your child's
understanding of coordinate geometry.

5

1. Find the sum. Write what comes next in the pattern.

$$\begin{array}{r} 22 \\ +\ 9 \\ \hline \end{array} \qquad \begin{array}{r} 33 \\ +\ 9 \\ \hline \end{array} \qquad \begin{array}{r} 44 \\ +\ 9 \\ \hline \end{array} \qquad \begin{array}{r} \\ +\ \_\_ \\ \hline \end{array}$$

2. Double each amount.
   ✔ the number sentence if you regrouped pennies as dimes.

   3 dimes 4 pennies   ___ + ___ = ___   ___

   2 dimes 6 pennies   ___ + ___ = ___   ___

3. Find the sum mentally. Tell your strategy.

   $22 + 30 + 30 =$ ___ $19 + 1 + 23 =$ ___

   PORTFOLIO Choose 1 of these projects.
   Use a separate sheet of paper.

4. You have $60 to spend. List three items you can buy.

Item	Cost
Book	$ 8
Video	$19
Game	$32
Headphones	$15
CD	$14

5. Spin twice for each addend. Record sums to fill in a table like the one below. Tell if the sum is even or odd and less than or greater than 50.

Addends		Sum	
11	23	34	even, less than 50

ASSESSMENT

This page provides a variety of informal assessment opportunities in order to measure your child's understanding of Chapter 5.

5   two hundred five **205**

Find the sum. Regroup where needed.

1. $\begin{array}{r} 44 \\ +\ 9 \\ \hline \end{array}$ $\begin{array}{r} 34 \\ +20 \\ \hline \end{array}$ $\begin{array}{r} 82 \\ +\ 5 \\ \hline \end{array}$ $\begin{array}{r} 19¢ \\ +49¢ \\ \hline \end{array}$ $\begin{array}{r} 47¢ \\ +26¢ \\ \hline \end{array}$

2. $\begin{array}{r} 58 \\ +24 \\ \hline \end{array}$ $\begin{array}{r} 73 \\ +\ 9 \\ \hline \end{array}$ $\begin{array}{r} 45 \\ +\ 8 \\ \hline \end{array}$ $\begin{array}{r} 18¢ \\ +72¢ \\ \hline \end{array}$ $\begin{array}{r} 19¢ \\ +20¢ \\ \hline \end{array}$

3. $\begin{array}{r} 25 \\ 15 \\ +31 \\ \hline \end{array}$ $\begin{array}{r} 47 \\ 2 \\ +35 \\ \hline \end{array}$ $\begin{array}{r} 16 \\ 22 \\ +37 \\ \hline \end{array}$ $\begin{array}{r} 43¢ \\ 11¢ \\ +19¢ \\ \hline \end{array}$ $\begin{array}{r} 38¢ \\ 40¢ \\ +\ 8¢ \\ \hline \end{array}$

Compare. Write $<$, $=$, or $>$.

4. $62 + 20$ ___ $62 + 2$ 5. $32 + 5$ ___ $35 + 2$

6. $20 + 4 + 2$ ___ $20 + 40 + 2$

 PROBLEM SOLVING   Use a strategy you have learned.

7. Jess was tenth in the ticket line. 12 people were behind him. How many people in all were in the ticket line?

8. 29 scouts are at camp. A dozen more come. How many scouts are at camp now?

_____ people   _____ scouts

5

Name _____

Mark the ◯ for your answer.

Listening Section

Ⓐ

cannot
yes   no   tell
◯   ◯   ◯

Ⓑ   9 ✈
         7 ✈

                              cannot
add   subtract   tell
◯   ◯   ◯

1.

2 tens 7 ones + 3 ones = _____

◯ 5 tens 7 ones   ◯ 3 tens 7 ones
◯ 3 tens 3 ones   ◯ 3 tens 0 ones

2.

30 + 30 = _____

                          not
30   60   50   given
◯   ◯   ◯   ◯

3. What comes next?

20, 30, 40, _____

◯ 50   ◯ 40
◯ 60   ◯ 10

4.

50 + 4

◯ 5
◯ 40
◯ 45
◯ 54

5.

☐ + 5 = 10

◯ 4   ◯ 10
◯ 5   ◯ 6

6.

  17
− 8

◯ 11
◯ 8
◯ 9
◯ 7

7.

   70
 +20

◯ 50
◯ 60
◯ 72
◯ 90

8.

   34
   12
 +33

◯ 69
◯ 46
◯ 79
◯ 76

9.

  15¢
 +40¢

◯ 55¢
◯ 25¢
◯ 45¢
◯ 60¢

10.

   53
 +27

◯ 70
◯ 79
◯ 80
◯ 81

11.

◯ 46¢
◯ 41¢
◯ 55¢
◯ 36¢

ASSESSMENT

Mark the ◯ for your answer.

12.

◯ 40¢
◯ 60¢
◯ 55¢
◯ 65¢

13.

◯ 10:30
◯ 10:03
◯ 10:15
◯ 3:50

14.

◯ 2:20
◯ 4:10
◯ 2:04
◯ 2:40

15.

19
+22
⬚1

◯ 2
◯ 3
◯ 4
◯ 5

16.

36¢
+ 8¢

◯ 32¢
◯ 34¢
◯ 44¢
◯ 54¢

17. Which numbers are odd?

◯ 6 and 60
◯ 17 and 12
◯ 14 and 45
◯ 13 and 31

18. What is the amount?

◯ $1.23
◯ $1.30
◯ $1.32
◯ $1.50

19. How much time passed?

started   and   finished

◯ 15 minutes   ◯ 30 minutes
◯ 1 hour   ◯ 2 hours

20. What is not a fair trade for 🪙🪙?

◯ 25 pennies
◯ 10 nickels
◯ 5 dimes
◯ 1 half dollar

21. What does not belong to the same fact family?

◯ 7 + 3   ◯ 10 − 7
◯ 3 + 7   ◯ 7 − 3

22. Ty earned 1 dollar for chores. He found 3 nickels. How much does he have now?

$5.30 $1.30 $1.15 $1.05
◯   ◯   ◯   ◯

23. Today is Wednesday, February 11. In 1 week it will be February ____.

17   18   19   20
◯   ◯   ◯   ◯

Subtraction of Two-Digit Numbers

CRITICAL THINKING

You are in a line of 18 people to buy tickets. 6 people are ahead of you. How many people are behind you?

For more information about Chapter 6, visit the Family Information Center at **www.sadlier-oxford.com**

Internet

Dear Family,

Today your child began Chapter 6. As he/she studies subtraction of 2-digit numbers, you may want to read the poem below, which was read in class. Have your child talk about some of the math ideas pictured on page 209.

Look for the 🏠 at the bottom of each skills lesson. The suggestion on the page gives you an opportunity to improve your child's understanding of math and to reinforce his/her math language. You may want to have dimes, pennies and other countables available to your child to use throughout this chapter.

Home Activity

Raccoon Mystery Theater

Try this activity with your child. Tell a subtraction story like the one below. Ask your child to use stacks of ten pennies or blocks to act out the story. Have him/her write the problem and then solve it.

Dad made 40 cookies. When he went upstairs, a raccoon came in and took 10 cookies. How many cookies were left by the raccoon?

Home Reading Connection

On with the Show

The time has come! Today's the day!
The second grade presents a play.
We've learned our lines and know each part.
We'll tell a tale with skill and art.

Twenty costumes all in a row,
Twenty hats are set to go.
The stage is set, the lights are on.
The scenes are painted, curtains drawn.

Ninety tickets have been sold
To friends and family, young and old.
We'll make them happy, make them proud.
They'll laugh and clap and cheer out loud.

Twenty students wait to start.
Waiting is the hardest part.
The time has come! Today's the day!
The second grade presents a play.

Marie A. Cooper

Name

Complete the increasing patterns.

Complete the repeating patterns.

CONNECTIONS

Complete an increasing pattern and a repeating pattern.

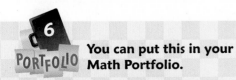

People all over the world use masks for celebrations and shows.

Draw and color to make the right half match the left half.

C O N N E C T I O N S

You can put this in your Math Portfolio.

Name _____

Subtract: 56 − 24 = ?

| 1st | Subtract ones. |

tens	ones
5	6
− 2	4
	2

| 2nd | Subtract tens. |

tens	ones
5	6
− 2	4
3	2

```
  56
− 24
  32
```

Record. Subtract. Use models to check.

1. 62 − 51

tens	ones
6	2
− 5	1
1	1

2. 38 − 25

tens	ones
3	8
−	

3. 29 − 14

tens	ones
2	9
−	

Subtract.

4.
```
  49      75      99      48      55      66
− 32    − 53    − 28    − 17    − 35    − 43
  17
```

5.
```
  88      97      44      99      86      77
− 58    − 43    − 20    − 67    − 13    − 37
```

6.
```
  69      54      74      97      63      86
− 33    − 43    − 64    − 51    − 42    − 73
```

Do not write 0 when no tens are left.

Find the difference.

1.
$$73 - 71 = 2$$ $$44 - 41$$ $$68 - 62$$ $$97 - 95$$

2.
$$19 - 14$$ $$36 - 15$$ $$56 - 10$$ $$65 - 24$$ $$79 - 26$$ $$68 - 14$$

3.
$$86 - 22$$ $$37 - 34$$ $$99 - 15$$ $$76 - 46$$ $$84 - 32$$ $$85 - 42$$

4.
$$77 - 32$$ $$85 - 62$$ $$78 - 10$$ $$95 - 31$$ $$78 - 16$$ $$58 - 53$$

PROBLEM SOLVING

5. There were 96 singers on stage. There were 75 dancers on stage. How many more singers than dancers were on stage?

_____

CRITICAL THINKING

6. Sean had 36 flyers. Anna had 14 flyers. After Sean gave Anna some of his flyers, they each had the same number. How many flyers did Sean give Anna?

_____

"Of Thee We Sing"

Main Auditorium
Grade 2 Players
Friday at 2 P.M.

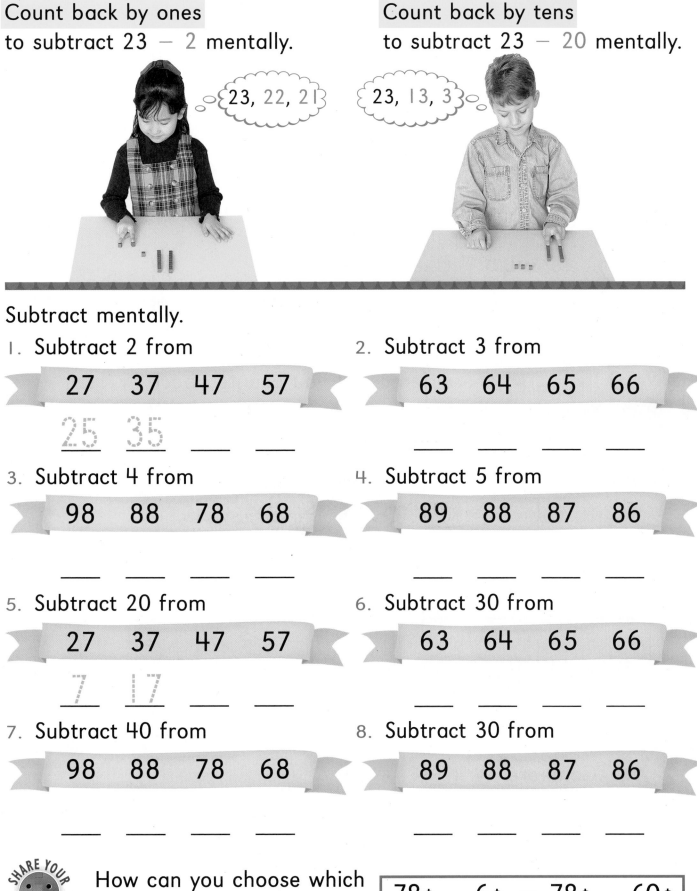

Name _____

Count back by ones
to subtract 23 − 2 mentally.

23, 22, 21

Count back by tens
to subtract 23 − 20 mentally.

23, 13, 3

Subtract mentally.

1. Subtract 2 from

27   37   47   57

25   35   ___   ___

2. Subtract 3 from

63   64   65   66

___   ___   ___   ___

3. Subtract 4 from

98   88   78   68

___   ___   ___   ___

4. Subtract 5 from

89   88   87   86

___   ___   ___   ___

5. Subtract 20 from

27   37   47   57

7   17   ___   ___

6. Subtract 30 from

63   64   65   66

___   ___   ___   ___

7. Subtract 40 from

98   88   78   68

___   ___   ___   ___

8. Subtract 30 from

89   88   87   86

___   ___   ___   ___

SHARE YOUR THINKING

How can you choose which
is less without subtracting?

| 78¢ − 6¢ or 78¢ − 60¢ |

6-2 Tell your child to describe how
counting back by ones or tens
makes it easy to subtract mentally.

two hundred fifteen **215**

Subtract.

1.
49	58	64	35	78	75
− 7	− 6	− 2	− 5	− 8	− 4
42					

2.
48	55	77	56	69	28
−20	−30	−40	−20	−50	−20
28					

3.
82	49	54	38	94	43
−20	− 6	−50	−20	− 4	−30

 In 1–3 ✔ the difference if you counted back by 1s. Ring the difference if you counted back by 10s.

Complete. Write the missing difference.

4. *27*

70 80
60 · 10
97−
50 · 20
40 30

5. *90*

7 8
6 · 1
98−
5 · 2
4 3

Compare. Write <, =, or >.

6. 50 − 10 ___ 70 − 30   7. 30 − 10 ___ 60 − 20

8. 90 − 60 ___ 80 − 40   9. 75 − 15 ___ 75 − 25

10. 74 − 40 ___ 84 − 40   11. 68 − 3 ___ 58 − 3

Name _____

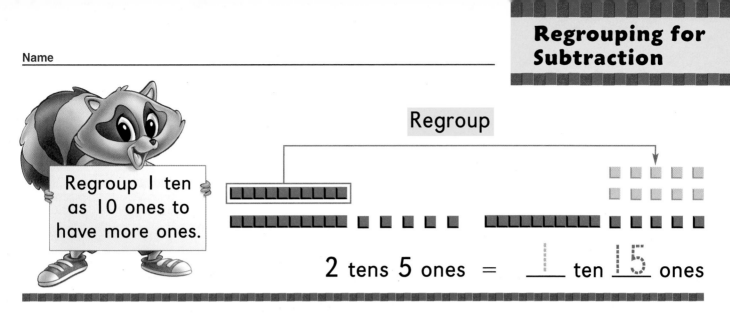

Regroup 1 ten as 10 ones to have more ones.

Regroup

2 tens 5 ones = ___ ten 15 ones

Model as you regroup 1 ten as 10 ones.
Write the missing numbers.

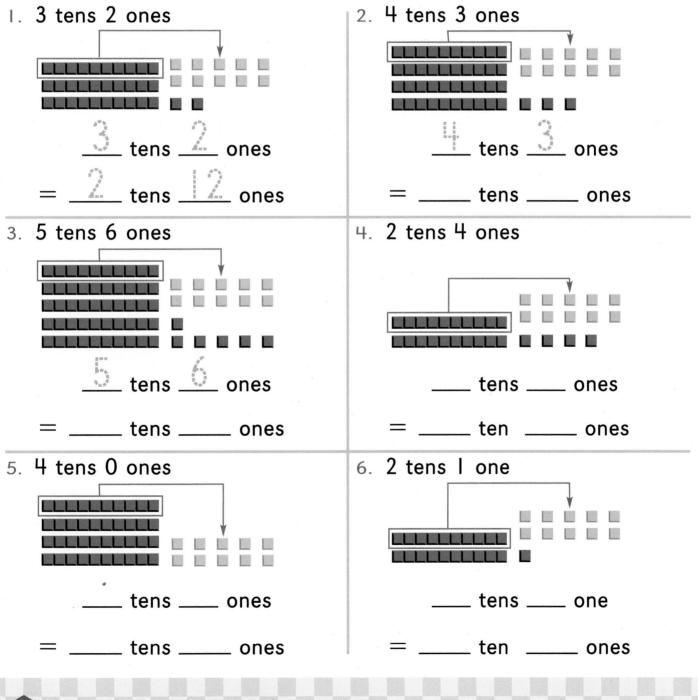

1. 3 tens 2 ones

___ tens ___ ones

= ___ tens ___ ones

2. 4 tens 3 ones

___ tens ___ ones

= ___ tens ___ ones

3. 5 tens 6 ones

___ tens ___ ones

= ___ tens ___ ones

4. 2 tens 4 ones

___ tens ___ ones

= ___ ten ___ ones

5. 4 tens 0 ones

___ tens ___ ones

= ___ tens ___ ones

6. 2 tens 1 one

___ tens ___ one

= ___ ten ___ ones

6-3 Invite your child to model subtracting
4 ones from 2 tens 1 one by regrouping.

two hundred seventeen **217**

2 tens 3 ones − 6 ones = ?

tens	ones

Then subtract.

tens	ones

Not enough ones.
Regroup 1 ten
as 10 ones.

2 tens 3 ones
= __1__ ten __13__ ones

1 ten   13 ones
−   6 ones
1 ten   7 ones

Subtract. Regroup 1 ten to have enough ones.

1. 3 tens 0 ones − 5 ones

Regroup. __2__ tens __10__ ones
−   __5__ ones
__2__ tens __5__ ones

2. 4 tens 7 ones − 9 ones

Regroup. __3__ tens ____ ones
−   __9__ ones
____ tens ____ ones

3. 5 tens 1 one − 4 ones

Regroup. ____ tens ____ ones
−   __4__ ones
____ tens ____ ones

4. 3 tens 6 ones − 8 ones

Regroup. ____ tens ____ ones
−   ____ ones
____ tens ____ ones

5. 2 tens 8 ones − 9 ones

Regroup. ____ ten ____ ones
−   ____ ones
____ ten ____ ones

6. 4 tens 2 ones − 7 ones

Regroup. ____ tens ____ ones
−   ____ ones
____ tens ____ ones

MATH JOURNAL

7. You have 5 ▭▭▭▭▭▭▭▭▭ 4 ▫. For which rules will you need to regroup?

−3 ones −7 ones
−9 ones −1 one −6 ones

Name

Subtract 17 from 31.

Not enough ones to subtract 7 from 1.

Then subtract.

tens	ones
3	1
– 1	7

Regroup 1 ten as 10 ones.

3 tens 1 one
= 2 tens 11 ones

tens	ones
²3̷	¹¹1̷
– 1	7
	1 4

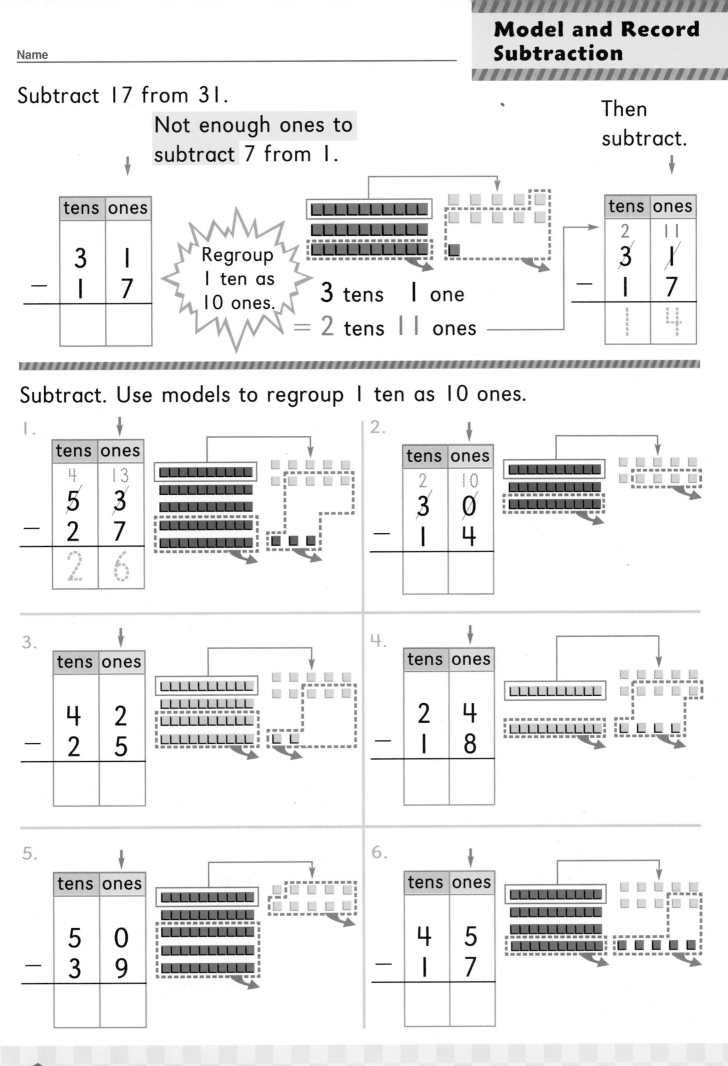

Subtract. Use models to regroup 1 ten as 10 ones.

1.

tens	ones
⁴5̷	¹³3̷
– 2	7
2	6

2.

tens	ones
²3̷	¹⁰0̷
– 1	4

3.

tens	ones
4	2
– 2	5

4.

tens	ones
2	4
– 1	8

5.

tens	ones
5	0
– 3	9

6.

tens	ones
4	5
– 1	7

6-4 Have your child tell how the picture of subtraction with regrouping in exercises 1 and 2 relates to the subtraction.

two hundred nineteen **219**

Record the regrouping. Find the difference.

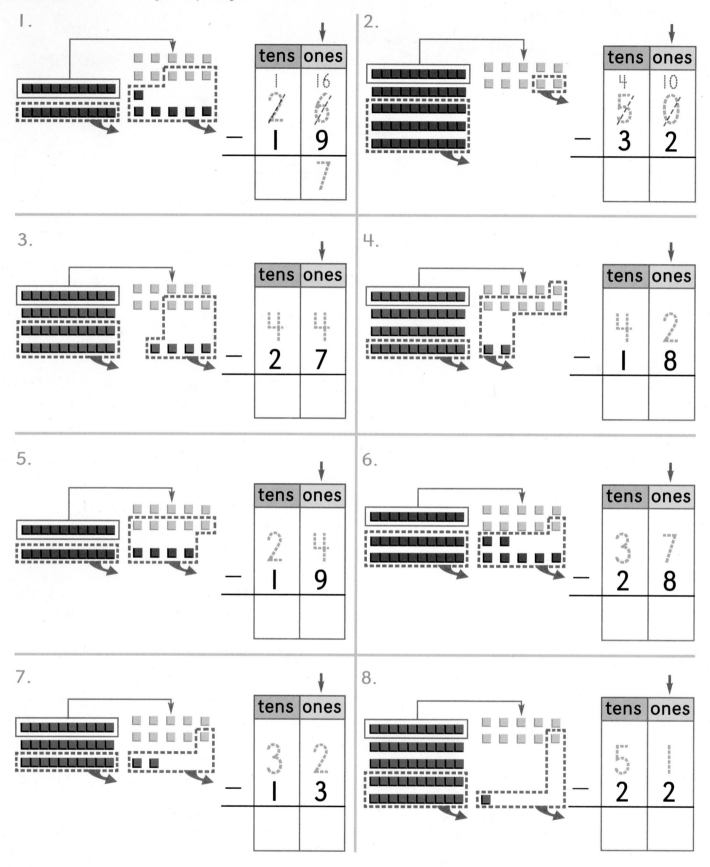

1.

tens	ones
1 2̶	16 6̶
− 1	9
	7

2.

tens	ones
4 5̶	10 0̶
− 3	2

3.

tens	ones
4	4
− 2	7

4.

tens	ones
4	2
− 1	8

5.

tens	ones
2	4
− 1	9

6.

tens	ones
3	7
− 2	8

7.

tens	ones
3	2
− 1	3

8.

tens	ones
5	1
− 2	2

Look at the differences in 1–8. Why do some
of the answers have no tens?

Subtract 6 from 34.

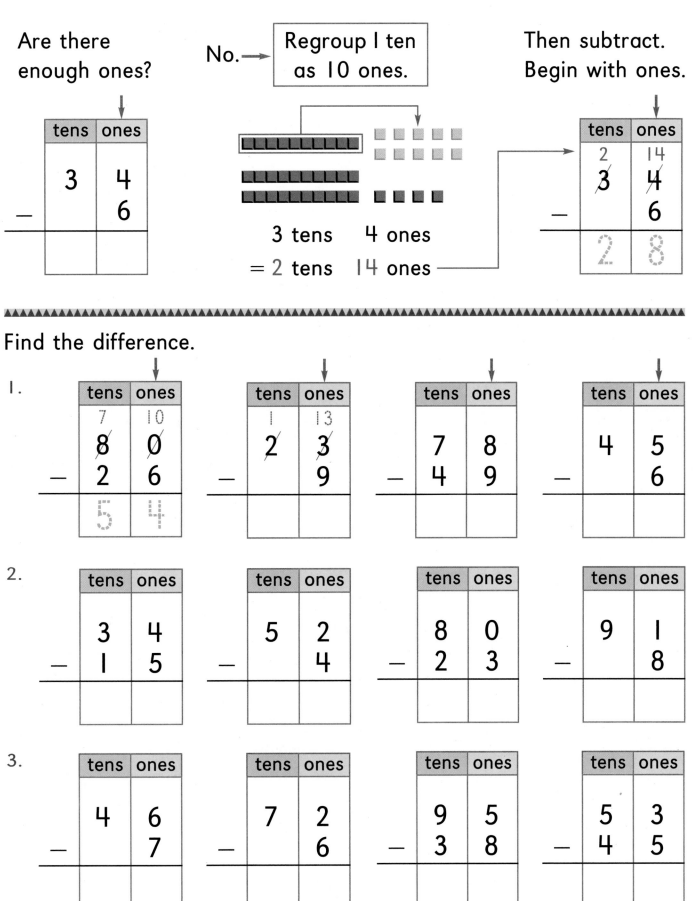

Are there enough ones?

No. → Regroup 1 ten as 10 ones.

Then subtract. Begin with ones.

tens	ones
3	4
−	6

3 tens   4 ones
= 2 tens   14 ones

tens	ones
2	14
3̷	4̷
−	6
2	8

Find the difference.

1.

tens	ones
7	10
8̷	0̷
− 2	6
5	4

tens	ones
1	13
2	3̷
−	9

tens	ones
7	8
− 4	9

tens	ones
4	5
−	6

2.

tens	ones
3	4
− 1	5

tens	ones
5	2
−	4

tens	ones
8	0
− 2	3

tens	ones
9	1
−	8

3.

tens	ones
4	6
−	7

tens	ones
7	2
−	6

tens	ones
9	5
− 3	8

tens	ones
5	3
− 4	5

6-5 Prompt your child to subtract 3 from 28 and 13 from 28 and to explain how these exercises are alike and different.

two hundred twenty-one **221**

Find the difference. Regroup where needed.

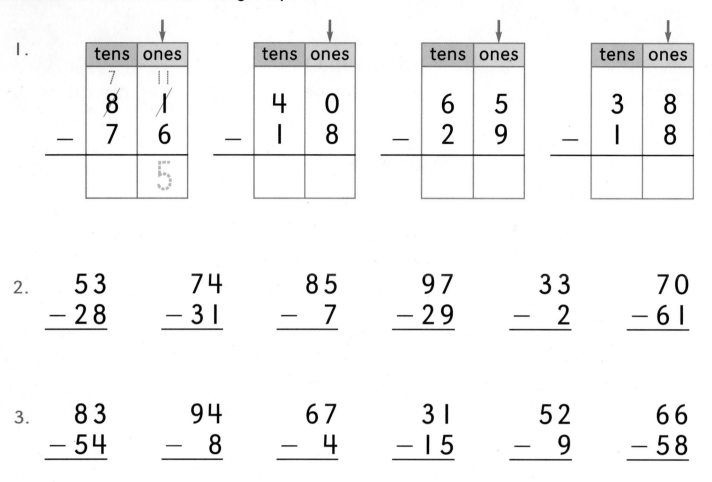

1.

tens	ones
⁷8̸	¹¹1̸
− 7	6
	5

tens	ones
4	0
− 1	8

tens	ones
6	5
− 2	9

tens	ones
3	8
− 1	8

2.
$$53 - 28 \qquad 74 - 31 \qquad 85 - 7 \qquad 97 - 29 \qquad 33 - 2 \qquad 70 - 61$$

3.
$$83 - 54 \qquad 94 - 8 \qquad 67 - 4 \qquad 31 - 15 \qquad 52 - 9 \qquad 66 - 58$$

PROBLEM SOLVING

4. There are 45 🎵 on a shelf.
There are 29 children in the band.
Each child takes a 🎵.
How many 🎵 are left
on the shelf?

_____ 🎵 are left.

5. Ned counted 27 📯 and 45 🪘.
How many more 🪘 than 📯
did he count?

_____ 🪘 more

6. Ben had 38 🎟. He sold
18 of them. How many 🎟
were left?

_____ 🎟 were left.

Name

Julio has 5 🪙 and 7 🪙. He spends
28¢. How much does he have left?

Subtract pennies first.

Not enough pennies to subtract.
Regroup I dime as 10 pennies.

5 dimes   7 pennies
= 4 dimes 17 pennies

dimes	pennies
4	17
5̸	7̸
− 2	8
2	9

```
  4 17
  5̸7̸¢
− 2 8¢
  2 9¢
```

Find the difference. Regroup where needed.

1.
```
 7 15
 8̸5̸¢
−4 6¢
 3 9¢
```
```
 6 10
 7̸0̸¢
−  4¢
```
```
 4 0¢
−2 7¢
```
```
 3 4¢
−1 0¢
```
```
 6 1¢
−3 2¢
```

2.
```
 6 4¢
−4 7¢
```
```
 7 8¢
−2 9¢
```
```
 5 8¢
−1 8¢
```
```
 3 6¢
−2 9¢
```
```
 2 3¢
−  7¢
```

3.
```
 7 3¢
−2 4¢
```
```
 8 6¢
−  4¢
```
```
 9 4¢
−5 9¢
```
```
 6 2¢
−1 7¢
```
```
 9 1¢
−2 9¢
```

4.
```
 6 3¢
−2 6¢
```
```
 6 0¢
−3 1¢
```
```
 7 7¢
−1 7¢
```
```
 4 6¢
−3 8¢
```
```
 8 2¢
−7 3¢
```

SHARE YOUR THINKING Which one in each row can you
solve by subtracting mentally? Why?

Subtract. Regroup where needed.

1.
$$\begin{array}{r} \overset{6}{\cancel{7}}\overset{15}{5}¢ \\ - 8¢ \\ \hline 67¢ \end{array}$$

$$\begin{array}{r} 52¢ \\ -25¢ \\ \hline \end{array}$$

$$\begin{array}{r} 62¢ \\ -48¢ \\ \hline \end{array}$$

$$\begin{array}{r} 63¢ \\ -15¢ \\ \hline \end{array}$$

$$\begin{array}{r} 79¢ \\ -30¢ \\ \hline \end{array}$$

2.
$$\begin{array}{r} 75¢ \\ -21¢ \\ \hline \end{array}$$

$$\begin{array}{r} 91¢ \\ -23¢ \\ \hline \end{array}$$

$$\begin{array}{r} 86¢ \\ -47¢ \\ \hline \end{array}$$

$$\begin{array}{r} 50¢ \\ -2¢ \\ \hline \end{array}$$

$$\begin{array}{r} 82¢ \\ -38¢ \\ \hline \end{array}$$

3.
$$\begin{array}{r} 21¢ \\ -6¢ \\ \hline \end{array}$$

$$\begin{array}{r} 32¢ \\ -19¢ \\ \hline \end{array}$$

$$\begin{array}{r} 41¢ \\ -24¢ \\ \hline \end{array}$$

$$\begin{array}{r} 60¢ \\ -25¢ \\ \hline \end{array}$$

$$\begin{array}{r} 95¢ \\ -84¢ \\ \hline \end{array}$$

PROBLEM SOLVING

4. The 🎭 for my costume costs 91¢.
   I have only 75¢. How much more
   money do I need to buy it?   _____

5. Dan had 75¢. He bought a 🎀. **59¢**
   How much money did he have left?   _____

6. Joy had 60¢. Lyle had 1 quarter,
   1 dime, and 3 pennies. How much
   more money did Joy have than Lyle?   _____

CHALLENGE

7. A 🌼 costs a dime. What is the
   greatest number of 🌼 I can buy
   with 45¢? How much money
   will I have left?

   _____ 🌼 _____ left

To check subtraction, add.

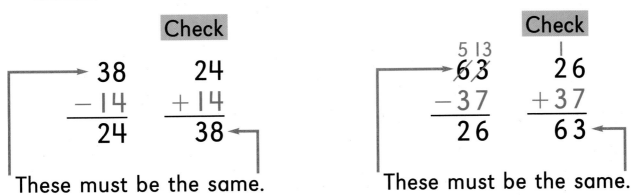

Check

$$38 \qquad 24$$
$$-14 \qquad +14$$
$$\overline{24} \qquad \overline{38}$$

These must be the same.

Check

$$\overset{5\ 13}{\cancel{63}} \qquad \overset{1}{26}$$
$$-37 \qquad +37$$
$$\overline{26} \qquad \overline{63}$$

These must be the same.

In each example, what number stands for the whole and what numbers stand for the parts? How are these examples different?

Subtract and check.

1.
$$\overset{3\ 12}{\cancel{42}} \qquad \overset{1}{16}$$
$$-26 \qquad +26$$
$$\overline{16} \qquad \overline{42}$$

2.
$$\overset{4\ 10}{\cancel{50}} \qquad 33$$
$$-17 \qquad +17$$
$$\overline{33} \qquad$$

3.
$$82$$
$$-47 \qquad +$$

4.
$$48$$
$$-\ 5 \qquad +$$

5.
$$63$$
$$-23 \qquad +$$

6.
$$91$$
$$-56 \qquad +$$

7.
$$27$$
$$-13 \qquad +$$

8.
$$33$$
$$-16 \qquad +$$

9.
$$78$$
$$-34 \qquad +$$

10.
$$75$$
$$-69 \qquad +$$

11.
$$90$$
$$-30 \qquad +$$

12.
$$60$$
$$-25 \qquad +$$

6-7 Direct your child to subtract and check 38−29, explaining which number is the whole and which numbers are the parts.

two hundred twenty-five **225**

Subtract and check.

1. $\begin{array}{r} \small 5\;14 \\ \cancel{6}4¢ \\ -27¢ \\ \hline 37¢ \end{array}$ $\begin{array}{r} 37¢ \\ +27¢ \\ \hline 64¢ \end{array}$

2. $\begin{array}{r} 76 \\ -59 \\ \hline \end{array}$

3. $\begin{array}{r} 82 \\ -69 \\ \hline \end{array}$

4. $\begin{array}{r} 75¢ \\ -26¢ \\ \hline \end{array}$

5. $\begin{array}{r} 40 \\ -22 \\ \hline \end{array}$

6. $\begin{array}{r} 53 \\ -48 \\ \hline \end{array}$

7. $\begin{array}{r} 61¢ \\ -\;\;4¢ \\ \hline \end{array}$

8. $\begin{array}{r} 93 \\ -34 \\ \hline \end{array}$

9. $\begin{array}{r} 74 \\ -36 \\ \hline \end{array}$

10. $\begin{array}{r} 25¢ \\ -18¢ \\ \hline \end{array}$

11. $\begin{array}{r} 81¢ \\ -26¢ \\ \hline \end{array}$

12. $\begin{array}{r} 57¢ \\ -39¢ \\ \hline \end{array}$

PROBLEM SOLVING   Is the change correct? Subtract and check.

13. I bought a 🎵 for 29¢. I gave the clerk 2 quarters. He gave me 2 dimes and 1 penny as change.

_____

14. I paid 3 quarters for a 🎵 that cost 67¢. The cashier gave me 1 nickel and 2 pennies.

_____

15. Write a problem like 13 and 14 in your Math Journal. Solve it.

Name _____

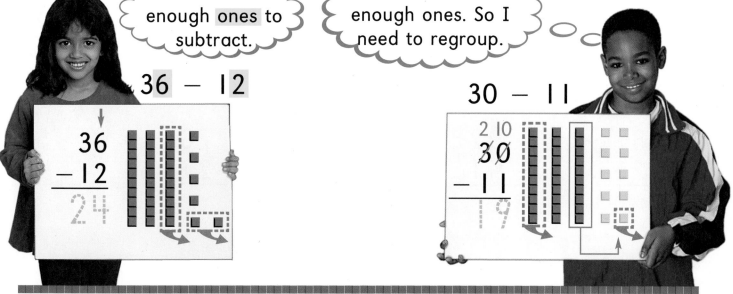

There are enough **ones** to subtract.

36 − 12

```
  36
− 12
  24
```

There are not enough ones. So I need to regroup.

30 − 11

```
  2 10
  3 0
− 1 1
  1 9
```

Ring the ones ⟨ green ⟩ if there are enough ones.
Subtract. Regroup if needed.

1.
```
  36        50        17        43        20        85
− 24      − 35      −  3      − 37      − 12      − 29
  12
```

2.
```
  94        32        69        27        82        46
− 36      −  6      − 59      − 11      − 44      − 28
```

3.
```
  48        63        67        56        94        25
− 23      − 18      − 49      − 13      − 47      −  8
```

4.
```
  91        39        79        52        81        44
− 27      − 32      − 24      − 45      − 73      − 18
```

6-8   Ask your child to fill in the missing digits in 6 □ − 2 □ and to tell whether or not regrouping is needed.

two hundred twenty-seven **227**

Subtract. Regroup where needed.

1.
$$\begin{array}{r} \overset{6\ 12}{\cancel{7}\cancel{2}}¢ \\ -53¢ \\ \hline 19¢ \end{array}$$
$$\begin{array}{r} 47¢ \\ -28¢ \\ \hline \end{array}$$
$$\begin{array}{r} 94¢ \\ -75¢ \\ \hline \end{array}$$
$$\begin{array}{r} 36¢ \\ -17¢ \\ \hline \end{array}$$
$$\begin{array}{r} 68¢ \\ -49¢ \\ \hline \end{array}$$

2.
$$\begin{array}{r} 87¢ \\ -26¢ \\ \hline \end{array}$$
$$\begin{array}{r} 92¢ \\ -41¢ \\ \hline \end{array}$$
$$\begin{array}{r} 54¢ \\ -13¢ \\ \hline \end{array}$$
$$\begin{array}{r} 78¢ \\ -27¢ \\ \hline \end{array}$$
$$\begin{array}{r} 45¢ \\ -\ 4¢ \\ \hline \end{array}$$

3.
$$\begin{array}{r} 99¢ \\ -15¢ \\ \hline \end{array}$$
$$\begin{array}{r} 87¢ \\ -15¢ \\ \hline \end{array}$$
$$\begin{array}{r} 75¢ \\ -15¢ \\ \hline \end{array}$$
$$\begin{array}{r} 63¢ \\ -15¢ \\ \hline \end{array}$$
$$\begin{array}{r} 51¢ \\ -15¢ \\ \hline \end{array}$$

 Describe the pattern in each row above.
Name another example that belongs to each.

 4. Sam and Lori each had 75¢.
Each bought a snack at the
show. Sam got 18¢ change
and Lori got 8¢ change.
How much did each spend?   Sam _____ Lori _____

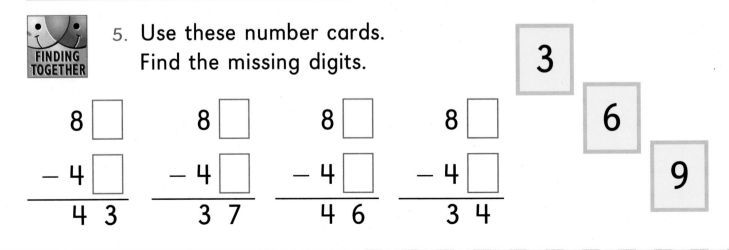

5. Use these number cards.
Find the missing digits.

$$\begin{array}{r} 8\ \square \\ -4\ \square \\ \hline 4\ 3 \end{array}$$
$$\begin{array}{r} 8\ \square \\ -4\ \square \\ \hline 3\ 7 \end{array}$$
$$\begin{array}{r} 8\ \square \\ -4\ \square \\ \hline 4\ 6 \end{array}$$
$$\begin{array}{r} 8\ \square \\ -4\ \square \\ \hline 3\ 4 \end{array}$$

3   6   9

Name

Work from left to right.
Add or subtract.

First count on:
49, 50, 51.

$$49 + 2 - 21$$
$$51 \quad - 21 = 30$$

Add or subtract from left to right.

1. $68 + 11 - 7 =$
   $79 - 7 = 72$

2. $35 - 4 + 12 =$
   $31 + 12 =$

3. $25 - 15 + 9 =$

4. $73 + 5 - 22 =$

5. $49 + 1 + 20 =$

6. $50 + 17 - 11 =$

7. $92 - 22 + 8 =$

8. $60 + 32 - 3 =$

9. $40 - 5 - 32 =$

10. $18 + 2 - 15 =$

PROBLEM SOLVING

11. Fifty-seven singers were on stage.
    17 left the stage, 9 came back
    as dancers. How many were on
    the stage then?

_____

Name _____

Add. Regroup where needed. Color each sum.

has 0 ones — blue
has 4 tens — orange
greater than 90 — purple
between 50 and 60 — red
less than 40 — green
between 70 and 80 — brown

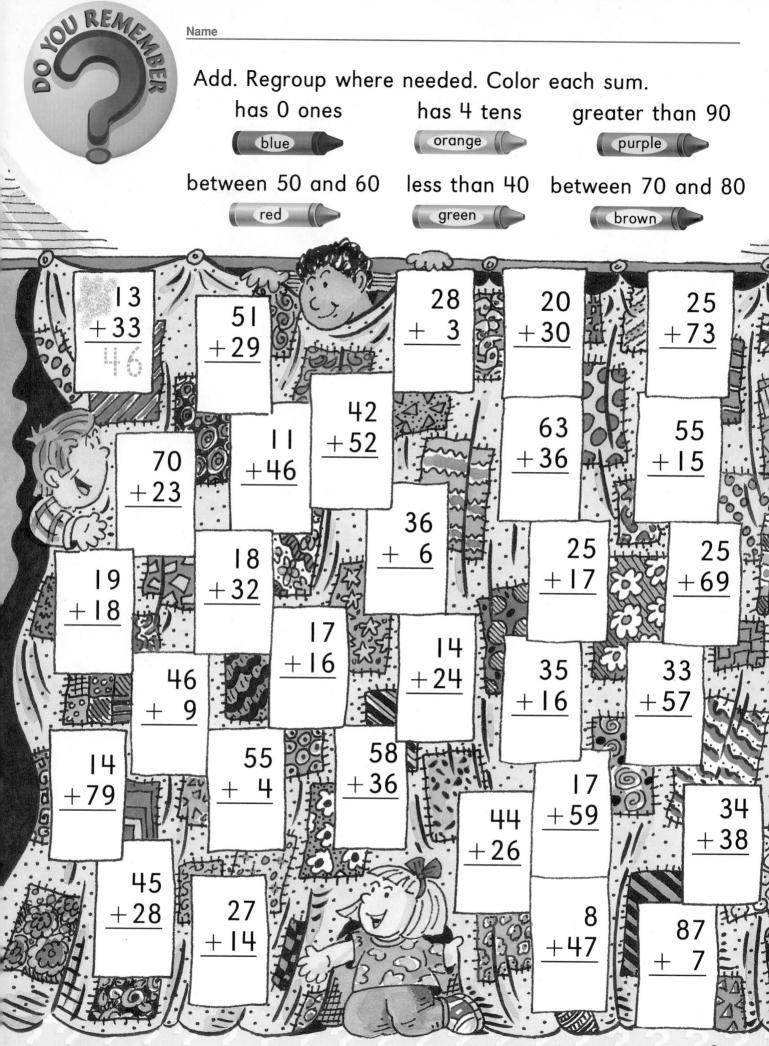

$$13 + 33 = 46$$

$$51 + 29$$

$$28 + 3$$

$$20 + 30$$

$$25 + 73$$

$$42 + 52$$

$$63 + 36$$

$$55 + 15$$

$$70 + 23$$

$$11 + 46$$

$$36 + 6$$

$$25 + 17$$

$$25 + 69$$

$$19 + 18$$

$$18 + 32$$

$$17 + 16$$

$$14 + 24$$

$$35 + 16$$

$$33 + 57$$

$$46 + 9$$

$$14 + 79$$

$$55 + 4$$

$$58 + 36$$

$$44 + 26$$

$$17 + 59$$

$$34 + 38$$

$$45 + 28$$

$$27 + 14$$

$$8 + 47$$

$$87 + 7$$

This page reviews the mathematical content presented in Chapter 5.

6

Name

You can estimate sums and differences by rounding the numbers to the nearest ten.

57 − 32 is about?   32 + 57 is about?

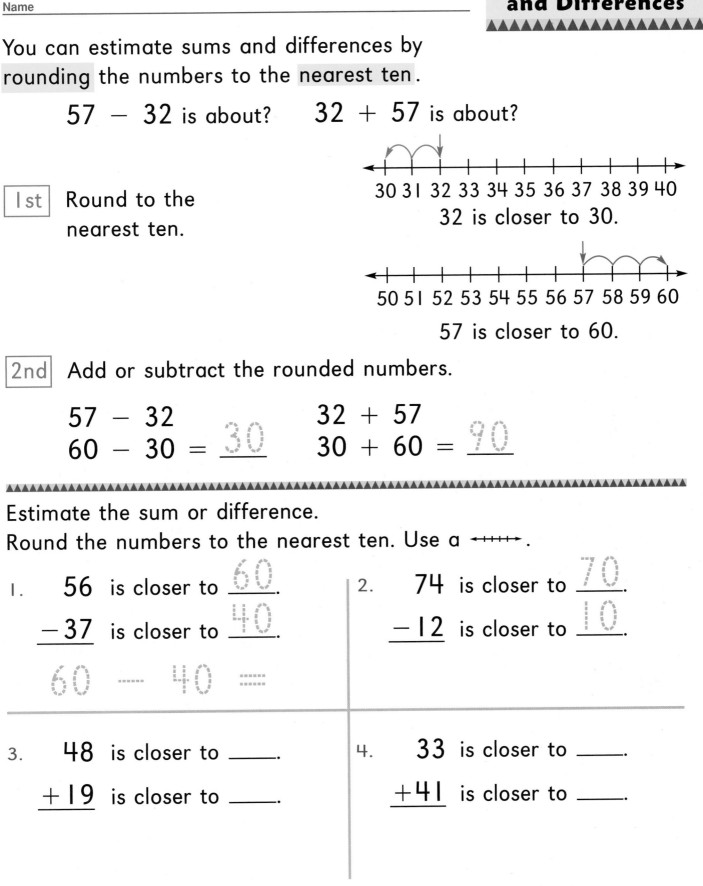

| 1st | Round to the nearest ten. |

30 31 32 33 34 35 36 37 38 39 40
32 is closer to 30.

50 51 52 53 54 55 56 57 58 59 60
57 is closer to 60.

| 2nd | Add or subtract the rounded numbers. |

57 − 32   32 + 57
60 − 30 = 30   30 + 60 = 90

Estimate the sum or difference.
Round the numbers to the nearest ten. Use a ⟵┼┼┼┼⟶.

1.   56 is closer to 60.
   − 37 is closer to 40.

   60 ── 40 ═══

2.   74 is closer to 70.
   − 12 is closer to 10.

3.   48 is closer to ____.
   + 19 is closer to ____.

4.   33 is closer to ____.
   + 41 is closer to ____.

Did you round the numbers up or down to the nearest ten in 1−4?

6-10 Give your child 2 two-digit numbers and have her/him estimate their sum and their difference.

Round numbers to the nearest ten.
Then estimate the sum or the difference.

1.
$74 \rightarrow 70$
$+23 \rightarrow +20$
about 90

$38 \rightarrow$
$+18 \rightarrow +$___
about

$46 \rightarrow$
$+ 9 \rightarrow +$___
about

2.
$32 \rightarrow$
$+34 \rightarrow +$___
about

$47 \rightarrow$
$+29 \rightarrow +$___
about

$41 \rightarrow$
$+36 \rightarrow +$___
about

3.
$75 \rightarrow 80$
$-27 \rightarrow -30$
about 50

$84 \rightarrow$
$-71 \rightarrow -$___
about

$53 \rightarrow$
$- 8 \rightarrow -$___
about

4.
$92 \rightarrow$
$-51 \rightarrow -$___
about

$68 \rightarrow$
$-39 \rightarrow -$___
about

$61 \rightarrow$
$-46 \rightarrow -$___
about

PROBLEM SOLVING

5. About how much more does the ![saxophone] cost than the ![tambourine] ?

58¢

23¢

¢ $\rightarrow$
¢ $\rightarrow$ ___
about

6. About how much money would you need to buy both toys?

¢ $\rightarrow$
¢ $\rightarrow$ ___
about

MAKE UP YOUR OWN

7. Write a problem like 5 and 6 in your Math Journal. Solve it.

Name _____

Find the sum.

1.
```
   21        18        67        80       34¢       19¢
 +32       +48       +24       +10      +14¢      +13¢
  53        66
```

2.
```
   37        55        93        27       27¢       48¢
 +35       +30       + 6       +27      +43¢      +42¢
```

3.
```
   75        68        36        59       50¢       46¢
 + 5       + 9       +57       +32      +47¢      + 6¢
```

Find the difference.

4.
```
  8 15
   95        48        20        62       75¢       70¢
 -27       -17       - 9       -24      -22¢      -63¢
  68        31
```

5.
```
   65        70        58        90       98¢       85¢
 -43       -12       -14       -76      -40¢      - 8¢
```

6.
```
   33        98        45        82       64¢       50¢
 - 9       -11       -26       -43      -58¢      -27¢
```

Add or subtract. Watch for + and −.

1.
$$\begin{array}{r} \overset{7\;10}{\cancel{8}\,\cancel{0}} \\ -25 \\ \hline 55 \end{array}$$
$$\begin{array}{r} 46 \\ -\;9 \\ \hline \end{array}$$
$$\begin{array}{r} 59¢ \\ +15¢ \\ \hline \end{array}$$
$$\begin{array}{r} 85¢ \\ +\;8¢ \\ \hline \end{array}$$
$$\begin{array}{r} 38¢ \\ -12¢ \\ \hline \end{array}$$

2.
$$\begin{array}{r} 47 \\ +19 \\ \hline \end{array}$$
$$\begin{array}{r} 88 \\ -39 \\ \hline \end{array}$$
$$\begin{array}{r} 70¢ \\ -44¢ \\ \hline \end{array}$$
$$\begin{array}{r} 73¢ \\ +18¢ \\ \hline \end{array}$$
$$\begin{array}{r} 97¢ \\ -15¢ \\ \hline \end{array}$$

3.
$$\begin{array}{r} 51 \\ -47 \\ \hline \end{array}$$
$$\begin{array}{r} 32 \\ +49 \\ \hline \end{array}$$
$$\begin{array}{r} 84¢ \\ -17¢ \\ \hline \end{array}$$
$$\begin{array}{r} 28¢ \\ +40¢ \\ \hline \end{array}$$
$$\begin{array}{r} 61¢ \\ +19¢ \\ \hline \end{array}$$

PROBLEM SOLVING

4. In the class play, I saw 27 singers and 45 dancers. How many children were on stage in all?

_____ children

5. Sara had 87 lines to say. She said 86 of them. How many lines did Sara have left to say?

_____ line

6. ADMIT ONE to the show cost 60¢. Jim has 38¢. How much more money does Jim need to buy a ADMIT ONE ?

_____

7. At the show cost 22¢. and cost 35¢. How much would it cost to buy both?

_____

MENTAL MATH

8. Subtract. Are the differences odd or even?

20 −	19	17	15	13

Name _____

You can add or subtract using paper and pencil or mental math.

Sometimes I count on mentally.

Sometimes I use paper and pencil.

30 + 12 = ?
Think: 30 + 10 + 2
Count on: 30, 40, 41, 42
30 + 12 = 42

```
  8 13
  9 3
- 3 8
-----
  5 5
```

Choose ✏️ or Mental Math. Then add or subtract.

1. 79
   − 20
    59   Mental Math

2. 36
   + 57   Mental Math

3. 92
   − 34   Mental Math

4. 23
   + 23   Mental Math

5. 45
   − 11   Mental Math

6. 58
   + 29   Mental Math

7. 38¢ − 10¢ = _____   Mental Math

8. 58 + 12 + 27 = _____   Mental Math

9. 47 − 20 − 7 = _____   Mental Math

10. 90¢ − 38¢ = _____   Mental Math

6-12 Invite your child to explain how she/he decided which way to compute.

two hundred thirty-five **235**

Choose ✏ or (Mental Math). Then solve.

1.

Dancers	
Tap	41 children
Ballet	15 children

How many more tap dancers than ballet dancers are there?

✏

(Mental Math) _____ more

2.

Stage Decorations	
Tallest	72 inches
Shortest	10 inches

How many inches more is the tallest decoration than the shortest decoration?

✏

(Mental Math) _____ inches more

3. Mateo saw 28 band members, 20 singers, and 36 dancers on the stage. How many children is that in all?

✏

(Mental Math) _____ children

4. Jennifer has 91¢. She uses 65¢ to buy some beads for the show. How much money is left?

✏

(Mental Math) _____ ¢

Sula practiced 35 minutes for the show the first week and 22 minutes the second week.

5. How many more minutes did she practice the first week?

✏

(Mental Math) _____ more minutes

6. How many minutes did she practice in all?

✏

(Mental Math) _____ minutes

MATH JOURNAL

7. Write a problem you can solve using paper and pencil. Write another you can solve using mental math.

Add and Subtract with a Calculator

To add 2-digit numbers using a calculator,
- enter the tens digit
- then enter the ones digit.

tens	ones
4	5
+ 3	6

Add 45 + 36 = ___

Press ON/AC   4   5   +   3   6   =

Display $0.$ $4.$ $45.$ $45.$ $3.$ $36.$ $81.$

TALK IT OVER Which keys would you press to subtract 36 from 45?

Find the sum or difference. Use a 🧮.

1. 23 + 65 = _88_   72 − 24 = ___

2. 68 + 31 = ___   36 − 19 = ___

3. 47 + 33 = ___   40 − 5 = ___

4. 37 + 15 + 26 = ___   96 − 34 − 47 = ___

5. 51 + 40 + 7 = ___   47 − 23 − 19 = ___

6. 12 + 55 + 14 = ___   44 − 39 − 5 = ___

7. 81 + 3 + 5 = ___   60 − 27 − 11 = ___

TECHNOLOGY

6-13 This lesson teaches your child how to add and subtract 2-digit numbers using a calculator.

two hundred thirty-seven **237**

Use a 🖩 and the numbers in the box
to find sums of 88.

| 51 | 37 | 19 | 44 | 65 | 69 | 23 |

1. $\underline{51} + \underline{37} = 88$ 　　 2. _____ + _____ = 88

3. _____ + _____ = 88 　　 4. _____ + _____ = 88

Use a 🖩 and the numbers in the box
to find differences of 26.

| 44 | 49 | 26 | 31 | 23 | 57 | 52 | 18 |

5. _____ − _____ = 26 　　 6. _____ − _____ = 26

7. _____ − _____ = 26 　　 8. _____ − _____ = 26

FINDING TOGETHER Use a 🖩 to find doubles to complete
these sums and differences.

9. $66 = \underline{\quad} + \underline{\quad}$ 　　 10. $24 = \underline{\quad} + \underline{\quad}$

11. $58 = \underline{\quad} + \underline{\quad}$ 　　 12. $32 = \underline{\quad} + \underline{\quad}$

13. $\underline{\quad} = 86 - \underline{\quad}$ 　　 14. $\underline{\quad} = 44 - \underline{\quad}$

15. $\underline{\quad} = 50 - \underline{\quad}$ 　　 16. $\underline{\quad} = 72 - \underline{\quad}$

Name

1. **Read** Dan made 38 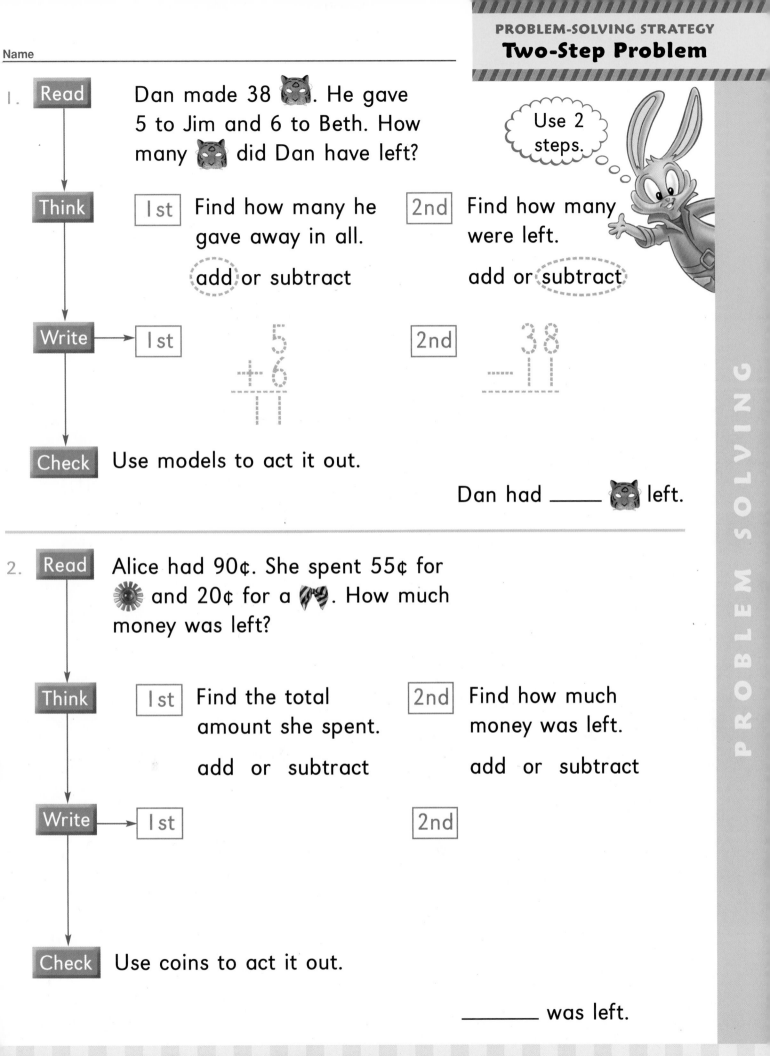. He gave 5 to Jim and 6 to Beth. How many 🐱 did Dan have left?

Use 2 steps.

Think | 1st | Find how many he gave away in all.

add or subtract

| 2nd | Find how many were left.

add or subtract

Write → | 1st |

$$\begin{array}{r} 5 \\ +6 \\ \hline 11 \end{array}$$

| 2nd |

$$\begin{array}{r} 38 \\ -11 \\ \hline \end{array}$$

Check Use models to act it out.

Dan had _____ 🐱 left.

2. **Read** Alice had 90¢. She spent 55¢ for 🌻 and 20¢ for a 🎀. How much money was left?

Think | 1st | Find the total amount she spent.

add or subtract

| 2nd | Find how much money was left.

add or subtract

Write → | 1st |

| 2nd |

Check Use coins to act it out.

_____ was left.

6-14 Your child solved problems involving two steps. Invite her/him to explain how to solve one of the problems above.

two hundred thirty-nine **239**

Read → Think → Write → Check

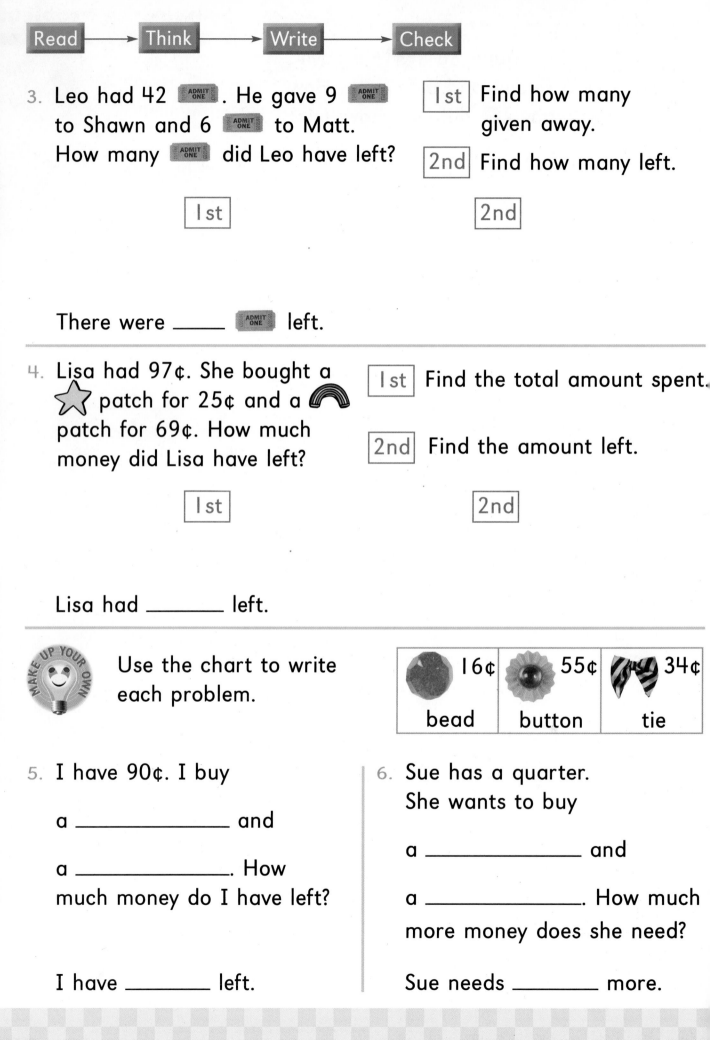

3. Leo had 42 [ADMIT ONE]. He gave 9 [ADMIT ONE] to Shawn and 6 [ADMIT ONE] to Matt. How many [ADMIT ONE] did Leo have left?

1st Find how many given away.

2nd Find how many left.

1st

2nd

There were _____ [ADMIT ONE] left.

4. Lisa had 97¢. She bought a ⭐ patch for 25¢ and a 🌈 patch for 69¢. How much money did Lisa have left?

1st Find the total amount spent.

2nd Find the amount left.

1st

2nd

Lisa had _____ left.

MAKE UP YOUR OWN   Use the chart to write each problem.

16¢	55¢	34¢
bead	button	tie

5. I have 90¢. I buy

a _____ and

a _____. How much money do I have left?

I have _____ left.

6. Sue has a quarter. She wants to buy

a _____ and

a _____. How much more money does she need?

Sue needs _____ more.

PROBLEM SOLVING

1. **Read** Sue made 28 posters.
Pam made 43 posters.
About how many did
they make in all?

About how
many tens?

Think Round to estimate the sum.

Sue Pam

Ring Sue made about | 20 ⟨30⟩ |.

Pam made about | ⟨40⟩ 50 |.

Write 30 ⊕ 40 ▭ ____   about ____ posters

2. **Read** Kim had 37 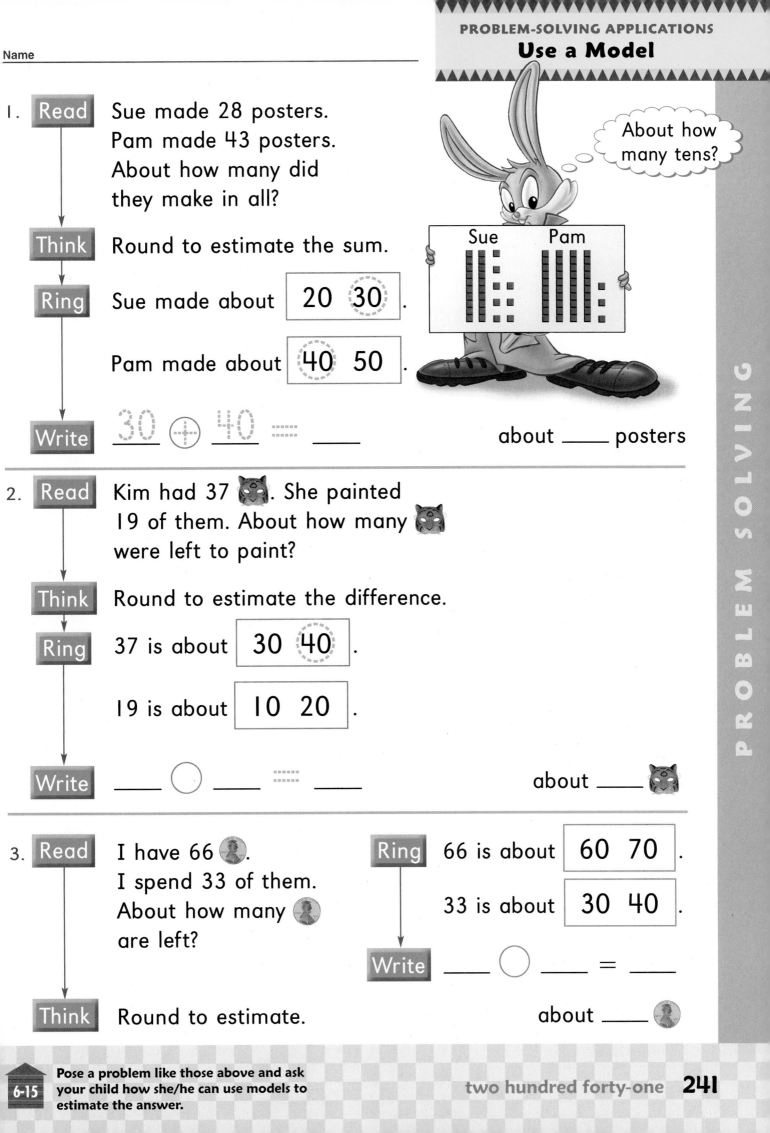. She painted
19 of them. About how many
were left to paint?

Think Round to estimate the difference.

Ring 37 is about | 30 ⟨40⟩ |.

19 is about | 10 20 |.

Write ____ ◯ ____ ▭ ____   about ____

3. **Read** I have 66 .
I spend 33 of them.
About how many
are left?

Ring 66 is about | 60 70 |.

33 is about | 30 40 |.

Write ____ ◯ ____ = ____

Think Round to estimate.   about ____

PROBLEM SOLVING

6-15   Pose a problem like those above and ask
your child how she/he can use models to
estimate the answer.   two hundred forty-one   **241**

Use a strategy you have learned.

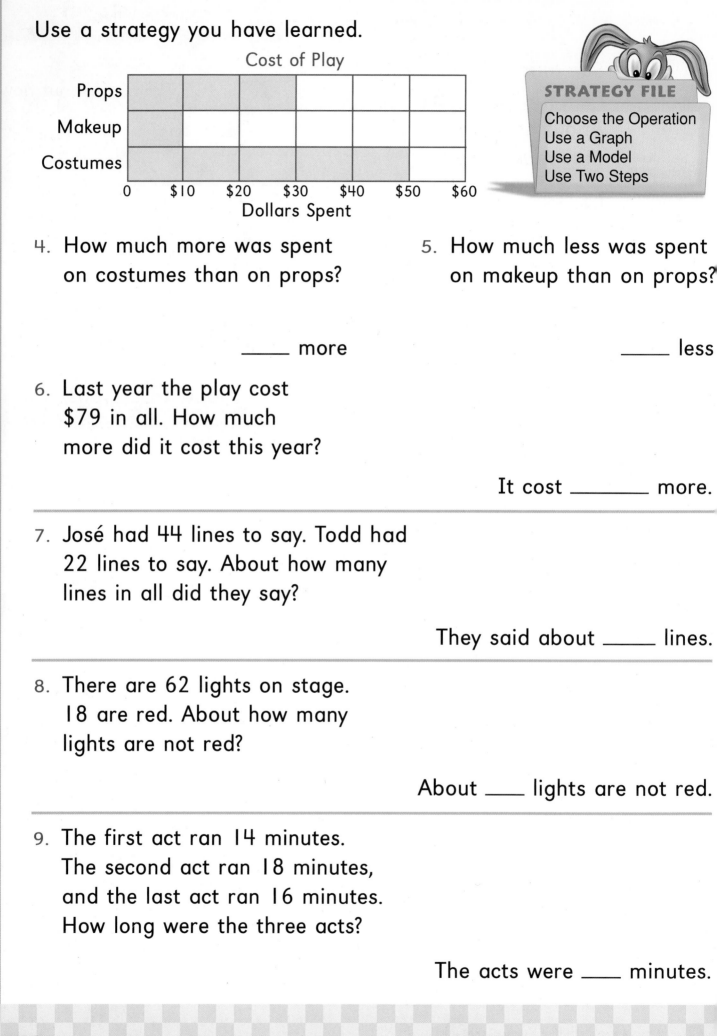

Cost of Play

Props

Makeup

Costumes

0   $10 $20 $30 $40 $50 $60

Dollars Spent

STRATEGY FILE

Choose the Operation
Use a Graph
Use a Model
Use Two Steps

4. How much more was spent on costumes than on props?

_____ more

5. How much less was spent on makeup than on props?

_____ less

6. Last year the play cost $79 in all. How much more did it cost this year?

It cost _____ more.

7. José had 44 lines to say. Todd had 22 lines to say. About how many lines in all did they say?

They said about _____ lines.

8. There are 62 lights on stage. 18 are red. About how many lights are not red?

About _____ lights are not red.

9. The first act ran 14 minutes. The second act ran 18 minutes, and the last act ran 16 minutes. How long were the three acts?

The acts were _____ minutes.

PROBLEM SOLVING

Subtract. Use models to check.

1.
$$76 - 30$$
$$34 - 4$$
$$49 - 42$$
$$44 - 5$$
$$62 - 38$$
$$91 - 48$$

2.
$$84 - 38$$
$$93 - 85$$
$$56 - 27$$
$$91 - 44$$
$$63 - 57$$
$$84 - 77$$

3.
$$79¢ - 8¢$$
$$61¢ - 46¢$$
$$80¢ - 7¢$$
$$37¢ - 19¢$$
$$90¢ - 68¢$$

Subtract. Add to check.

4.
$$93 - 9$$ + ___

5.
$$95 - 22$$ + ___

6.
$$40¢ - 24¢$$ + ___

PROBLEM SOLVING

7. There are 58 beads on a costume. 12 are red. About how many beads are not red?

   58 is about ___.

   12 is about ___.

   ___ ◯ ___ = ___

   about ___ beads

8. Marci spent 64¢ on props. Tito spent 27¢. About how much did they spend altogether?

   64¢ is about ___.

   27¢ is about ___.

   ___ ◯ ___ = ___

   about ___

REINFORCEMENT

This page reviews the mathematical content presented in Chapter 6.

Follow each path. Use the code.
Watch the + and − signs.

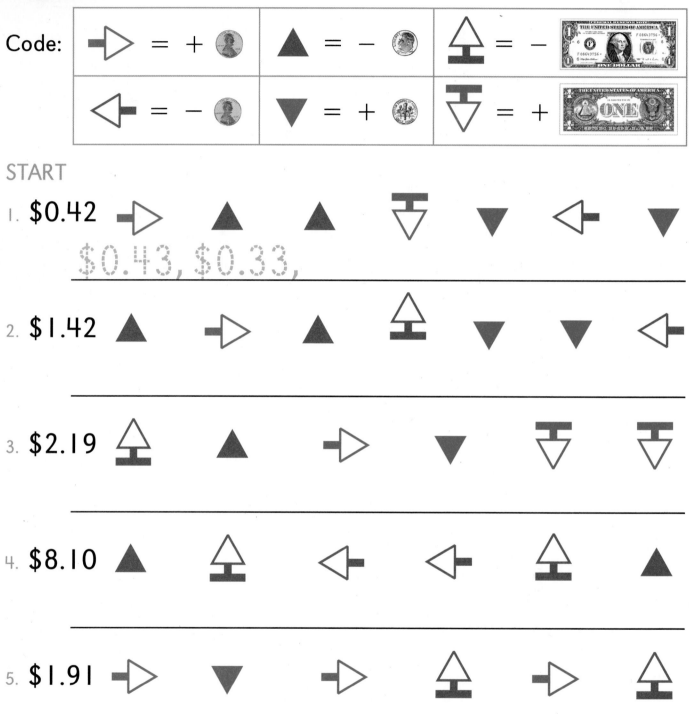

Code:

▷ = + 🪙	▲ = − 🪙	🔺 = − $1 bill
◁ = − 🪙	▼ = + 🪙	🔻 = + $1 bill

START

1. $0.42 ▷ ▲ ▲ 🔻 ▼ ◁ ▼

$0.43, $0.33,

2. $1.42 ▲ ▷ ▲ 🔺 ▼ ▼ ◁

3. $2.19 🔺 ▲ ▷ ▼ 🔻 🔻

4. $8.10 ▲ 🔺 ◁ ◁ 🔺 ▲

5. $1.91 ▷ ▼ ▷ 🔺 ▷ 🔺

What pattern do you see in 1 and 2?
In your Math Journal write the opposite
code for another exercise.

This page extends your child's understanding
of adding and subtracting money.

6

1. Use models to show how you solve each.

$25 + 8 - 13 =$ _____

$47 - 30 + 13 =$ _____

2. Use one of the numbers in the box to estimate the sum or difference. Tell how to estimate.

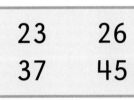

23	26
37	45

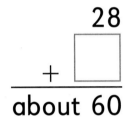
28
+ ☐
about 60

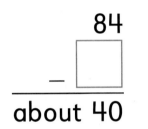
84
– ☐
about 40

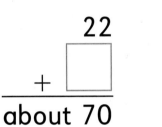
22
+ ☐
about 70

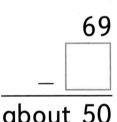
69
– ☐
about 50

PORTFOLIO

Choose 1 of these projects.
Use a separate sheet of paper.

3. Complete a table for each Rule. Make up 3 amounts for IN. Find the amounts for OUT.

Rule: – 1 quarter	
IN	OUT

Rule: – 2 dimes 8 pennies	
IN	OUT

4. You have 90¢ to buy letters to spell your name. Vowels cost 2¢ each. Consonants cost 1¢ each. How much do the letters in your name cost to buy? How much change will you get?

6

This page provides a variety of informal assessment opportunities in order to measure your child's understanding of Chapter 6.

two hundred forty-five **245**

Name _____

Subtract. Use models to check.

1.
$$\begin{array}{r} 67 \\ -27 \\ \hline \end{array}\qquad \begin{array}{r} 57 \\ -\ 5 \\ \hline \end{array}\qquad \begin{array}{r} 70 \\ -41 \\ \hline \end{array}\qquad \begin{array}{r} 81 \\ -39 \\ \hline \end{array}\qquad \begin{array}{r} 72 \\ -\ 6 \\ \hline \end{array}\qquad \begin{array}{r} 38 \\ -17 \\ \hline \end{array}$$

2.
$$\begin{array}{r} 40 \\ -\ 2 \\ \hline \end{array}\qquad \begin{array}{r} 61 \\ -52 \\ \hline \end{array}\qquad \begin{array}{r} 30 \\ -19 \\ \hline \end{array}\qquad \begin{array}{r} 65 \\ -49 \\ \hline \end{array}\qquad \begin{array}{r} 62 \\ -35 \\ \hline \end{array}\qquad \begin{array}{r} 85 \\ -28 \\ \hline \end{array}$$

3.
$$\begin{array}{r} 95¢ \\ -70¢ \\ \hline \end{array}\qquad \begin{array}{r} 83¢ \\ -24¢ \\ \hline \end{array}\qquad \begin{array}{r} 80¢ \\ -23¢ \\ \hline \end{array}\qquad \begin{array}{r} 44¢ \\ -16¢ \\ \hline \end{array}\qquad \begin{array}{r} 72¢ \\ -\ 3¢ \\ \hline \end{array}$$

Subtract. Add to check.

4.
$$\begin{array}{r} 63 \\ -28 \\ \hline \end{array}\quad +\underline{}$$

5.
$$\begin{array}{r} 75 \\ -\ 7 \\ \hline \end{array}\quad +\underline{}$$

6.
$$\begin{array}{r} 56¢ \\ -48¢ \\ \hline \end{array}\quad +\underline{}$$

PROBLEM SOLVING

7. Ralph made 32 posters for the show. Suni made 17. About how many posters were made in all?

32 is about _____.

17 is about _____.

_____ + _____ = _____

about _____ posters

8. A prop cost 36¢. Yuka has 19¢. About how much more does Yuka need?

36¢ is about _____.

19¢ is about _____.

_____ − _____ = _____

about _____

This page is a formal assessment of your child's understanding of the content presented in Chapter 6.

6

7

CRITICAL THINKING

Rectangles, circles, and squares cost 10¢ each. Triangles cost 3¢ each. Draw the front of a house that costs between 50¢ and $1.00.

For more information about Chapter 7, visit the Family Information Center at www.sadlier-oxford.com

Dear Family,

Today your child began Chapter 7. As he/she studies geometry, fractions and probability, you may want to read the poem below, which was read in class. Have your child talk about some of the math ideas pictured on page 247.

Look for the 🏠 at the bottom of each skills lesson. The suggestion on the page gives you an opportunity to improve your child's understanding of math and to reinforce his/her math language. As your child progresses through this chapter, point out and ask him/her to identify shapes, solid figures, and fractional parts in your home, in natural surroundings, and in stores.

Home Activity

Solid-Figure Safari

Take your child on a solid-figure safari in your local hardware or building supply store. Explain that he/she should look for spheres, cylinders, cones, rectangular prisms, cubes, and pyramids. When you return home, help your child categorize the objects he/she considers to be solid figures. Then use the data to create a bar graph to record the number of each solid figure seen. (See list and graph below.)

cylinder
paint can

rectangular prism
door
window pane

Solid Figures

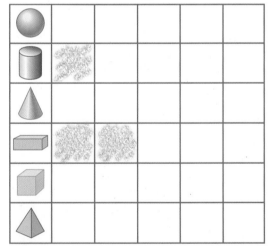

Home Reading Connection

HOUSES

Houses are faces
(haven't you found?)
with their hats in the air,
and their necks in the ground.

Windows are noses,
windows are eyes,
and doors are the mouths
of a suitable size.

And a porch—or the place
where porches begin—
is just like a mustache
shading the chin.

Aileen Fisher

This is half full.

This is more than half full.

This is less than half full.

Color the houses.

1. half 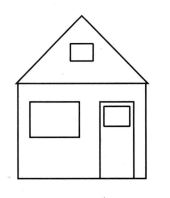 red

2. less than half green

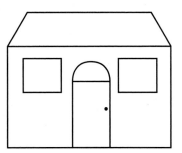

3. less than half blue

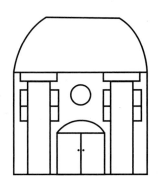

4. half yellow

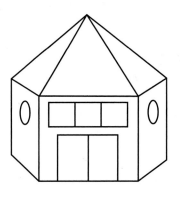

5. more than half purple

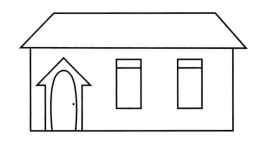

6. more than half brown

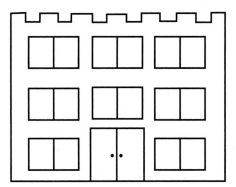

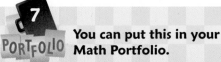

CONNECTIONS

Color the parts of this stained-glass window.

2 equal parts 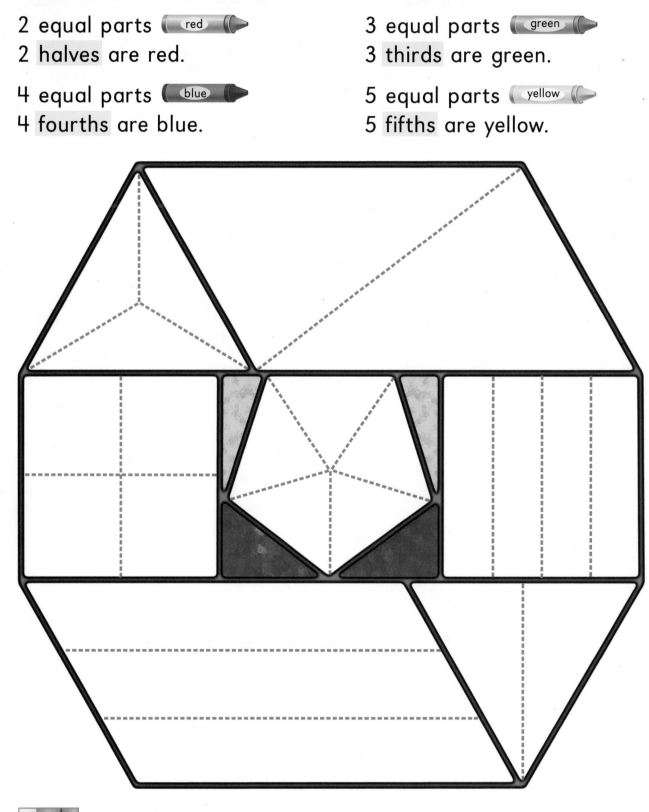 red
2 halves are red.

3 equal parts green
3 thirds are green.

4 equal parts blue
4 fourths are blue.

5 equal parts yellow
5 fifths are yellow.

FINDING TOGETHER List how many squares, triangles, and rectangles are in the window. Make a graph.

You can put this in your
Math Portfolio. PORTFOLIO

Name _____

Solids that cannot roll
have all flat surfaces.

Solids that roll have
a curved surface.

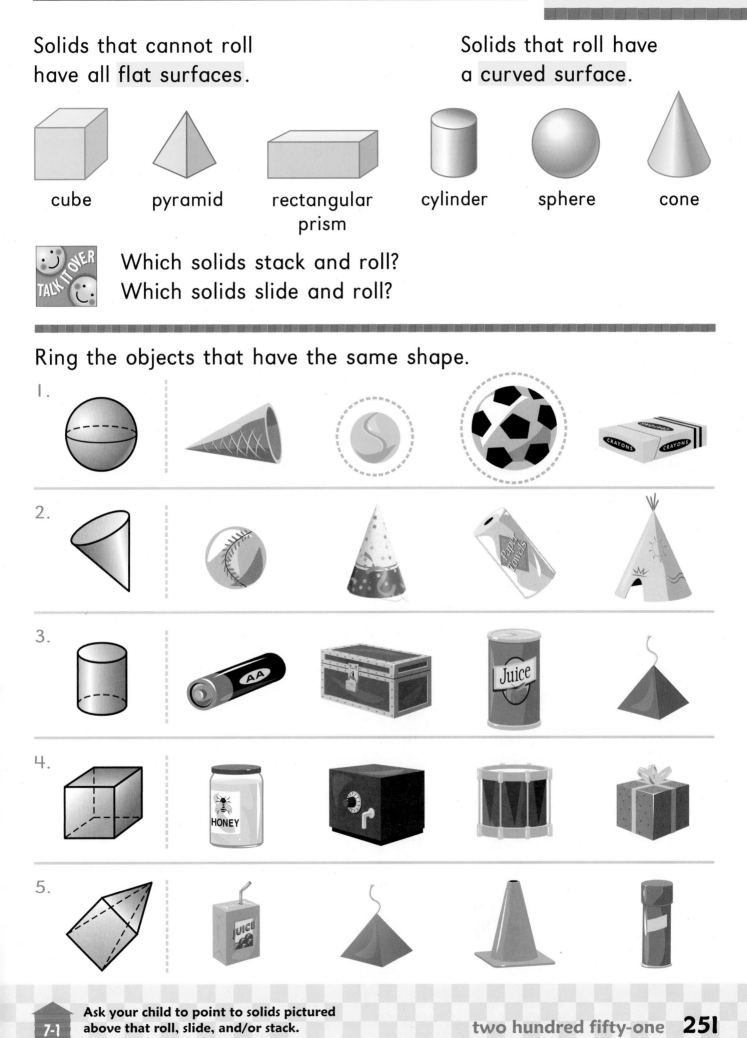

cube   pyramid   rectangular   cylinder   sphere   cone
                                prism

TALK IT OVER Which solids stack and roll?
Which solids slide and roll?

Ring the objects that have the same shape.

1.

2.

3.

4.

5.

7-1 Ask your child to point to solids pictured
above that roll, slide, and/or stack.

two hundred fifty-one **251**

cube   cone   pyramid   rectangular prism   cylinder   sphere

Sort the space figures. Write their names.

1. It rolls and has 2 flat surfaces. _____cylinder_____

2. It rolls and has 1 flat surface. _____

3. It has 6 flat surfaces. _____

   or _____

4. It has no flat surfaces. _____

5. It can be built from these 6 squares. _____

6. It has 5 flat surfaces. _____

PROBLEM SOLVING Use the space figures above.
Write Yes or No.

7. Can you stack a pyramid on top of a cone? _____

8. Can you stack a cone on top of a cylinder? _____

CRITICAL THINKING ✗ out what does not belong.

9.

space figures

10.

figures that slide

Name _____

Space figures that cannot roll have faces, corners, and edges.

The faces are flat surfaces.

The blue dots show some corners.

The red lines show some edges.

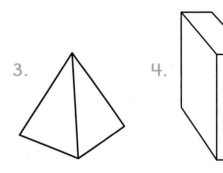

face →
corner →
edge →

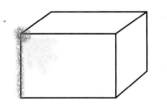 Each face of the number cube has a number from 1 to 6. What numbers are hidden?

Color the faces you see (yellow).

Color the edges you see (red).

Color the corners you see (blue).

1.

2.

3.

4.

5. Ring the space figures that have no corners.

6. Ring the space figures that have any square faces.

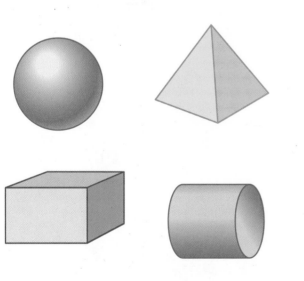

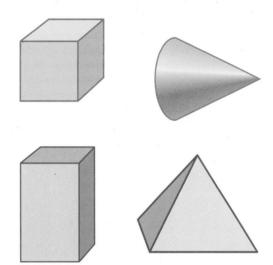

7-2 Ask your child to tell you about the faces, edges, and corners seen on the figures in exercises 1–4.

two hundred fifty-three **253**

1. Which two space figures have the same number of corners? Color them yellow. Name them.

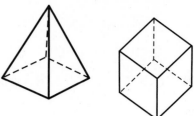

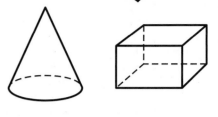

_____

2. Which two space figures have no edges and no faces? Color them blue. Name them.

_____

3. Look at the space figures above. Count and graph the number of corners in each kind of space figure.

TALK IT OVER

Which space figures in the graph have the most corners?

Corners of Space Figures

rectangular prism									
cube									
cylinder									
cone									
pyramid									

    0 1 2 3 4 5 6 7 8

Number of Corners

MAKE UP YOUR OWN

4. Tally the number of faces on each kind of space figure and show the results on a pictograph.

Name of Space Figure	Tally

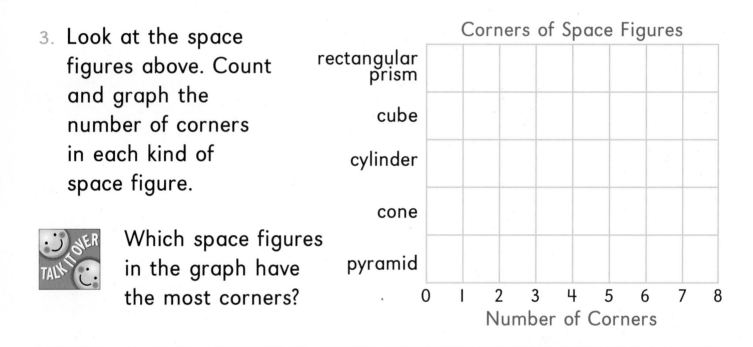

Jamal traced the 2 flat sufaces of a solid.
Which solid did he use?

| 1st | Trace each flat surface of the solids. |

| 2nd | Complete the table. |

Solid Figure	Number of Flat Surfaces	Shapes of Surfaces Traced
1. pyramid	5	△ △ △ △ ▢
2. rectangular prism		
3. cylinder		

4. Jamal traced the 2 flat surfaces of the _____.

 Name the shapes you traced. Were any of the flat surfaces the same shape and the same size? Which ones?

Complete the table below.

Solid Figure	Number of Flat Surfaces	Shapes of Surfaces Traced
5. cone		
6. cube		

7-3   Trace the flat surface of a familiar solid.
Have your child find the solid you used.

two hundred fifty-five **255**

Use these to cover the flat surfaces of one
of these solids. Color to match.

1. 2. 3. 4. 5.

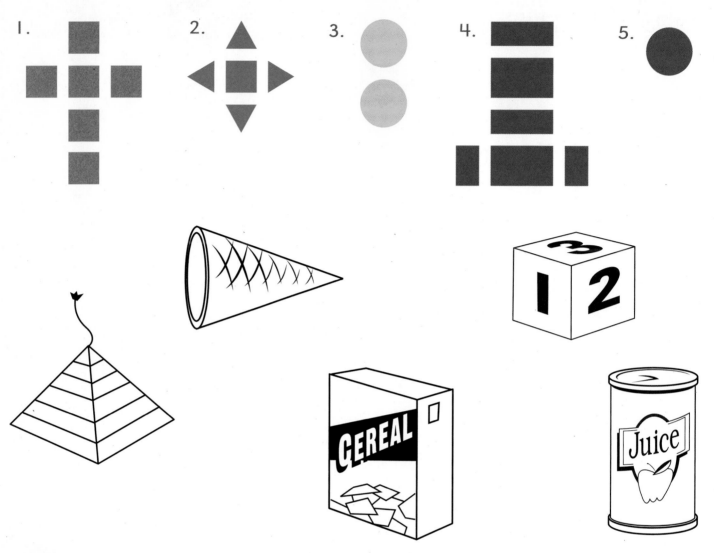

Can you cover the surface of a sphere?
Why or why not?

PROBLEM SOLVING

6. Jan needs to trace a circle.
   Which solid figures may
   she use? _____

7. Pablo needs to trace a triangle.
   What solid figure may he use? _____

8. Alan wants to trace a square.
   What solid figures may
   he use? _____

Name _____

These plane figures are closed figures.

These plane figures have curved lines.

These plane figures have straight lines.

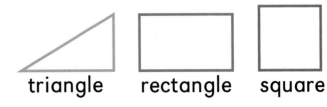

circles

triangle   rectangle   square

side ⟶

corner ⟶

A triangle has 3 sides and 3 corners.

Draw a line to make a closed figure. Write its name.

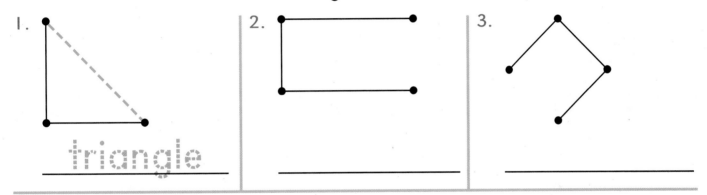

1.

triangle
_____

2.

_____

3.

_____

Draw a line to make a closed figure.
Write how many sides and corners.

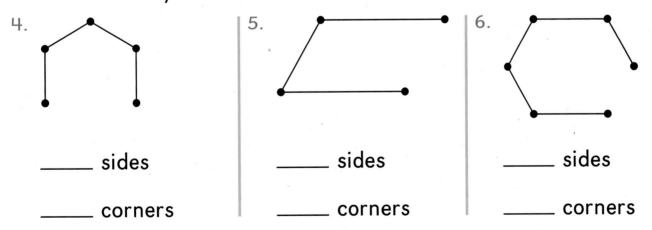

4.

_____ sides

_____ corners

5.

_____ sides

_____ corners

6.

_____ sides

_____ corners

How are squares and rectangles alike?
How are they different? Can a closed plane
figure have more sides than corners? Explain.

7-4   Draw three different shapes. Have your
child tell which shape has the most
corners and which the fewest sides.   two hundred fifty-seven   **257**

Draw each. Then color.

Figure	Color
circle	blue
rectangle	yellow
square	red
triangle	green
5 sides	purple

1. 0 sides

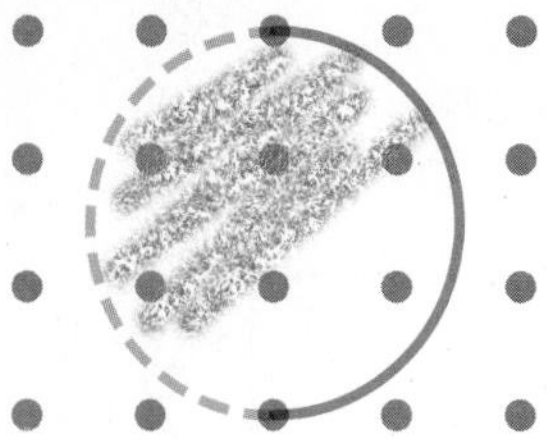

2. 5 corners

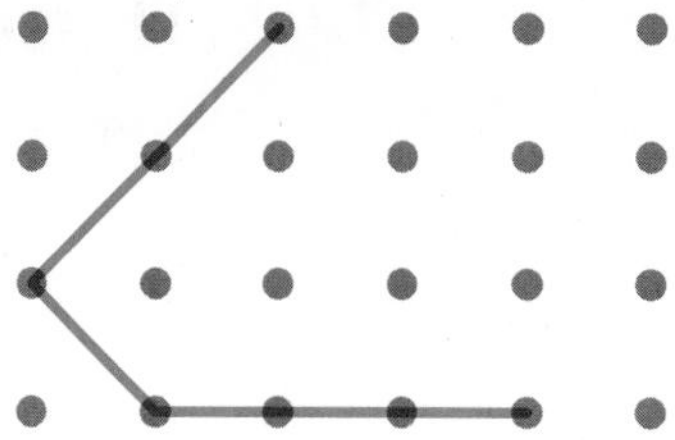

3. 4 sides

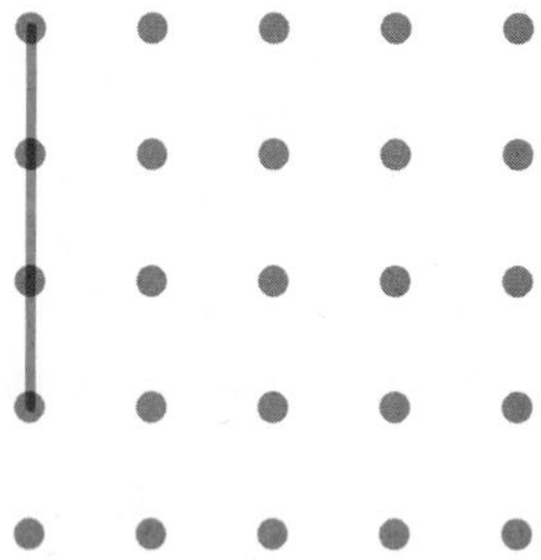

4. 3 corners

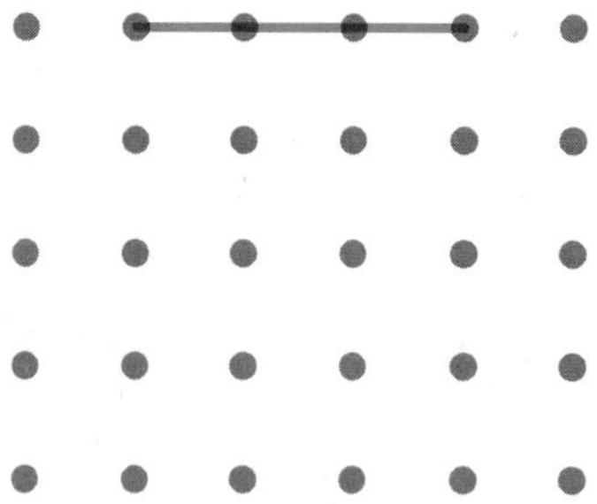

5. 4 corners

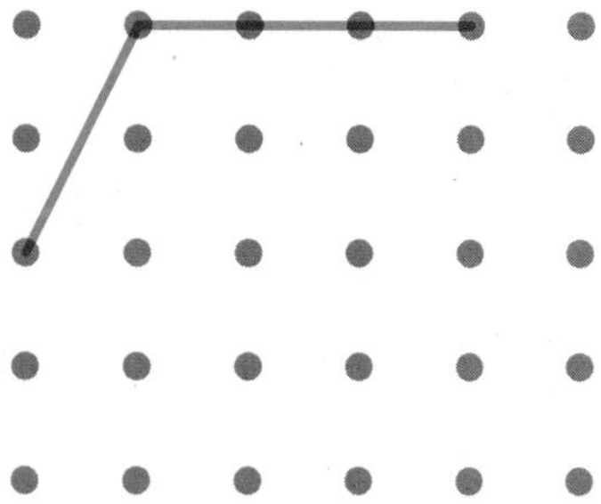

6. 3 sides

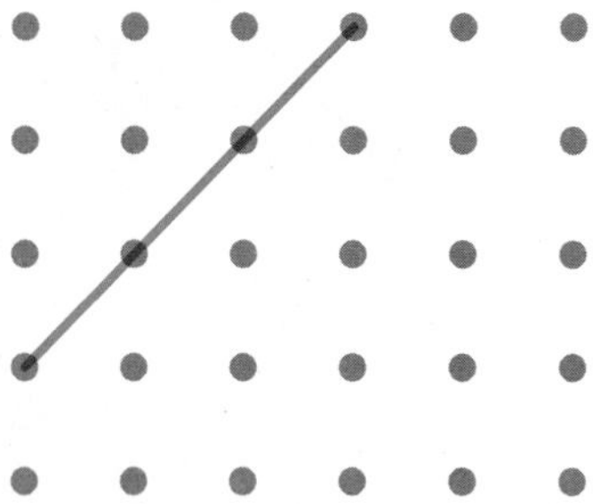

7. 4 sides

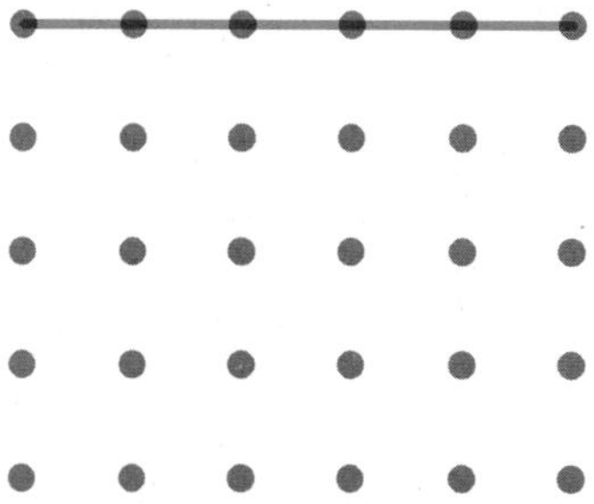

8. 0 corners

PROBLEM SOLVING

9. Chris's house has more than 3 sides.
Pilar's house is a rectangle.
Color Chris's house red and
Pilar's blue.

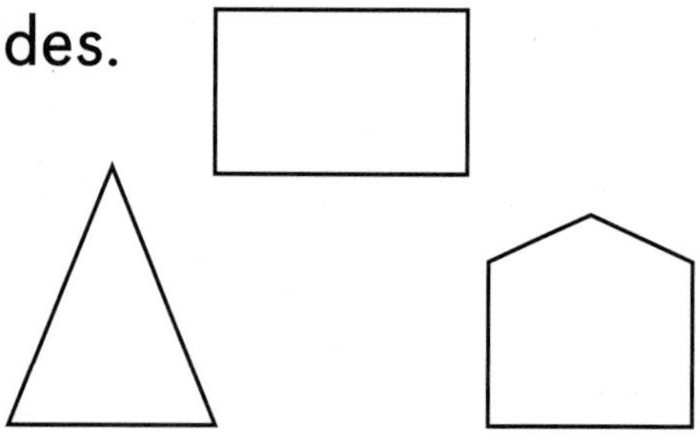

CRITICAL THINKING

Write All or No.

10. _____ squares and rectangles have 4 sides.

11. _____ triangles have 4 sides that are equal.

12. _____ circles have sides and corners.

Name

Figures that are congruent have the same shape and the same size.

 1st   Name the figures.

 2nd   Match the sides exactly.
Use the ●——● to help.

Are figures A and B congruent?
Are figures C and D congruent?

A   B

rectangles

C

D

Color figures that are congruent.

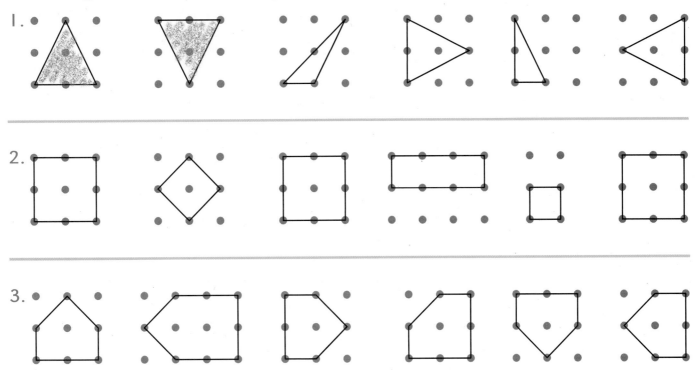

1.

2.

3.

 4. Draw 2 figures that are the same shape but not the same size. Are they congruent?

Draw a figure that is the same shape and size.

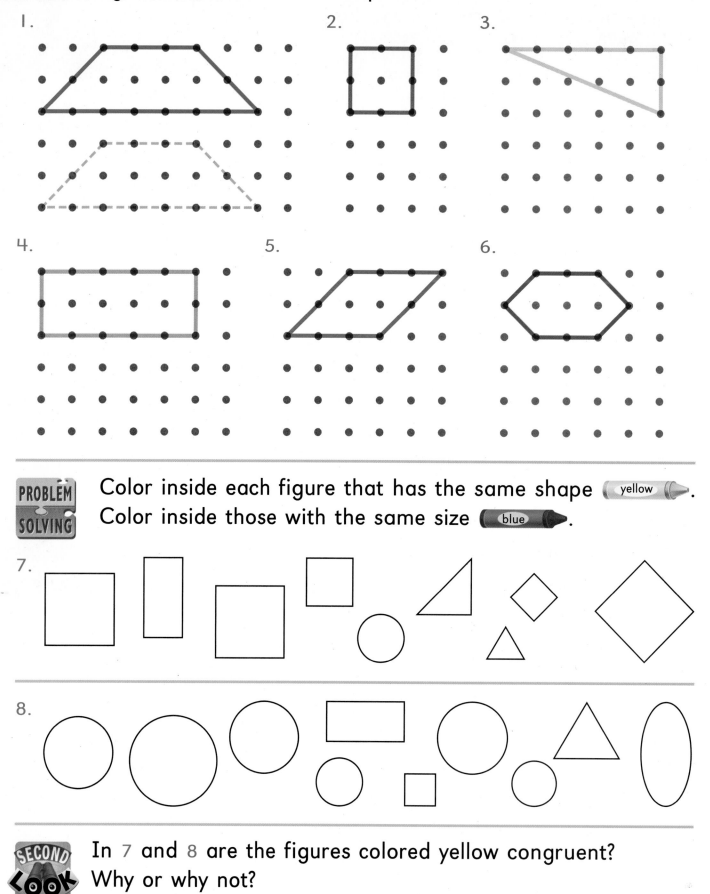

1.

2.

3.

4.

5.

6.

PROBLEM SOLVING Color inside each figure that has the same shape yellow.
Color inside those with the same size blue.

7.

8.

SECOND LOOK In 7 and 8 are the figures colored yellow congruent?
Why or why not?

Are the figures colored both yellow and blue congruent?
Why or why not?

Name _____

The two parts match.
The fold line is a **line of symmetry**.

This figure has
2 lines of symmetry.

Do the two parts match? Write Yes or No.

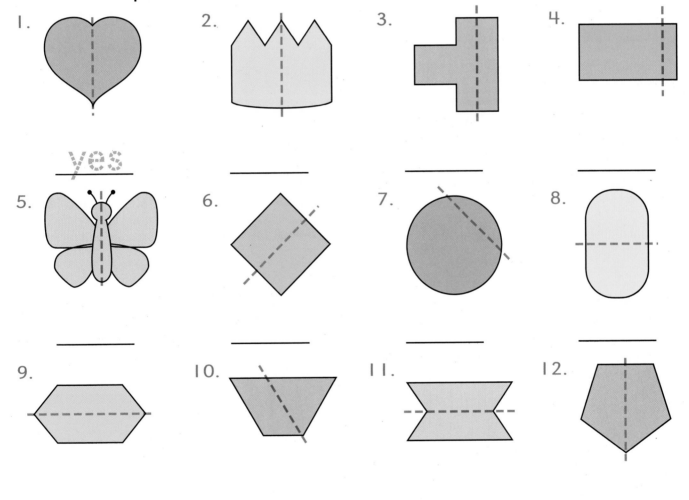

1. _yes_

2. _____

3. _____

4. _____

5. _____

6. _____

7. _____

8. _____

9. _____

10. _____

11. _____

12. _____

TALK IT OVER

Which shapes in 1–12 have a second
line of symmetry? How can you show it?

Draw one line of symmetry.

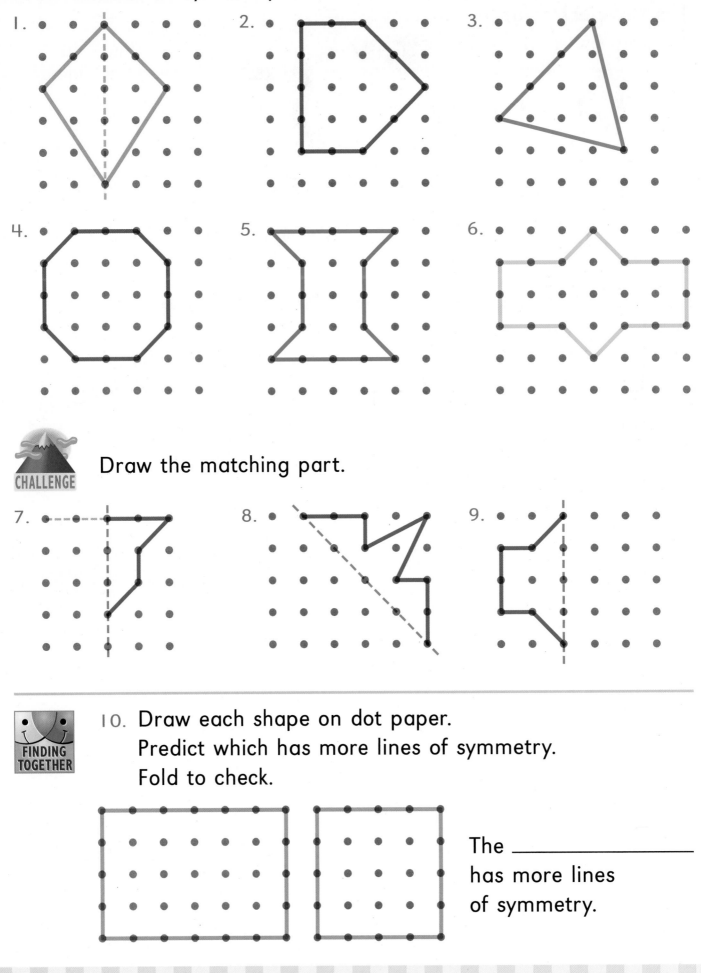

1.

2.

3.

4.

5.

6.

CHALLENGE Draw the matching part.

7.

8.

9.

10. Draw each shape on dot paper.
Predict which has more lines of symmetry.
Fold to check.

FINDING TOGETHER

The _____
has more lines
of symmetry.

Here are two ways to move an object.

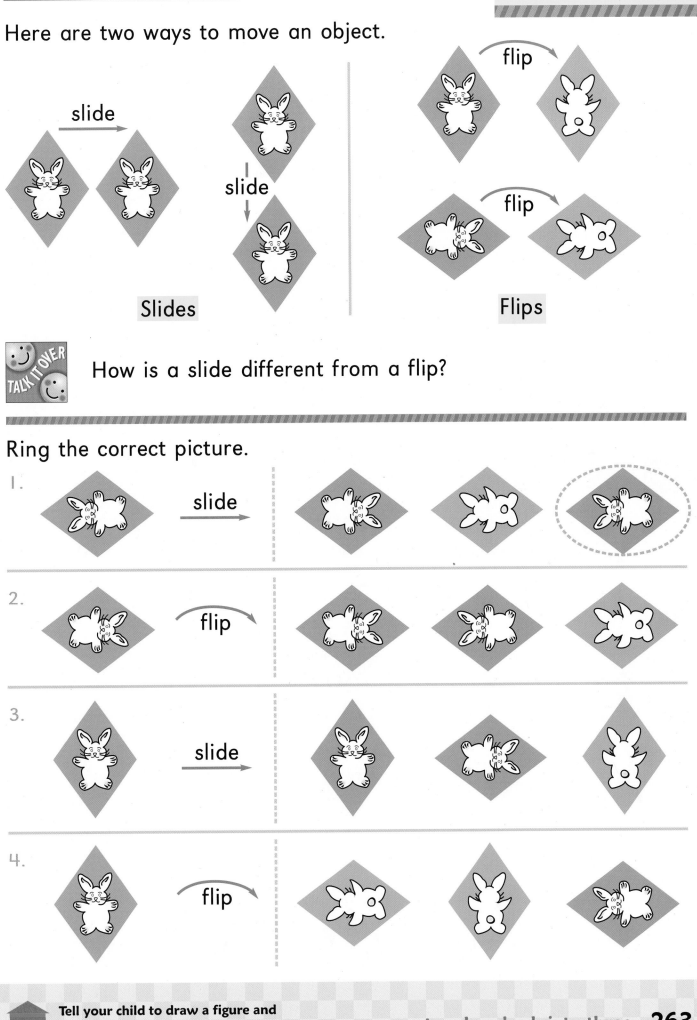

slide

Slides

flip

flip

Flips

TALK IT OVER

How is a slide different from a flip?

Ring the correct picture.

1.

slide

2.

flip

3.

slide

4.

flip

7-7 Tell your child to draw a figure and to make a pattern of alternating slides and flips with it.

two hundred sixty-three **263**

Name _____

| 1st | Find each difference.

| 2nd | Color the paths to the house.

less than 30 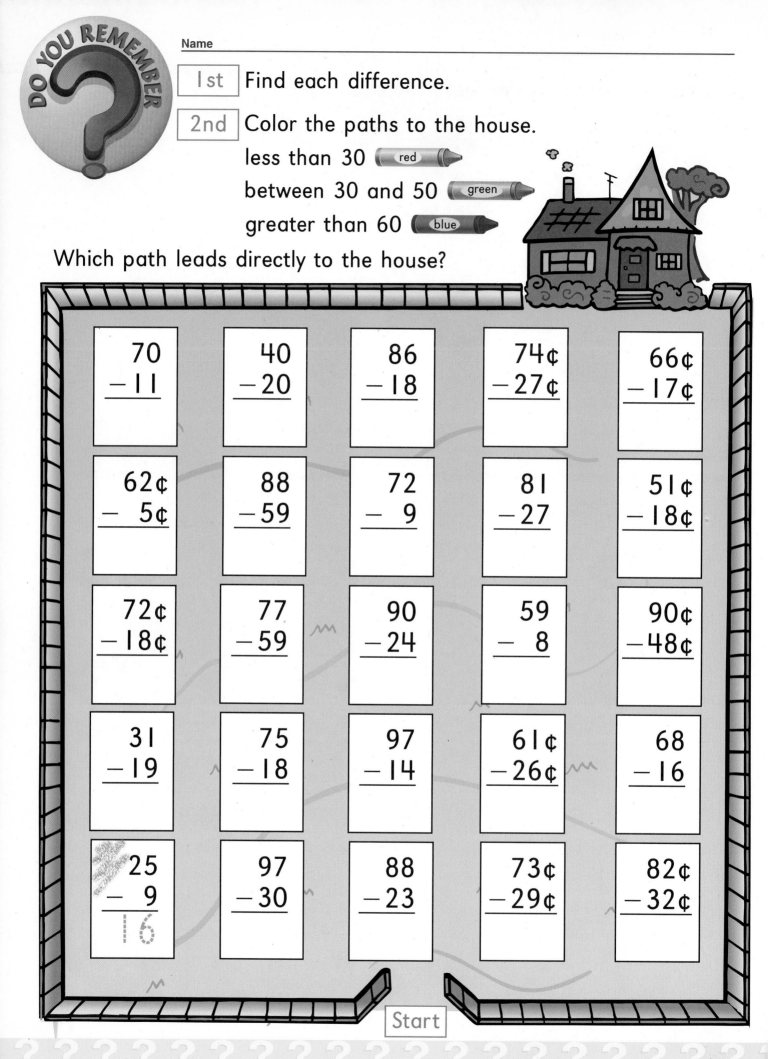 red

between 30 and 50 green

greater than 60 blue

Which path leads directly to the house?

$$70 - 11$$

$$40 - 20$$

$$86 - 18$$

$$74¢ - 27¢$$

$$66¢ - 17¢$$

$$62¢ - 5¢$$

$$88 - 59$$

$$72 - 9$$

$$81 - 27$$

$$51¢ - 18¢$$

$$72¢ - 18¢$$

$$77 - 59$$

$$90 - 24$$

$$59 - 8$$

$$90¢ - 48¢$$

$$31 - 19$$

$$75 - 18$$

$$97 - 14$$

$$61¢ - 26¢$$

$$68 - 16$$

$$25 - 9 = 16$$

$$97 - 30$$

$$88 - 23$$

$$73¢ - 29¢$$

$$82¢ - 32¢$$

Start

This page reviews the mathematical content presented in Chapter 6.

7

Name _____

You can separate figures into other figures.

| 1st | Draw a rectangle that has 3 dots inside. |

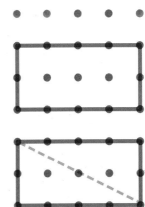

| 2nd | Draw a line to separate it into 2 triangles with square corners. |

| 3rd | Are the 2 parts congruent? _____ |

 How many ways can you separate a square into 2 triangles with square corners?

Separate each into congruent parts.

1. 2 squares

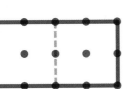

2. 3 triangles

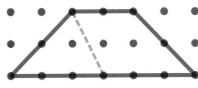

3. 4 triangles

4. 6 triangles

7-8 Tell your child to draw a figure, separate it into congruent parts, and name the parts.

two hundred sixty-five **265**

Make new figures.

1. Draw 2 triangles
to make a square.

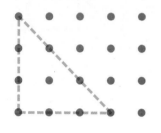

2. Draw 3 triangles
to make a rectangle.

SECOND LOOK Are the parts in 1 and 2 congruent?

CHALLENGE 3. Draw 1 triangle and
1 square to make a
figure with 5 sides
and 5 corners.

FINDING TOGETHER Use your punchouts to cover
the ⬡. Write how many.

4. _____ cover it.

5. _____ cover it.

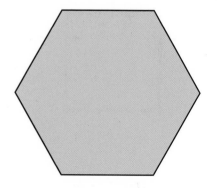

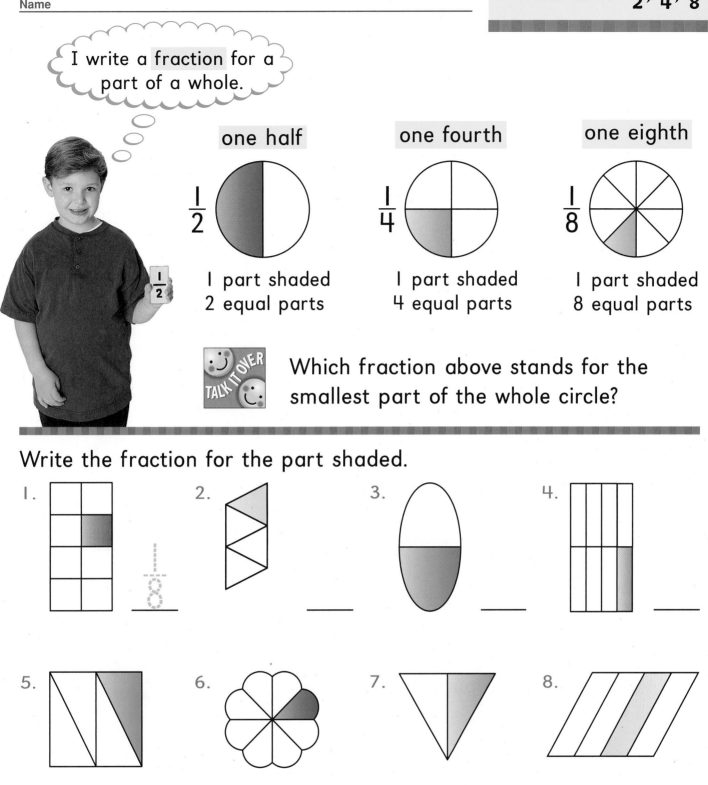

> I write a fraction for a part of a whole.

one half	one fourth	one eighth
$\frac{1}{2}$	$\frac{1}{4}$	$\frac{1}{8}$
1 part shaded 2 equal parts	1 part shaded 4 equal parts	1 part shaded 8 equal parts

TALK IT OVER Which fraction above stands for the smallest part of the whole circle?

Write the fraction for the part shaded.

1. $\frac{1}{8}$

2. ___

3. ___

4. ___

5. ___

6. ___

7. ___

8. ___

MATH JOURNAL

9. Draw 3 different rectangles, and show halves, fourths, and eighths. Color each part a different color. Describe the parts that make up each whole.

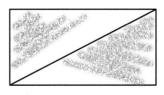

7-9   Provide your child with the outline of three different squares and have him/her shade $\frac{1}{4}$, $\frac{1}{8}$, and $\frac{1}{2}$.

two hundred sixty-seven **267**

Color 1 part of each whole.
Write the fraction for the part you colored.

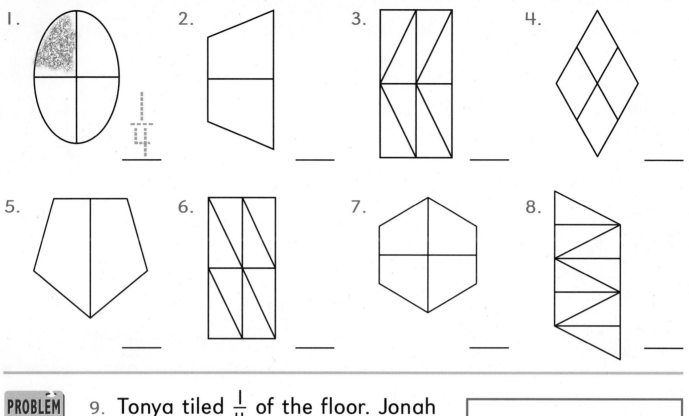

1. ____ 1/4

2. ____

3. ____

4. ____

5. ____

6. ____

7. ____

8. ____

PROBLEM SOLVING

9. Tonya tiled $\frac{1}{4}$ of the floor. Jonah tiled $\frac{1}{2}$ of the floor. Who tiled the greater amount?

_____

10. Daria and her 7 friends shared a pizza equally. What part of the pizza did Daria eat?

_____

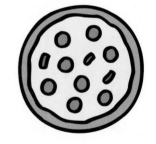

CHALLENGE Color the part of each shape to match the fraction.

11. $\frac{1}{2}$

12. $\frac{1}{2}$

13. $\frac{1}{4}$

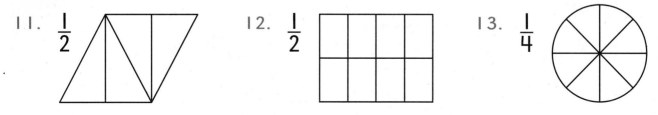

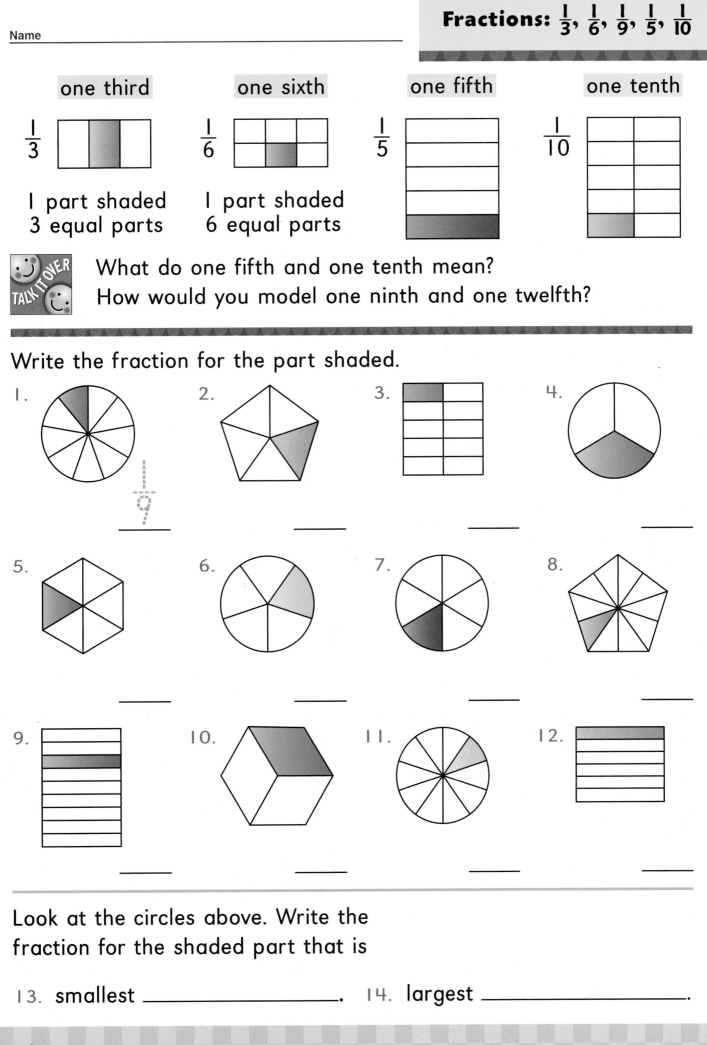

| one third | one sixth | one fifth | one tenth |

$\frac{1}{3}$ 1 part shaded   3 equal parts

$\frac{1}{6}$ 1 part shaded   6 equal parts

$\frac{1}{5}$

$\frac{1}{10}$

TALK IT OVER

What do one fifth and one tenth mean?
How would you model one ninth and one twelfth?

Write the fraction for the part shaded.

1. _____

2. _____

3. _____

4. _____

5. _____

6. _____

7. _____

8. _____

9. _____

10. _____

11. _____

12. _____

Look at the circles above. Write the
fraction for the shaded part that is

13. smallest _____.   14. largest _____.

7-10 After your child creates 3 different shapes of
6, 9, and 5 equal parts, have him/her shade
part of each and name the fractional part shaded.

Color 1 part of each whole.
Write the fraction for the part you colored.

1.

2. _____

3. _____

4. _____

5. _____

6. _____

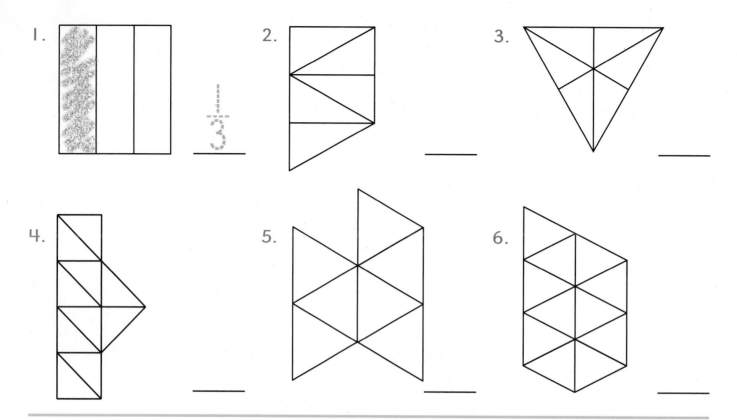

---

7. The stained-glass window in Rosa's house has 5 equal pieces of glass. Two are red and two are yellow. What part of the window is blue?

_____

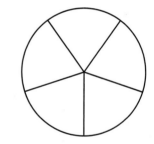

8. Leon made a brick wall. There were 9 red bricks and 1 white one. What part of the wall was white?

_____

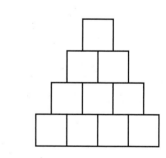

On dot paper make and color one half of a rectangle with:

9. 10 equal parts

10. 12 equal parts

Name _____

About $\frac{1}{3}$ of my house is red.

$\frac{1}{2}$

About $\frac{1}{4}$ of my house is brown.

about $\frac{1}{3}$

about $\frac{1}{2}$

about $\frac{1}{4}$

TALK IT OVER Suppose about one half of a floor is blue and about one fourth of it is white. Is the floor more blue or more white?

Write about what part of each is shaded.

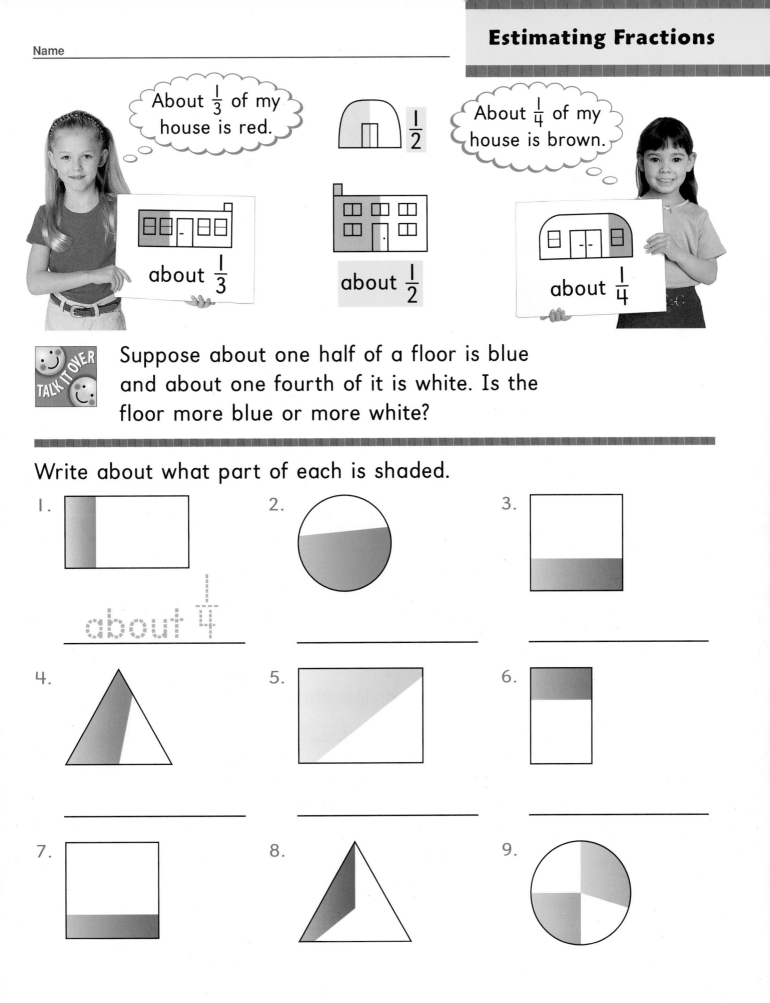

1. about $\frac{1}{4}$

2. _____

3. _____

4. _____

5. _____

6. _____

7. _____

8. _____

9. _____

7-11 Ask your child to shade each of 3 plane figures in a different amount and to estimate the amount shaded.

two hundred seventy-one **271**

Color each shape to match the estimate.

1. about one half

2. about one fourth
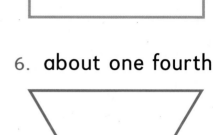

3. about one third

4. about one half

5. about one third
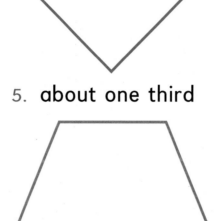

6. about one fourth
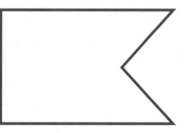

7. about one fourth

8. about one third

9. about one half

10. Sarah drank about $\frac{1}{2}$ glass of juice, Toy drank about $\frac{1}{4}$ glass, and Randi drank about $\frac{1}{3}$ glass. Match the girls with their glasses of juice and label each glass.

Sarah   Randi   Toy

11. _____ drank the most, _____ drank the least.

Name

The circle has 5 equal parts in all.

2 parts orange
5 equal parts is
two fifths.

3 parts blue
5 equal parts is
three fifths.

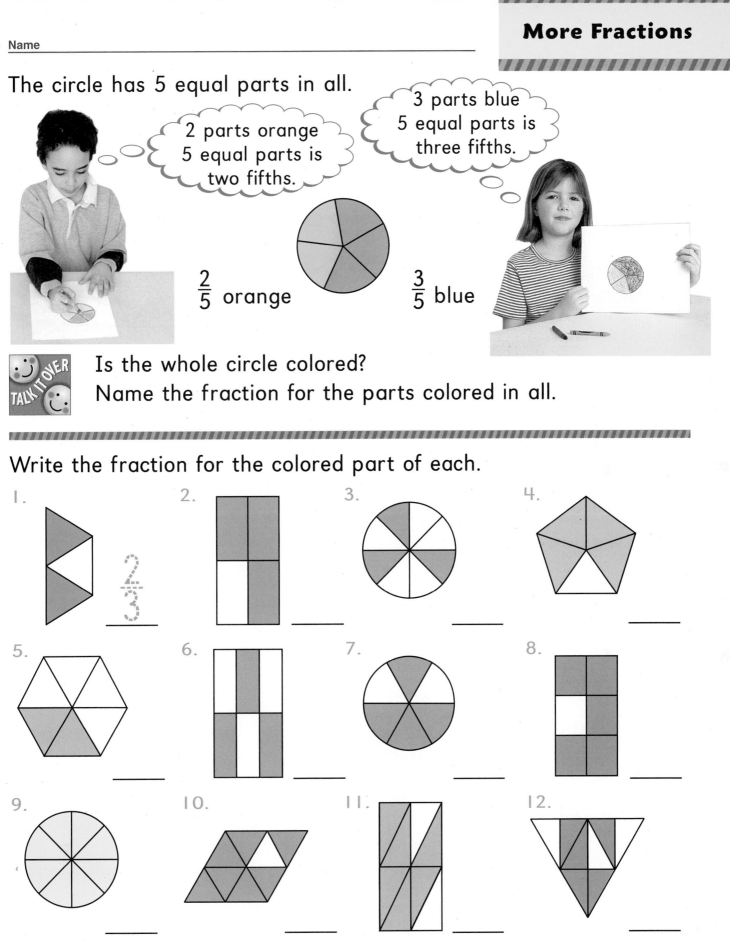

$\frac{2}{5}$ orange $\frac{3}{5}$ blue

TALK IT OVER

Is the whole circle colored?
Name the fraction for the parts colored in all.

Write the fraction for the colored part of each.

1.   2.   3.   4.

   $\frac{2}{3}$ ___   ___   ___

5.   6.   7.   8.

   ___   ___   ___   ___

9.   10.   11.   12.

   ___   ___   ___   ___

SHARE YOUR THINKING

Look at the fractions in each row. In which rows
is there a pattern? What comes next?

7-12 Tell your child to draw a circle, divide it
into 5 equal parts, and color it to
show $\frac{1}{5}$ orange and $\frac{4}{5}$ blue.

two hundred seventy-three **273**

Color the parts. Write the fraction.

1.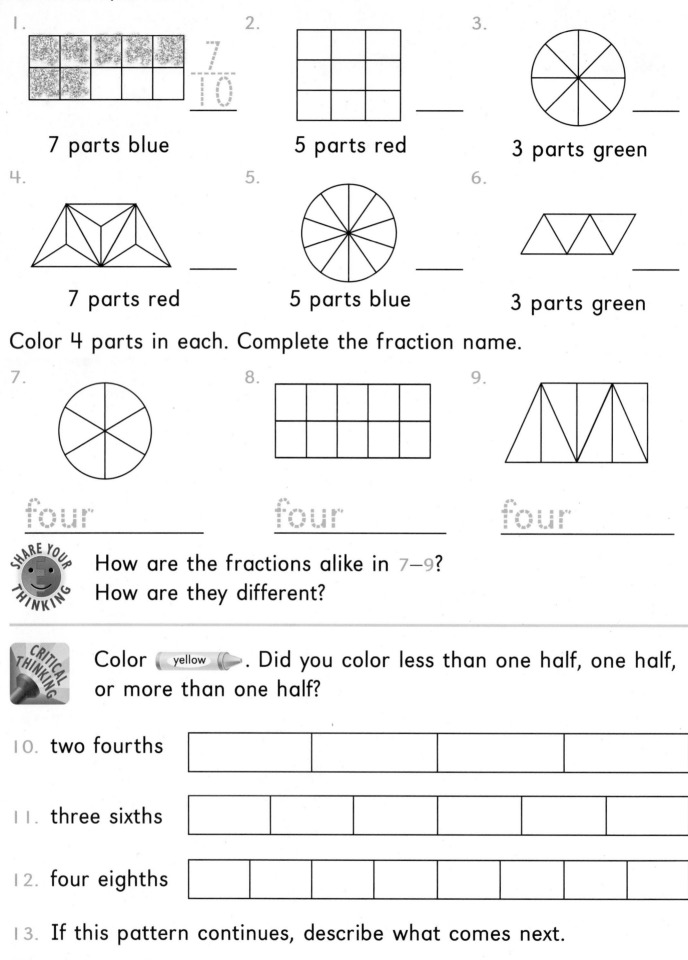
7 parts blue

$\dfrac{7}{10}$

2.
5 parts red

3.
3 parts green

4.
7 parts red

5.
5 parts blue

6.
3 parts green

Color 4 parts in each. Complete the fraction name.

7.

four

8.

four

9.

four

SHARE YOUR THINKING How are the fractions alike in 7–9?
How are they different?

CRITICAL THINKING Color yellow. Did you color less than one half, one half, or more than one half?

10. two fourths

11. three sixths

12. four eighths

13. If this pattern continues, describe what comes next.

Name

Part of each set of shapes is colored.

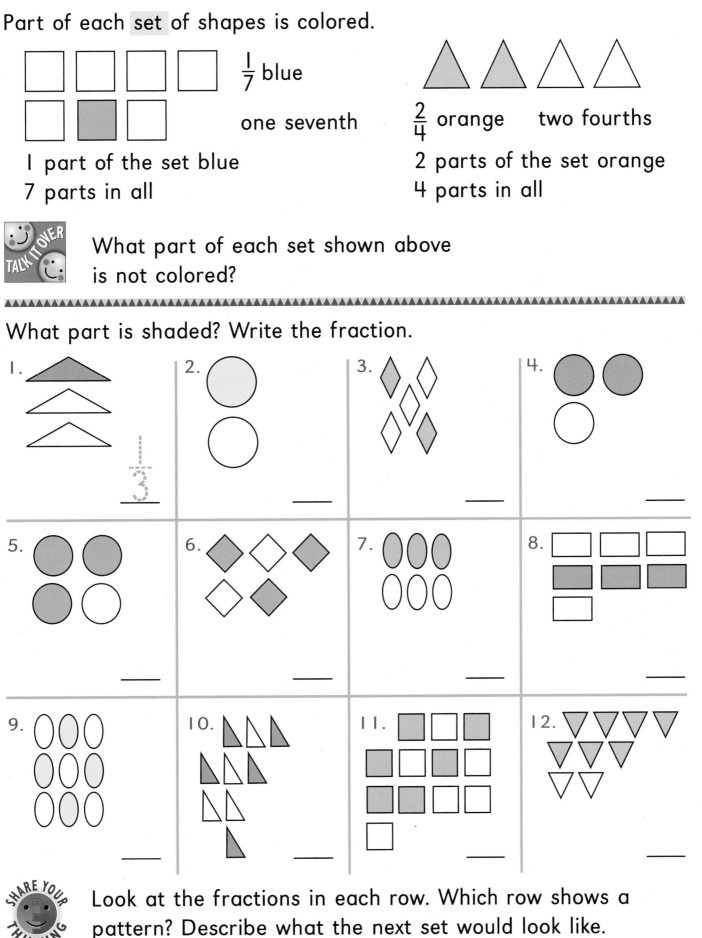

$\frac{1}{7}$ blue

one seventh

1 part of the set blue
7 parts in all

$\frac{2}{4}$ orange   two fourths

2 parts of the set orange
4 parts in all

TALK IT OVER

What part of each set shown above
is not colored?

What part is shaded? Write the fraction.

1. _____ $\frac{1}{3}$

2. _____

3. _____

4. _____

5. _____

6. _____

7. _____

8. _____

9. _____

10. _____

11. _____

12. _____

SHARE YOUR THINKING

Look at the fractions in each row. Which row shows a
pattern? Describe what the next set would look like.

7-13 Draw a set of objects. Have your child color part of the set and write a fraction for the part colored.

two hundred seventy-five **275**

Color the part of each set that matches the fraction.

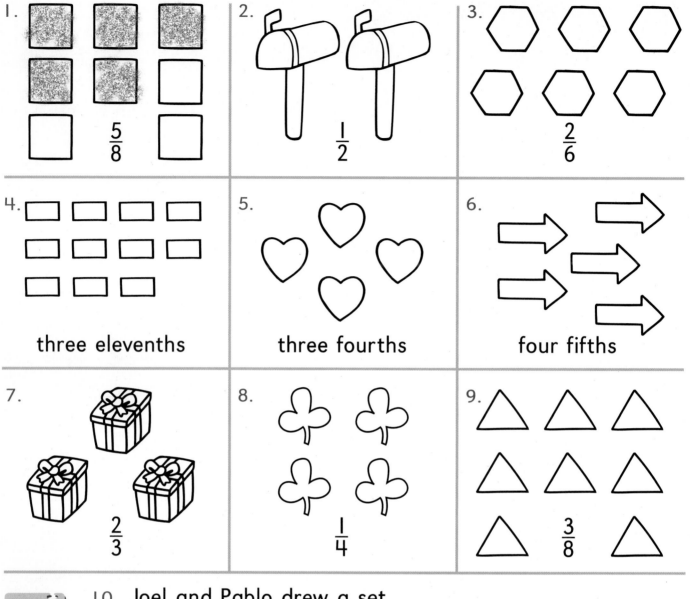

1. $\frac{5}{8}$

2. $\frac{1}{2}$

3. $\frac{2}{6}$

4. three elevenths

5. three fourths

6. four fifths

7. $\frac{2}{3}$

8. $\frac{1}{4}$

9. $\frac{3}{8}$

 PROBLEM SOLVING

10. Joel and Pablo drew a set of 12 circles. Joel colored 2 blue and Pablo colored 7 red. What part of the set is left to color?

_____ is left to color.

 CRITICAL THINKING

Ring one half of each set of stars.

11.

12.

13.

14.

Name

Both spinners have three outcomes.

A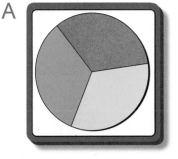

Outcomes
green
red
yellow

B

Outcomes
green
red
yellow

1st Predict the spinner that shows an equal chance of landing on each color.

2nd Use a ✏ and a 📎.
Spin each 12 times.
Color to record each outcome.

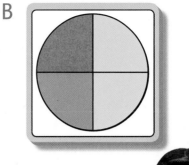

Outcomes

Spinner	1	2	3	4	5	6	7	8	9	10	11	12
A												
B												

 Compare your prediction with your results.
What do you notice?

List the outcomes. Ring the chance.

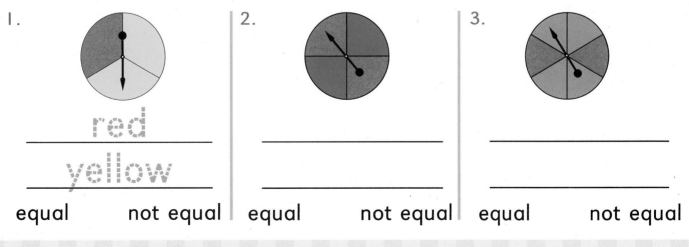

1.

red
yellow

equal   not equal

2.

equal   not equal

3.

equal   not equal

Certain and Impossible Outcomes

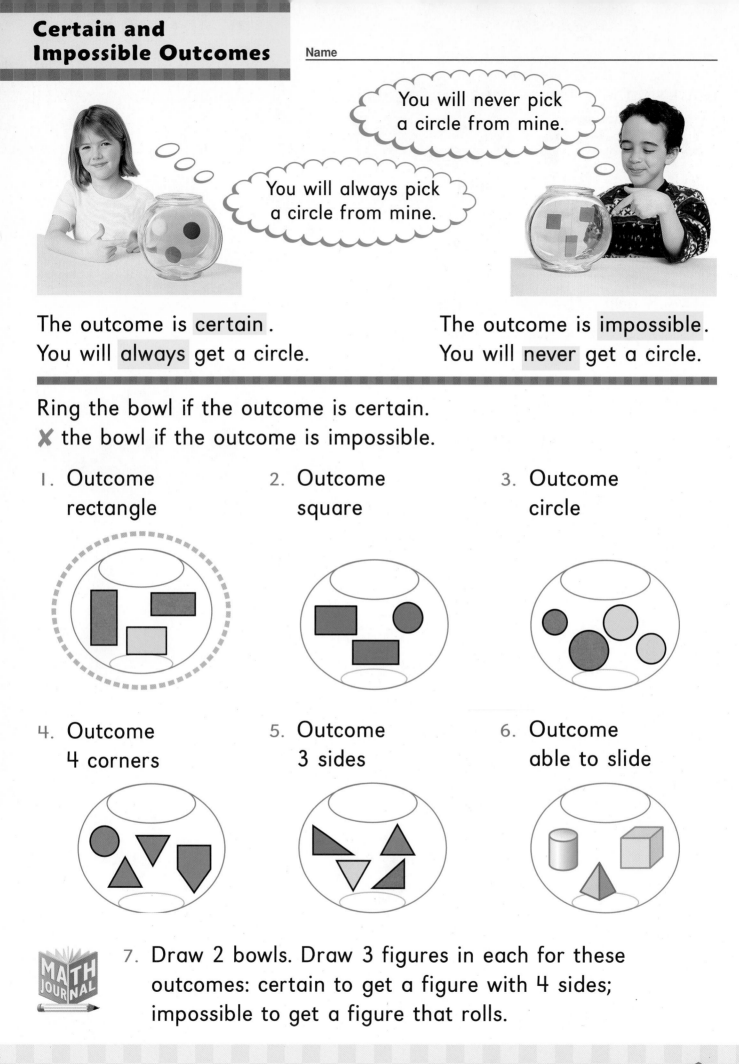

> You will never pick a circle from mine.

> You will always pick a circle from mine.

The outcome is **certain**.
You will **always** get a circle.

The outcome is **impossible**.
You will **never** get a circle.

Ring the bowl if the outcome is certain.
✗ the bowl if the outcome is impossible.

1. Outcome
   rectangle

2. Outcome
   square

3. Outcome
   circle

4. Outcome
   4 corners

5. Outcome
   3 sides

6. Outcome
   able to slide

MATH JOURNAL

7. Draw 2 bowls. Draw 3 figures in each for these outcomes: certain to get a figure with 4 sides; impossible to get a figure that rolls.

Ask your child to explain exercises 4–6.

 7-15

Name _____

The spinner I made is **more likely** to land on red.

Both spinners have **2 outcomes**.

The spinner I made is **less likely** to land on red.

Red	IIII I
Yellow	II

Red	III
Yellow	IIII II

Which color are you more likely to get?

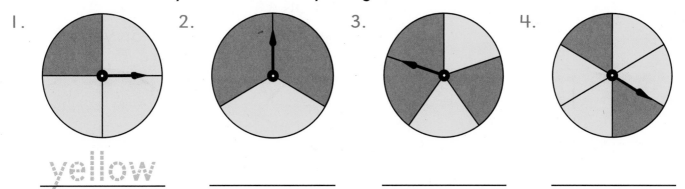

1. yellow

2. _____

3. _____

4. _____

Which color are you least likely to get?

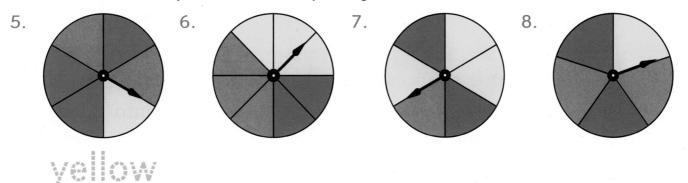

5. yellow

6. _____

7. _____

8. _____

 SHARE YOUR THINKING

How are the spinners in 1–4 alike?
How are the spinners in 5–8 alike?
How would you change the colors in 1, 4, and 5 to have an equal chance of landing on each color?

 7-16 Tell your child to create a spinner with 10 equal parts and to color it so there are equal chances of landing on 5 different colors.

Match each spinner with the more likely outcome.

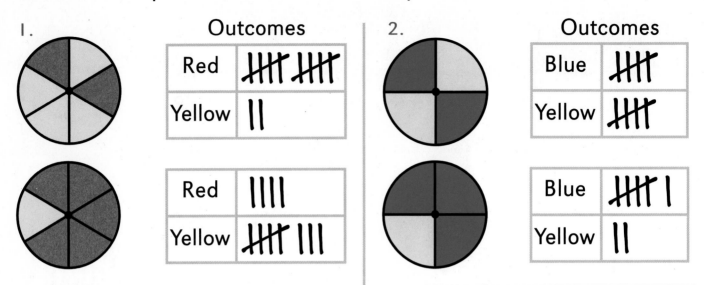

1.

Outcomes	
Red	卌 卌
Yellow	‖

Red	‖‖‖
Yellow	卌 ‖‖

2.

Outcomes	
Blue	卌
Yellow	卌

Blue	卌 ‖
Yellow	‖

Color each spinner.

3. Equal chance to get blue, red, or yellow

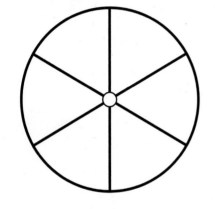

4. More likely to get red than blue or yellow

5. Less likely to get blue than red or yellow

6. Joan put 4 spheres and 2 cylinders into a bag. Which is she more likely to pick?

_____

7. Tom put 5 cones and 3 pyramids into a bag. Which is he less likely to pick?

_____

8. Make a tally table for each spinner in 3–5. Use a ⬭ and ——. Spin each 12 times and record the outcomes.

Name _____

Read → Name → Think → Write → Check

Use these steps.

Draw the 2 missing figures in each pattern.

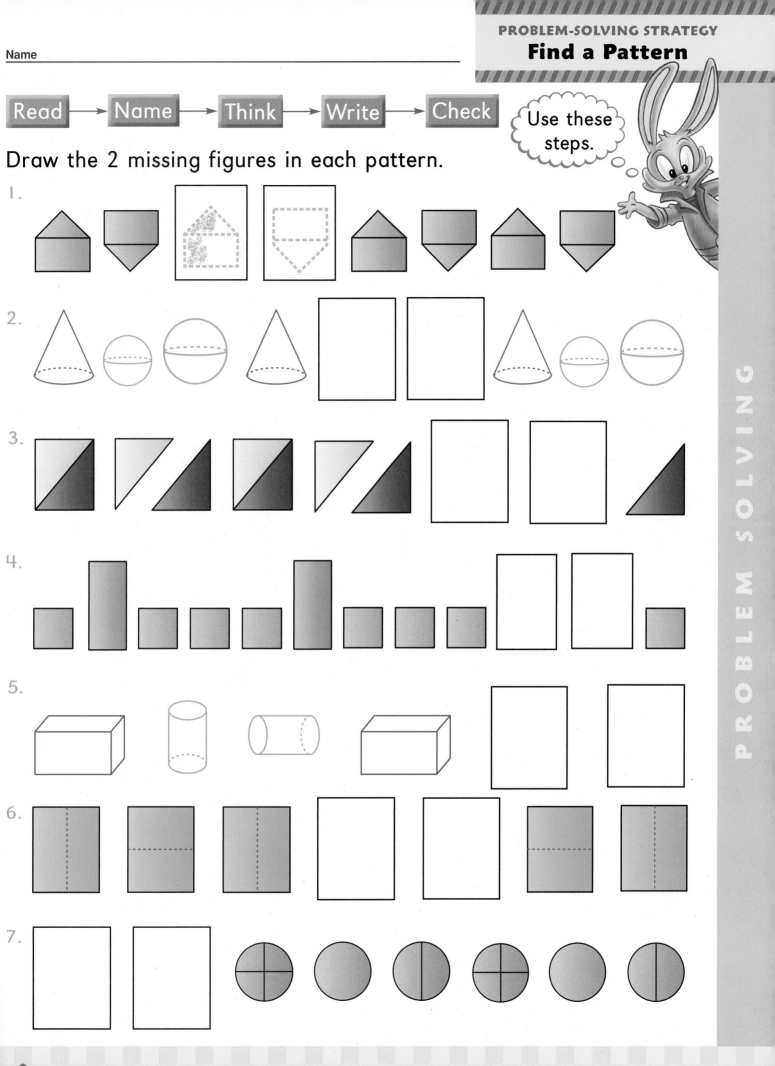

1.

2.

3.

4.

5.

6.

7.

PROBLEM SOLVING

Find the pattern. Color the next figures.

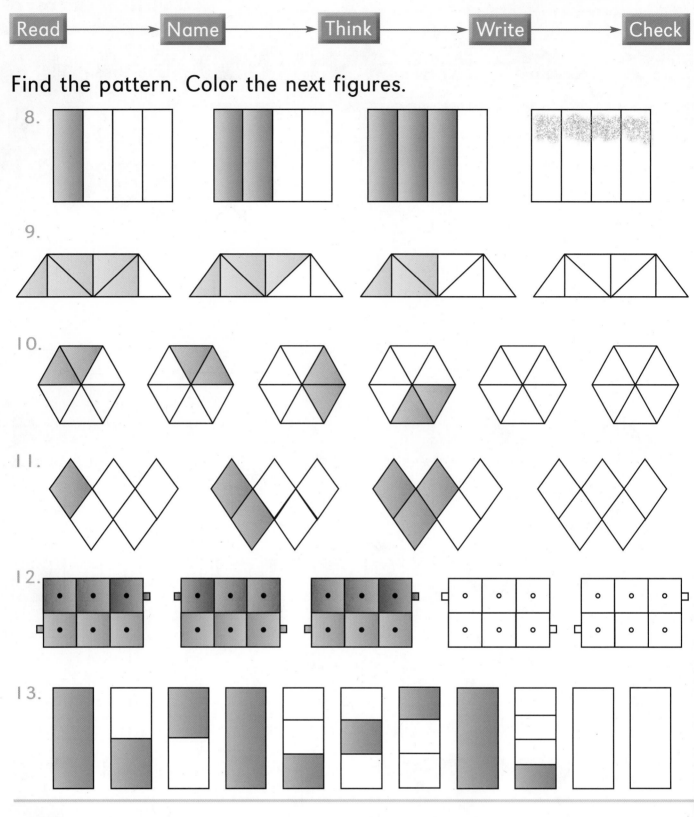

8.

9.

10.

11.

12.

13.

14. Use plane or space figures to draw
a pattern. Cover the first 2 figures.
Ask a classmate to draw them.

PROBLEM SOLVING

Name _____

Read → Draw → Think → Write → Check

1. Alex and his 3 friends share
   8 ● equally. Each of them
   gets only one fourth. How many
   ● does each one get?

   Use models to check.

   Each gets __2__ ●.

   $\frac{1}{4}$ of 8 = __2__

2. Trisha made 4 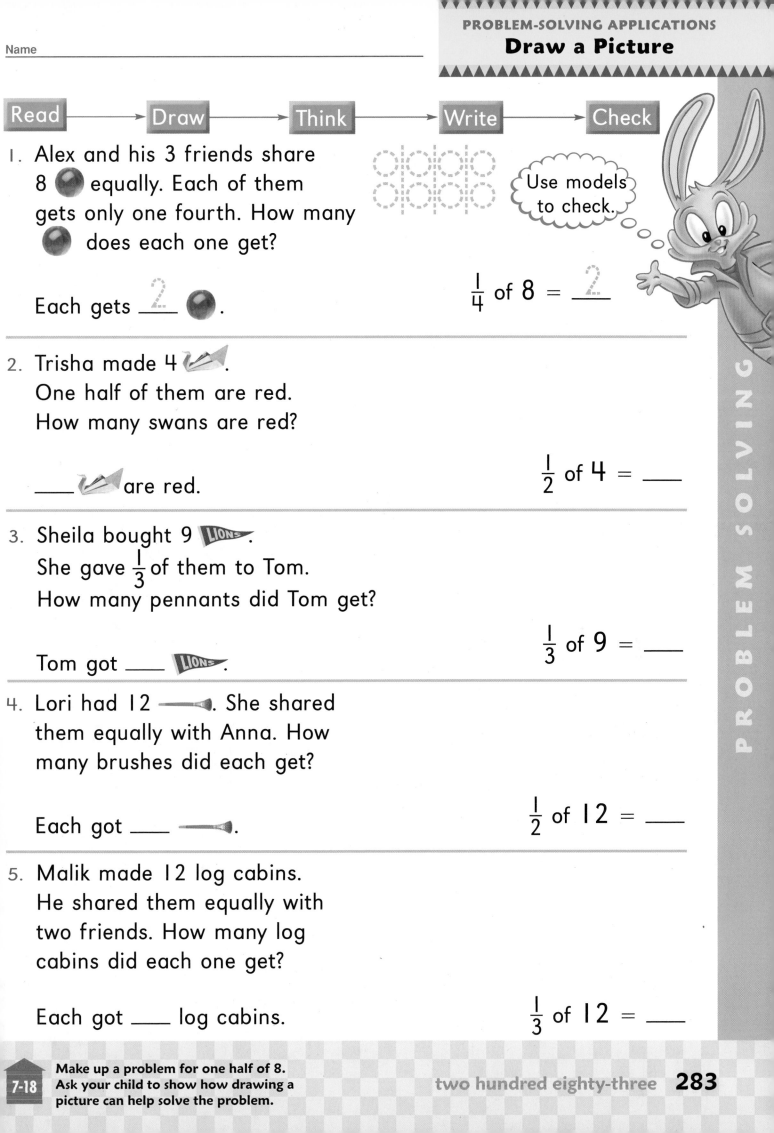.
   One half of them are red.
   How many swans are red?

   ____ are red.

   $\frac{1}{2}$ of 4 = ____

3. Sheila bought 9 LIONS.
   She gave $\frac{1}{3}$ of them to Tom.
   How many pennants did Tom get?

   Tom got ____ LIONS.

   $\frac{1}{3}$ of 9 = ____

4. Lori had 12 ——◀. She shared
   them equally with Anna. How
   many brushes did each get?

   Each got ____ ——◀.

   $\frac{1}{2}$ of 12 = ____

5. Malik made 12 log cabins.
   He shared them equally with
   two friends. How many log
   cabins did each one get?

   Each got ____ log cabins.

   $\frac{1}{3}$ of 12 = ____

PROBLEM SOLVING

7-18
Make up a problem for one half of 8.
Ask your child to show how drawing a
picture can help solve the problem.

two hundred eighty-three **283**

Use a strategy you know.

STRATEGY FILE

Use a Graph
Draw a Picture
Hidden Information
Logical Reasoning

6. Missy's quilt has 4 squares. Each square has 4 congruent triangles. 8 of the triangles are red. The rest are blue. How many are blue?

_____ triangles are blue.

7. Joel made a pattern using a dozen cylinders. The third, sixth, ninth, and last were yellow. How many were not yellow?

_____ were not yellow.

8. Tonya, Dora, and Leo each drew a different figure. Dora's had more than 3 corners. Tonya's had 5 sides and 1 line of symmetry. Color Leo's red, Dora's blue, and Tonya's yellow.

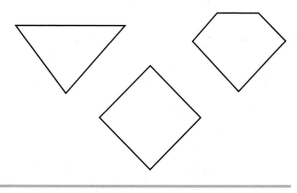

9. Two airplanes landed. What part of the group of airplanes landed?

_____ of the airplanes landed.

Vehicles Seen

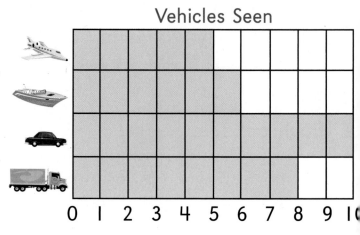

10. 3 of the cars are blue. 4 of the cars are white. The rest of the cars are black. What part of the group of cars are blue or white?

_____ of the cars are blue or white.

P R O B L E M   S O L V I N G

284   two hundred eighty-four

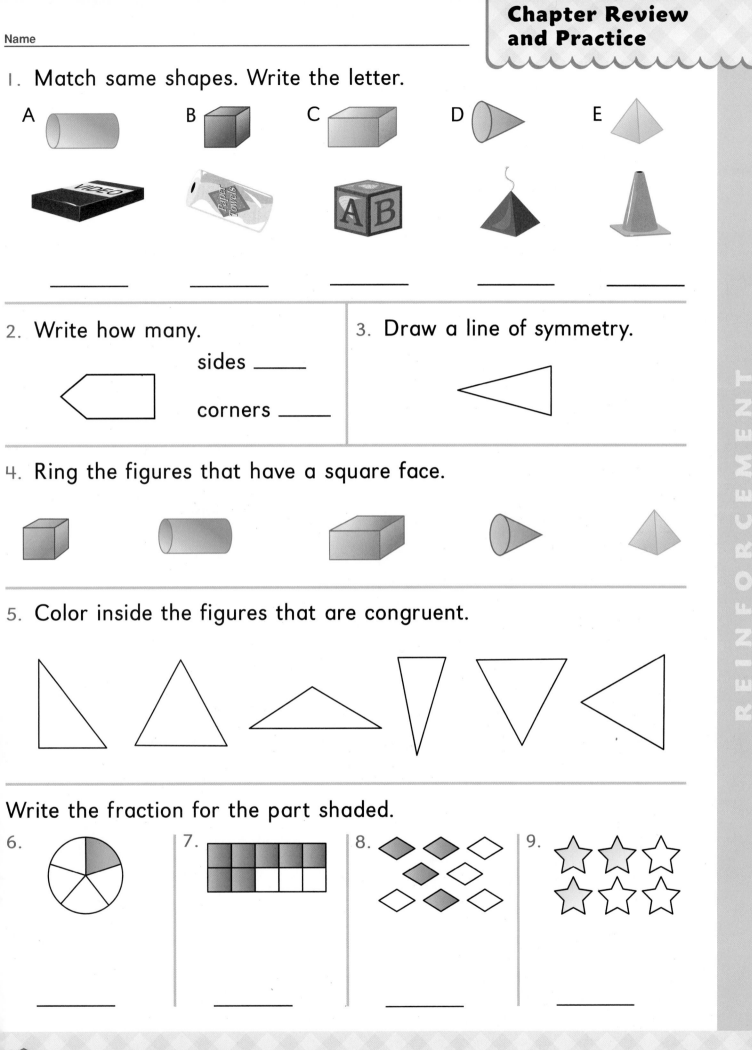

1. Match same shapes. Write the letter.

A   B   C   D   E

____   ____   ____   ____   ____

2. Write how many.

sides ____

corners ____

3. Draw a line of symmetry.

4. Ring the figures that have a square face.

5. Color inside the figures that are congruent.

Write the fraction for the part shaded.

6.   7.   8.   9.

____   ____   ____   ____

This page reviews the mathematical content presented in Chapter 7.

7

two hundred eighty-five **285**

Name _____

You can make patterns with slides , flips , and turns .
Name and color the next moves.

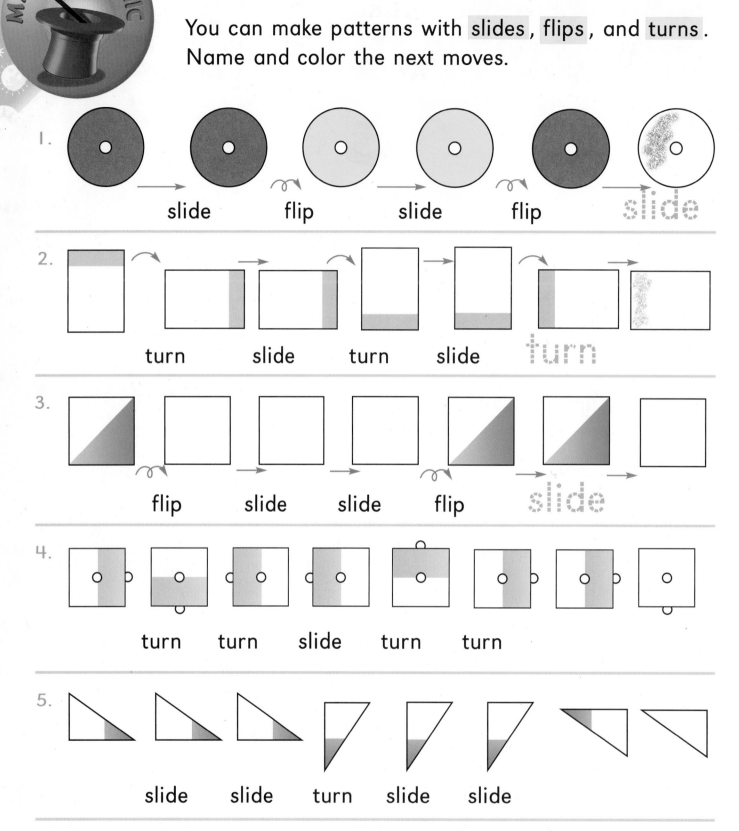

1. slide   flip   slide   flip   slide

2. turn   slide   turn   slide   turn

3. flip   slide   slide   flip   slide

4. turn   turn   slide   turn   turn

5. slide   slide   turn   slide   slide

6. Trace and cut out 10 ▱. Color one side of each red and the flip side yellow. Paste all of them on a separate sheet of paper to make up a slide, flip, and turn pattern.

This page extends your child's understanding of flips, slides, and turns.

7

1. Is one half of a set always
   the same number?   Yes   No
   Use models to check your answer.

$\frac{1}{2}$ of 6 = ___

$\frac{1}{2}$ of 10 = ___

$\frac{1}{2}$ of 1 dozen = ___ $\frac{1}{2}$ of 8 = ___

2. Use a separate sheet of paper.
   Draw figures for each one of these topics.
   • Two figures that have faces, corners, and edges.
   • A figure that has a line of symmetry.
   • Two congruent figures.
   • A figure that shows three fourths.
   • A figure that shows less than one half.

PORTFOLIO

Choose 1 of these projects.
Use a separate sheet of paper.

3. Draw 12 houses.
   Color some brown.
   Color some red.
   Color some blue.
   Write a fraction story
   about your houses.

> Houses
> More than one half
> of my houses are red.
> Only $\frac{1}{12}$ of them
> is brown.

4. Draw a spinner with space
   figures for each outcome.
   • impossible to get a figure
     with faces, corners, and
     edges
   • certain to get a figure
     that stacks

7

This page provides a variety of informal
assessment opportunities in order to measure
your child's understanding of Chapter 7.

two hundred eighty-seven **287**

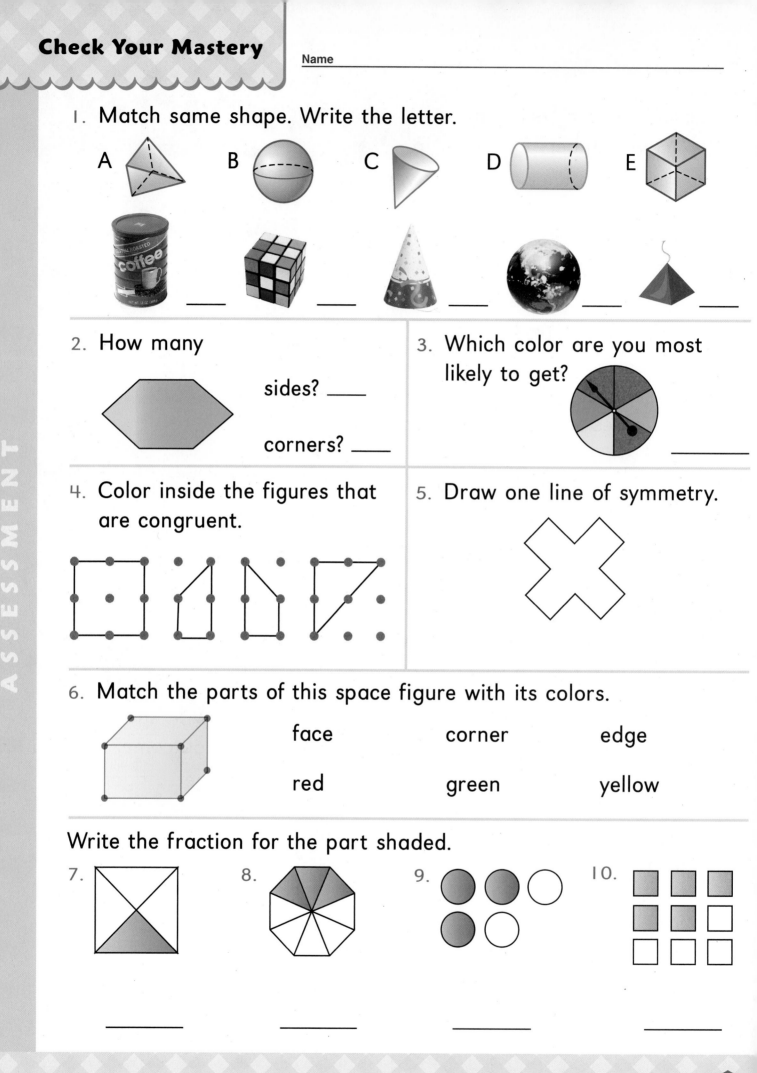

1. Match same shape. Write the letter.

A   B   C   D   E

coffee ____   ____   ____   ____   ____

2. How many

   sides? ____

   corners? ____

3. Which color are you most likely to get?

   _____

4. Color inside the figures that are congruent.

5. Draw one line of symmetry.

6. Match the parts of this space figure with its colors.

   face   corner   edge

   red   green   yellow

Write the fraction for the part shaded.

7.   8.   9.   10.

____   ____   ____   ____

This page is a formal assessment of your child's understanding of the content presented in Chapter 7.

7

Name _____

Mark the ◯ for your answer.
Listening Section

A   79 ◯ 84

Maria   Toby   cannot tell
◯   ◯   ◯

B

2   6   3
◯   ◯   ◯

---

1. Mark the time.

   ◯ 2:30
   ◯ 8:10
   ◯ 1:40
   ◯ 2:40

2. Find the sum.

   51
   $+35$

   ◯ 24
   ◯ 26
   ◯ 86
   ◯ not given

3. Add.

   $68¢$
   $+17¢$

   ◯ 51¢
   ◯ 75¢
   ◯ 89¢
   ◯ 85¢

4. Subtract.

   $79 - 71 = $ _____

   6   7   8   9
   ◯   ◯   ◯   ◯

5. Subtract.

   $83¢ - 12¢ = $ _____

   71¢   75¢   91¢   95¢
   ◯   ◯   ◯   ◯

6. Gus had 53 🦀. He gave 6 🦀 to Lupe and 10 to Darryl. How many did Gus have left?

   69   43   37   47
   ◯   ◯   ◯   ◯

7. What is the difference? Check by adding.

   $59 - 16 = $ ___

   ◯ 43 + 16   ◯ 59 + 16
   ◯ 43 + 59   ◯ 33 + 16

8. Estimate the sum.

   $39 \rightarrow \quad 40$
   $+18 \rightarrow \quad +20$
   about

   20   80   60   90
   ◯   ◯   ◯   ◯

9. Estimate the difference.

   $92 \rightarrow \quad 90$
   $-73 \rightarrow \quad -70$
   about

   80   20   30   60
   ◯   ◯   ◯   ◯

REINFORCEMENT

Mark the ○ for your answer.

10. How many corners?

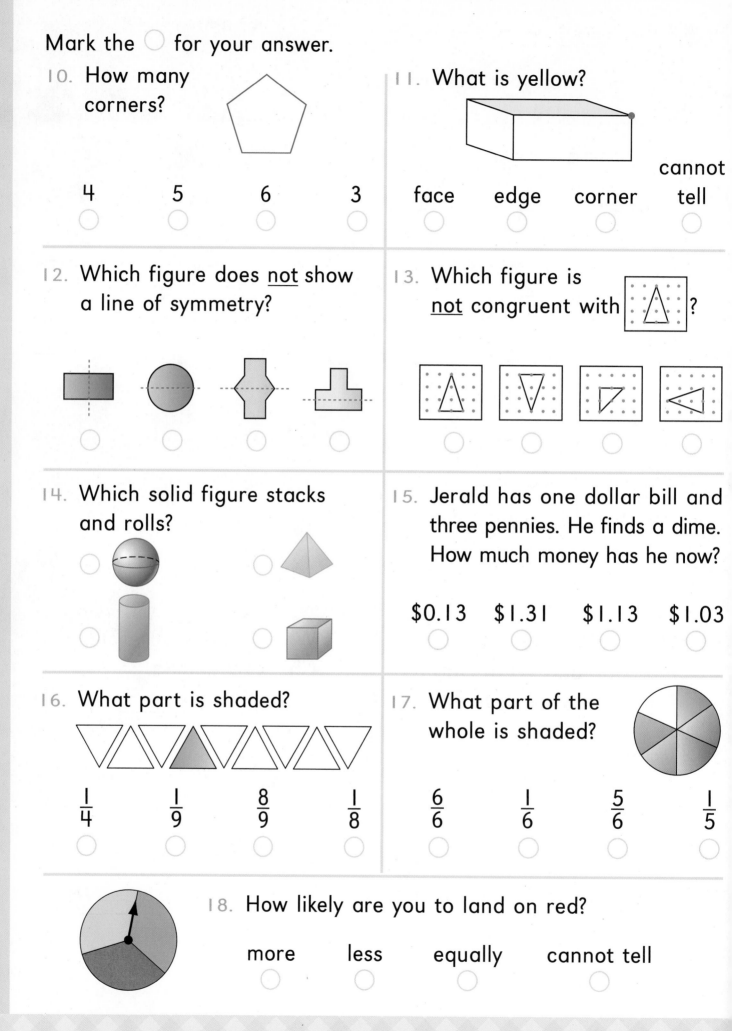

4	5	6	3
○	○	○	○

11. What is yellow?

| | | | cannot |
face	edge	corner	tell
○	○	○	○

12. Which figure does <u>not</u> show a line of symmetry?

○ ○ ○ ○

13. Which figure is <u>not</u> congruent with ?

○ ○ ○ ○

14. Which solid figure stacks and rolls?

○ ○

○ ○

15. Jerald has one dollar bill and three pennies. He finds a dime. How much money has he now?

$0.13	$1.31	$1.13	$1.03
○	○	○	○

16. What part is shaded?

$\frac{1}{4}$	$\frac{1}{9}$	$\frac{8}{9}$	$\frac{1}{8}$
○	○	○	○

17. What part of the whole is shaded?

$\frac{6}{6}$	$\frac{1}{6}$	$\frac{5}{6}$	$\frac{1}{5}$
○	○	○	○

18. How likely are you to land on red?

more	less	equally	cannot tell
○	○	○	○

Measurement

CRITICAL THINKING

The inchworm weighs the same as 1 connecting cube. How many cubes should be added to the right side to balance the scale?

Math Alive at Home

Dear Family,

Today your child began Chapter 8. As he/she studies measurement, you may want to read the poem below, which was read in class. Have your child talk about some of the math ideas pictured on page 291.

Look for the 🏠 at the bottom of each skills lesson. The suggestion on the page gives you an opportunity to improve your child's understanding of math and to reinforce his/her math language. You may want to have various measuring tools available for your child to use throughout this chapter.

Home Reading Connection

The Inchworm's Trip

Inch,
    by inch,
        by inch
            he crawls

through our classrooms,
    through our halls.
He's one inch long
    and will not stop,
Inching, inching,
    past the mop.
Inching, inching,
    up my chair.
Now he's inching
    through my hair—
His way of "walking's"
    fun to see—
Does it tickle him
    as much as me?

Sandra Liatsos

Home Activity

I Spy

For the first three lessons of this chapter, have a 12-inch ruler and a yardstick or tape measure available. Play the game "I Spy" by describing an object within your child's viewing range, including an estimation of its length or height. After your child has identified the object, ask him/her to use a ruler or yardstick to measure it to the nearest inch, half inch, then foot—as these skills are presented. Have him/her keep a log, recording the name of the object and its actual measurement. For example,

    I spy an object on the table.
    It is about 7 inches long and has
    2 silver and orange parts.

As your child progresses through this chapter, change the type of object, the unit of measure, and the measurement tools available.

Name _____

Bob measured the lengths
of some objects in spans.

teacher's desk	7 spans wide
board	10 spans long
textbook	2 spans long
bookcase	5 spans long

1 span

He made a pictograph of the measures.
Draw and color to complete the graph.

1.

Spans of Four Objects

teacher's desk	🖐🖐🖐🖐🖐🖐🖐
board	🖐🖐🖐🖐🖐🖐🖐🖐🖐🖐
textbook	
bookcase	
Key: Each 🖐 stands for 1 span.	

FINDING TOGETHER

Measure each classroom object in spans.
Then make your own graph.

2. teacher's desk ___ spans   3. board ___ spans

4. textbook ___ spans   5. bookcase ___ spans

TALK IT OVER

Why are your numbers different from those
in the pictograph above?

CONNECTIONS

Read each clue. Unscramble the word in the second column.
Then put the words in alphabetical order. Write 1st, 2nd, or 3rd.

1. how heavy   grkilamo   kilogram   _____

    how long   thleng   _____   _____

    [water bottle]   lerti   _____   _____

2. how high   ghthei   _____   _____

    [ruler]   toof   _____   _____

    more than 1 foot eeft   _____   _____

3. cold, warm, hot   pertemature   _____   _____

    opposite of
    shorter   allert   _____   _____

    [thermometer]   momthereter   _____   _____

4. [measuring cups] tipn   _____   _____

    unit of weight   pndou   _____   _____

    distance around   metperier   _____   _____

You can put this in your
Math Portfolio.

Name

This is 1 inch. |———|

The chalk is between
2 and 3 inches long.
It is closer to 2 inches.
It is about 2 inches long.

inches
| 1 | 2 | 3

TALK IT OVER How should you place your ruler when measuring?

Use your inch ruler. Estimate the length.

1. GLUE STICK Instant Adhesive NON-TOXIC WASHABLE Net Wt .28 oz (6 grams) SAFE

about __3__ inches

inches
| 1 | 2 | 3 | 4 | 5 | 6

2.

about ___ inches

3.

about ___ inches

4.

about ___ inches

8-1 Ask your child to explain how she/he uses a ruler to measure an object in inches.

two hundred ninety-five **295**

Estimate. Then use your ruler to measure.

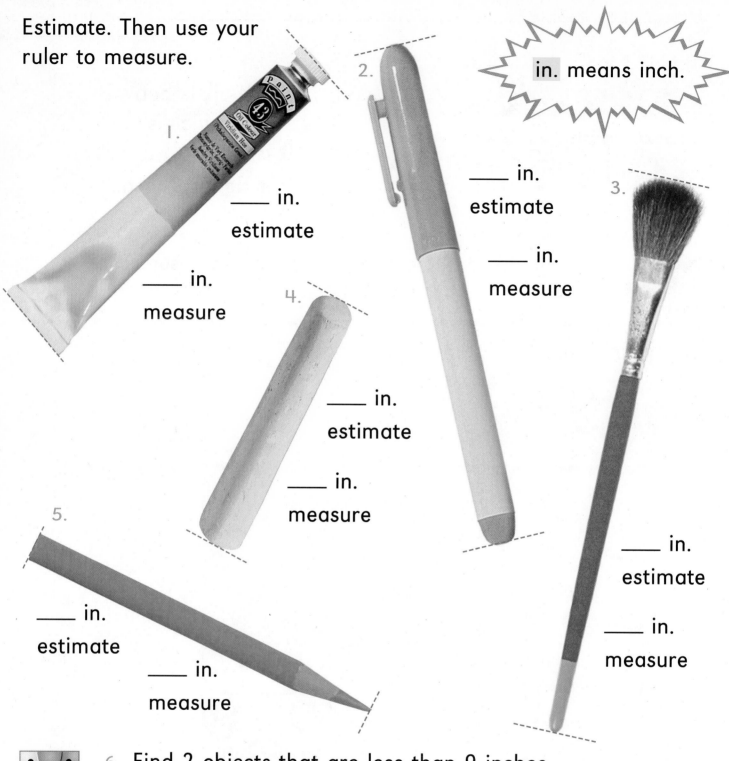

in. means inch.

1. ___ in. estimate
___ in. measure

2. ___ in. estimate
___ in. measure

3. ___ in. estimate
___ in. measure

4. ___ in. estimate
___ in. measure

5. ___ in. estimate
___ in. measure

FINDING TOGETHER

6. Find 2 objects that are less than 9 inches in width. Find 2 objects that are more than 9 inches in length.

Object	Width
	in.
	in.

Object	Length
	in.
	in.

Name

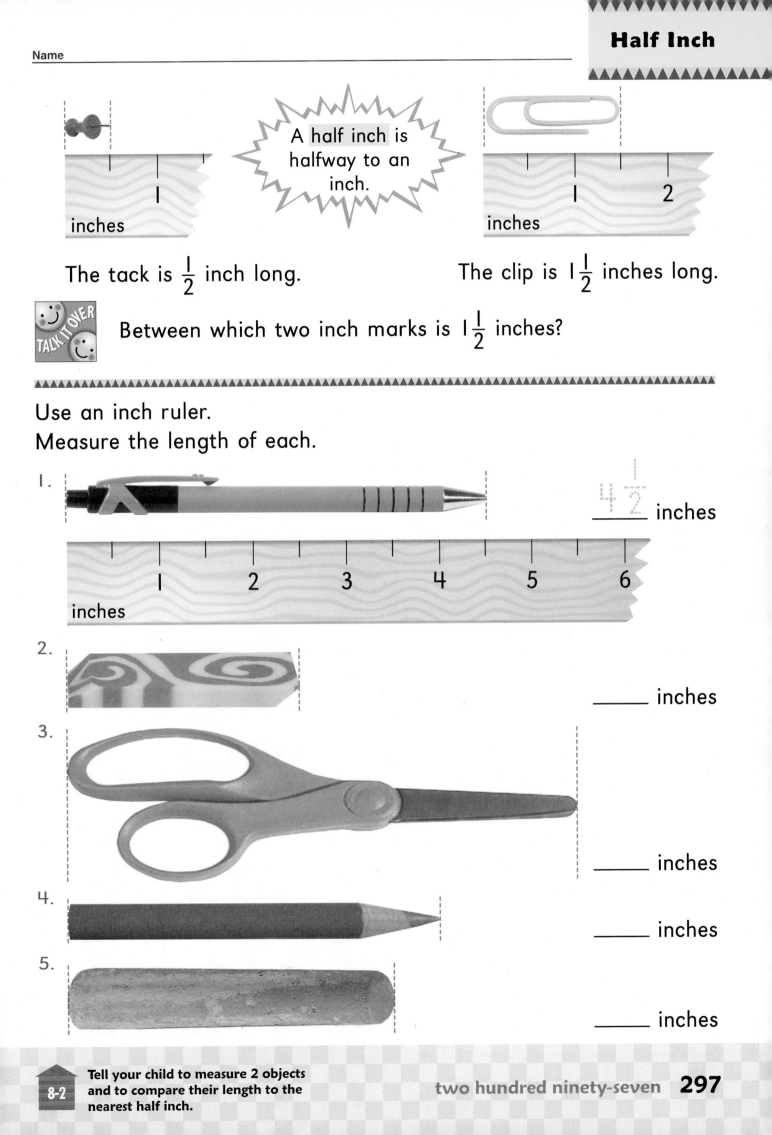

A half inch is halfway to an inch.

inches

inches

The tack is $\frac{1}{2}$ inch long.

The clip is $1\frac{1}{2}$ inches long.

TALK IT OVER

Between which two inch marks is $1\frac{1}{2}$ inches?

Use an inch ruler.
Measure the length of each.

1. $4\frac{1}{2}$ _____ inches

inches

2. _____ inches

3. _____ inches

4. _____ inches

5. _____ inches

8-2 Tell your child to measure 2 objects and to compare their length to the nearest half inch.

two hundred ninety-seven **297**

Start at the mark. Draw a line for each measure.

1. $2\frac{1}{2}$ inches

|

2. $5\frac{1}{2}$ inches

|

3. 4 inches

|

4. $3\frac{1}{2}$ inches

|

Color to show the same length.

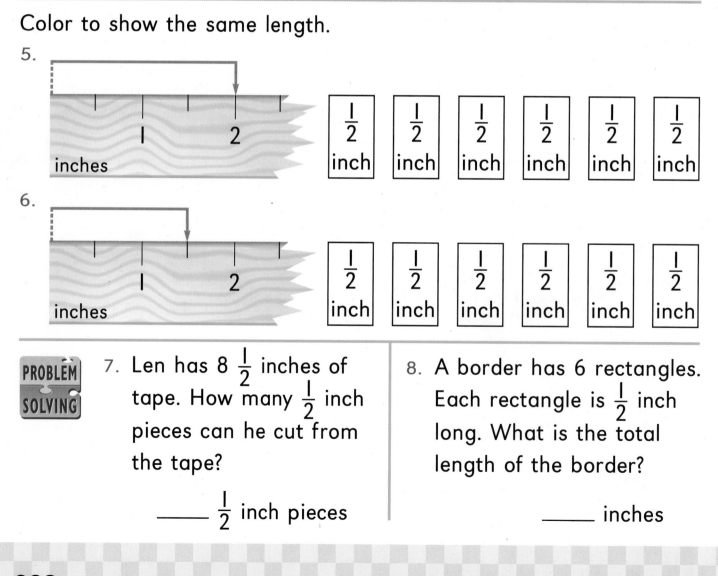

5.

6.

7. Len has $8\frac{1}{2}$ inches of tape. How many $\frac{1}{2}$ inch pieces can he cut from the tape?

_____ $\frac{1}{2}$ inch pieces

8. A border has 6 rectangles. Each rectangle is $\frac{1}{2}$ inch long. What is the total length of the border?

_____ inches

A foot is a unit
of measure.
A ruler shows 1 foot.

12 inches = 1 foot

A yard is a unit
of measure.
A yardstick shows 3 feet.

3 feet = 1 yard

Ring the best unit of measure.

1.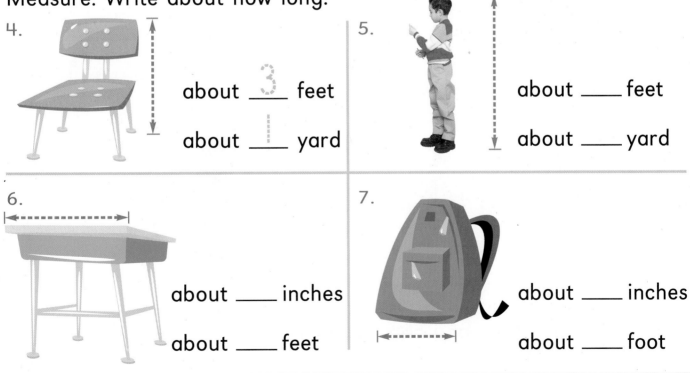

foot   (yard)

2.

foot   inch

3.

foot   yard

Predict which unit will give the greater number.
Measure. Write about how long.

4.

about __3__ feet

about __1__ yard

5.

about ___ feet

about ___ yard

6.

about ___ inches

about ___ feet

7.

about ___ inches

about ___ foot

8-3 Measure several objects in yards and have
your child measure the same objects in feet.

two hundred ninety-nine **299**

Ring the unit of measure. Estimate. Then measure.

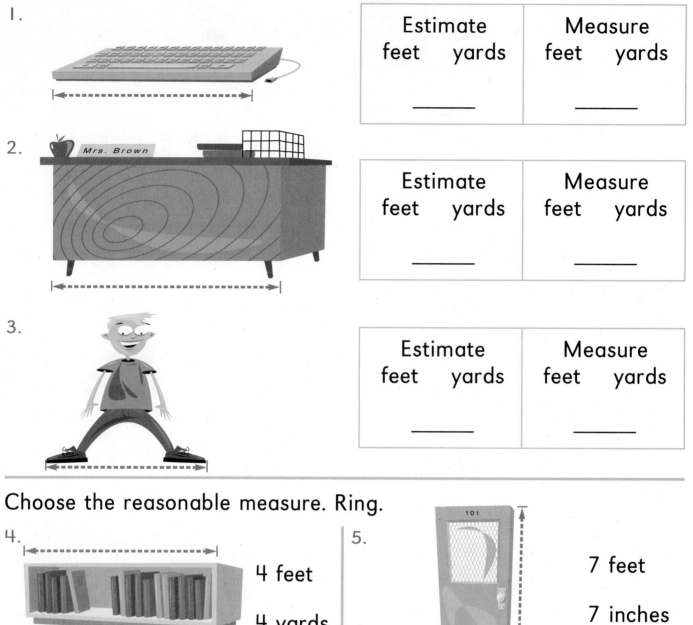

1.

Estimate feet yards	Measure feet yards
_____	_____

2.

Mrs. Brown

Estimate feet yards	Measure feet yards
_____	_____

3.

Estimate feet yards	Measure feet yards
_____	_____

Choose the reasonable measure. Ring.

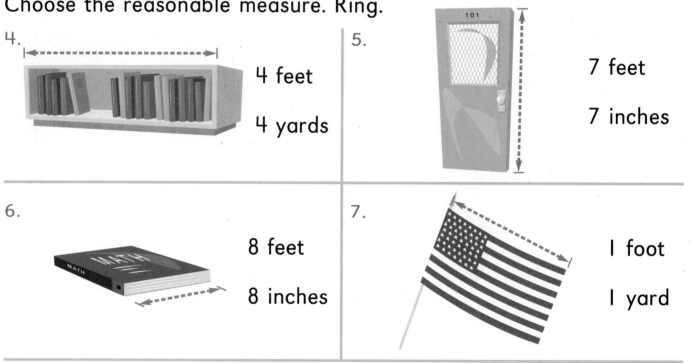

4.

4 feet

4 yards

5.

101

7 feet

7 inches

6.

MATH

8 feet

8 inches

7.

1 foot

1 yard

Find an object that is about 1 inch,
about 1 foot, and about 1 yard.

Name _____

You can measure liquids in cups, pints, or quarts.

2 cups = 1 pint

2 pints = 1 quart

Color to show the same amount. Complete.

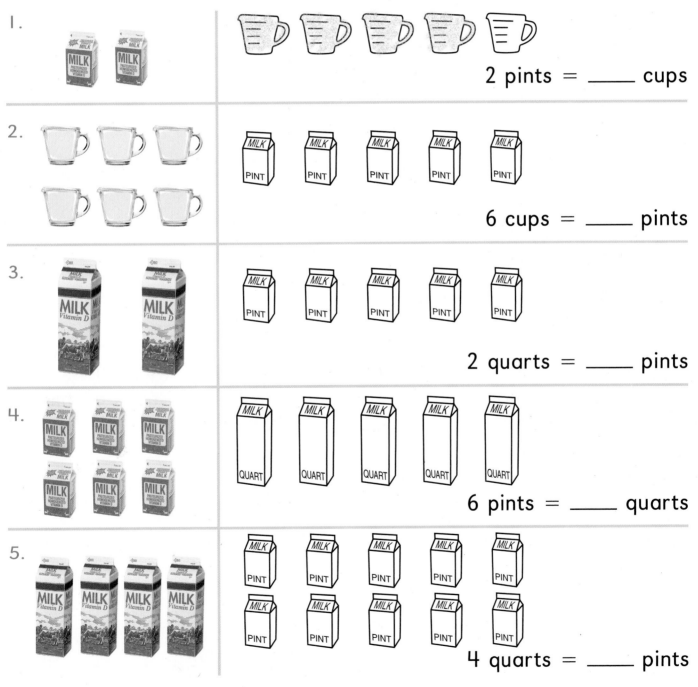

1.

2 pints = _____ cups

2.

6 cups = _____ pints

3.

2 quarts = _____ pints

4.

6 pints = _____ quarts

5.

4 quarts = _____ pints

Ring the better estimate.

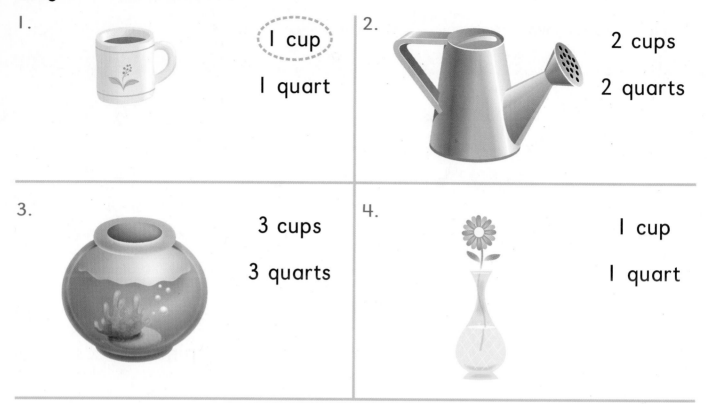

1. **1 cup**

   1 quart

2. 2 cups

   2 quarts

3. 3 cups

   3 quarts

4. 1 cup

   1 quart

Write R for reasonable or U for unreasonable.

5. Sara drinks 5 quarts of water at dinner. _____

6. Martin used 1 pint of water to wash the car. _____

7. Kate pours 1 quart of milk into 2 pint bottles. _____

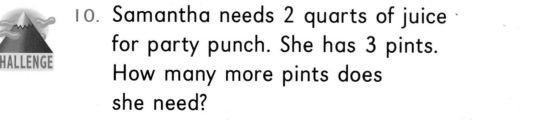

PROBLEM SOLVING

8. Josh has 4 pints of juice. Stan has 3 quarts of juice. Who has more?

   _____ has more.

9. Lee has 4 pints of milk. Josh has 5 cups of milk. Chen has 6 pints of milk.

   _____ has the least.

CHALLENGE

10. Samantha needs 2 quarts of juice for party punch. She has 3 pints. How many more pints does she need?

    She needs ____ pints.

4 quarts = 1 gallon

Color to show the same amount.

1.

MILK QUART   MILK QUART   MILK QUART   MILK QUART   MILK QUART   MILK QUART   MILK QUART   MILK QUART

____ gallons = ____ quarts

2.

MILK QUART   MILK QUART   MILK QUART   MILK QUART   MILK QUART   MILK QUART

MILK QUART   MILK QUART   MILK QUART   MILK QUART   MILK QUART   MILK QUART

____ gallons = ____ quarts

SHARE YOUR THINKING

3. There are 4 quarts in 1 gallon.
   How many quarts are in
   1 half gallon?   ____ quarts

4. Does $\frac{1}{2}$ gallon have more or
   less than 2 pints?   _____

MILK Vitamin D Half Gallon

8-5 Prompt your child to look at exercises 1-2 and to tell you how many more quarts are in 3 gallons than are in 2 gallons.

three hundred three **303**

Color red for Yes.   Color brown for No.

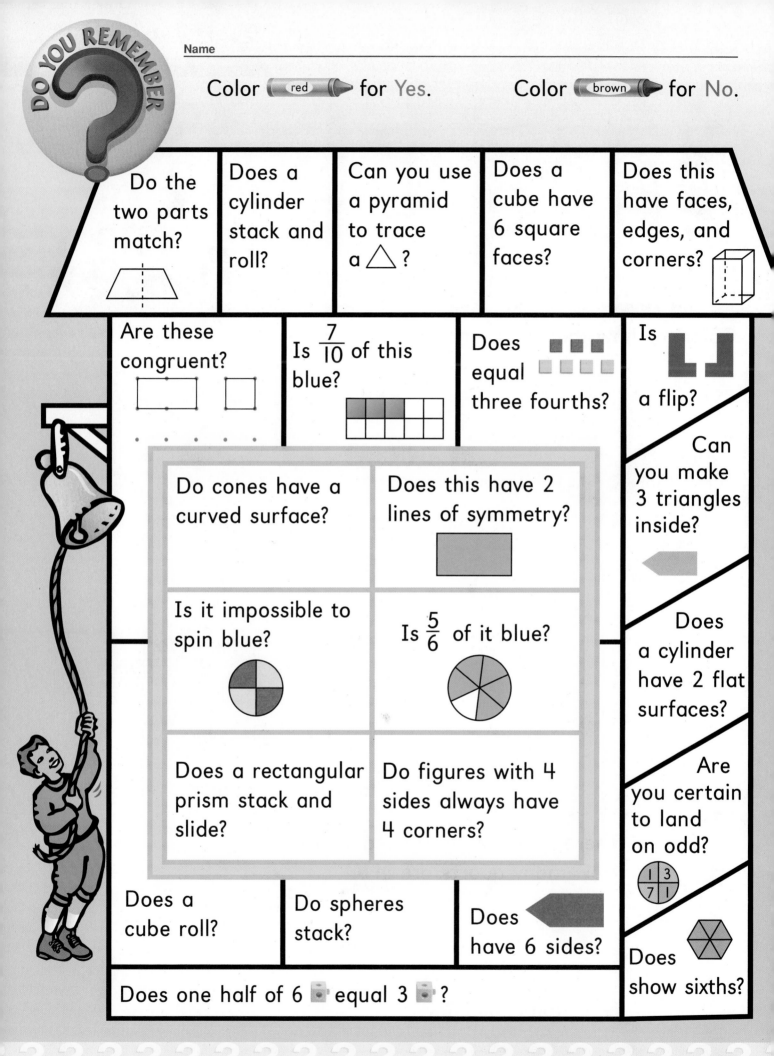

Do the two parts match?

Does a cylinder stack and roll?

Can you use a pyramid to trace a △ ?

Does a cube have 6 square faces?

Does this have faces, edges, and corners?

Are these congruent?

Is $\frac{7}{10}$ of this blue?

Does ▪▪▪ ▫▫▫▫ equal three fourths?

Is ⌐⌐ a flip?

Do cones have a curved surface?

Does this have 2 lines of symmetry?

Is it impossible to spin blue?

Is $\frac{5}{6}$ of it blue?

Does a rectangular prism stack and slide?

Do figures with 4 sides always have 4 corners?

Can you make 3 triangles inside?

Does a cylinder have 2 flat surfaces?

Are you certain to land on odd?

Does a cube roll?

Do spheres stack?

Does ◀ have 6 sides?

Does show sixths?

Does one half of 6 ▮ equal 3 ▮ ?

This page reviews the mathematical content in Chapter 7.

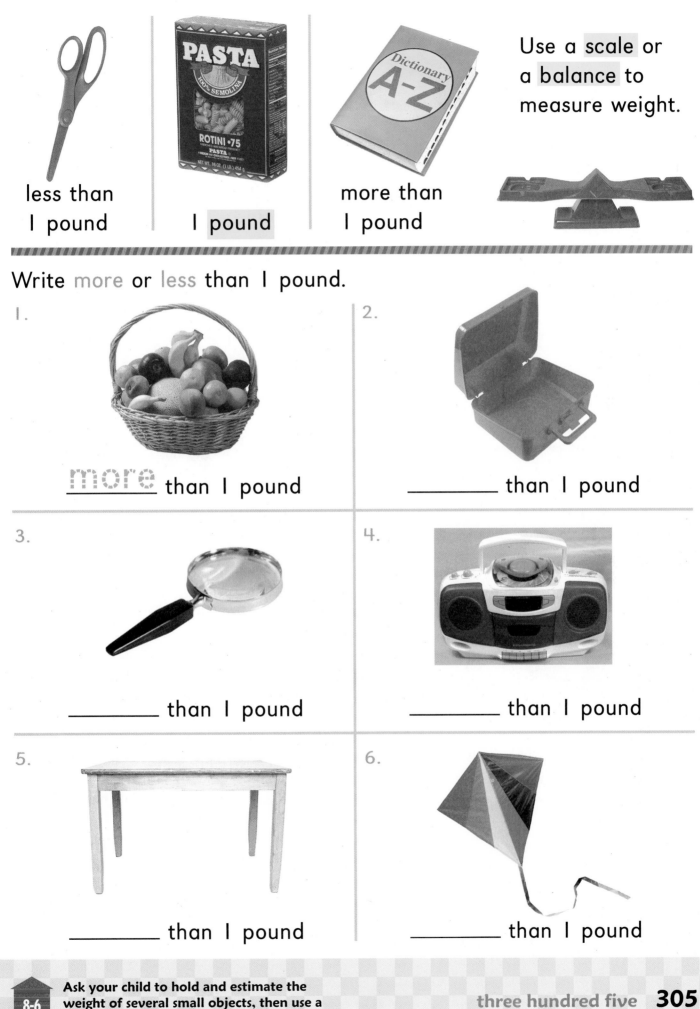

less than
1 pound

1 pound

more than
1 pound

Use a scale or
a balance to
measure weight.

Write more or less than 1 pound.

1. _more_ than 1 pound

2. _____ than 1 pound

3. _____ than 1 pound

4. _____ than 1 pound

5. _____ than 1 pound

6. _____ than 1 pound

8-6   **Ask your child to hold and estimate the weight of several small objects, then use a scale to weigh them and compare.**

three hundred five **305**

Write to the nearest pound.

1.

The watermelon weighs between 7 and 8 pounds.

about ___ pounds

2. _____ pounds

3. _____ pound

4. _____ pounds

5. _____ pound

6. _____ pounds

7. _____ pounds

Explain how you found each answer.

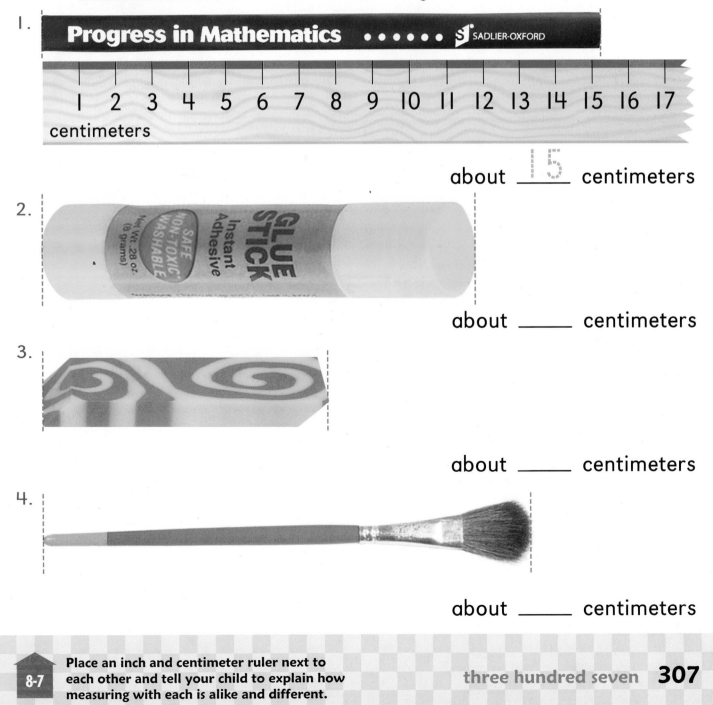

The chalk measures closer
to 10 centimeters.
It is about 10 centimeters.

The eraser measures
closer to 3 centimeters.
It is about 3 centimeters.

Use a centimeter ruler. Estimate the length.

1.

Progress in Mathematics • • • • • • **SADLIER-OXFORD**

| 1 | 2 | 3 | 4 | 5 | 6 | 7 | 8 | 9 | 10 | 11 | 12 | 13 | 14 | 15 | 16 | 17 |

centimeters

about ___15___ centimeters

2.

GLUE STICK
Instant Adhesive
SAFE NON-TOXIC WASHABLE
Net Wt. 28 oz. (6 grams)

about _____ centimeters

3.

about _____ centimeters

4.

about _____ centimeters

8-7   Place an inch and centimeter ruler next to
each other and tell your child to explain how
measuring with each is alike and different.

three hundred seven **307**

Estimate the length of each object.
Then use your ruler to measure.

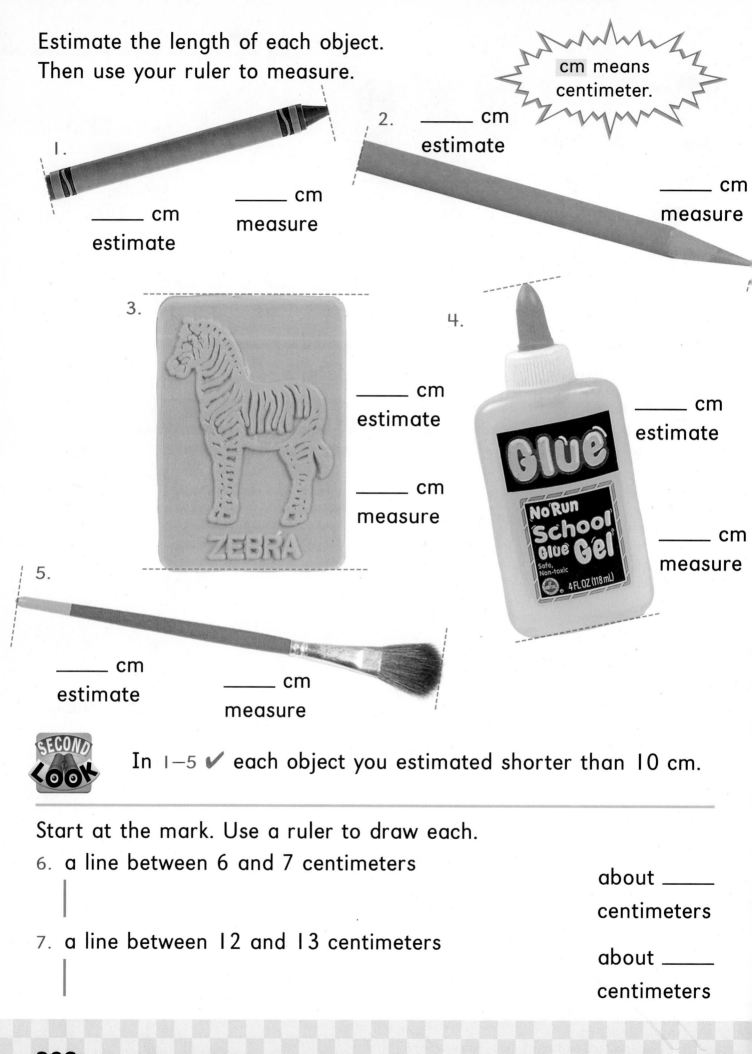

cm means
centimeter.

1. _____ cm
estimate

_____ cm
measure

2. _____ cm
estimate

_____ cm
measure

3. _____ cm
estimate

_____ cm
measure

4. _____ cm
estimate

_____ cm
measure

5. _____ cm
estimate

_____ cm
measure

SECOND LOOK   In 1–5 ✔ each object you estimated shorter than 10 cm.

Start at the mark. Use a ruler to draw each.

6. a line between 6 and 7 centimeters

about _____ centimeters

7. a line between 12 and 13 centimeters

about _____ centimeters

Name _____

How many centimeters equal 1 decimeter?

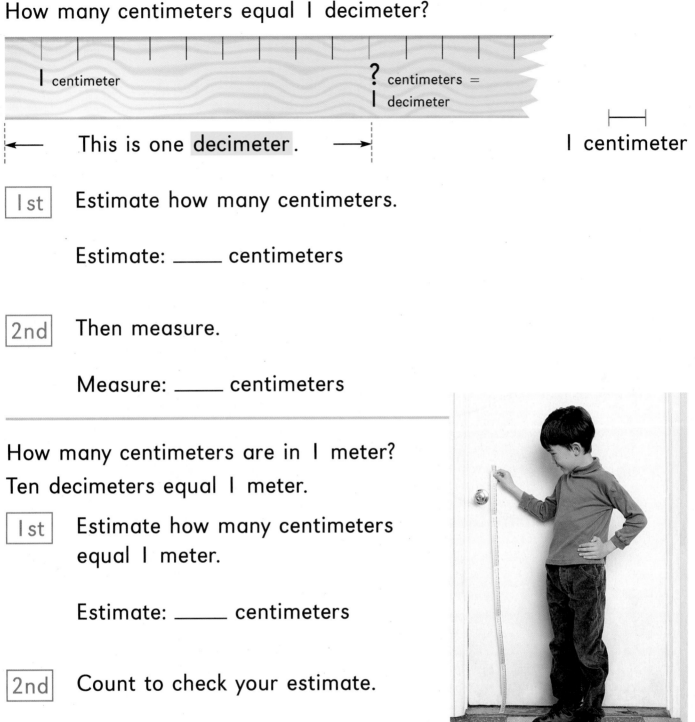

1 centimeter

? centimeters = 1 decimeter

This is one decimeter.

1 centimeter

| 1st | Estimate how many centimeters. |

Estimate: _____ centimeters

| 2nd | Then measure. |

Measure: _____ centimeters

How many centimeters are in 1 meter?
Ten decimeters equal 1 meter.

| 1st | Estimate how many centimeters equal 1 meter. |

Estimate: _____ centimeters

| 2nd | Count to check your estimate. |

Measure: _____ centimeters

FINDING TOGETHER

List 2 items that measure about 1 decimeter and 2 that measure about 1 meter.

Write less than or more than . Then measure.

1.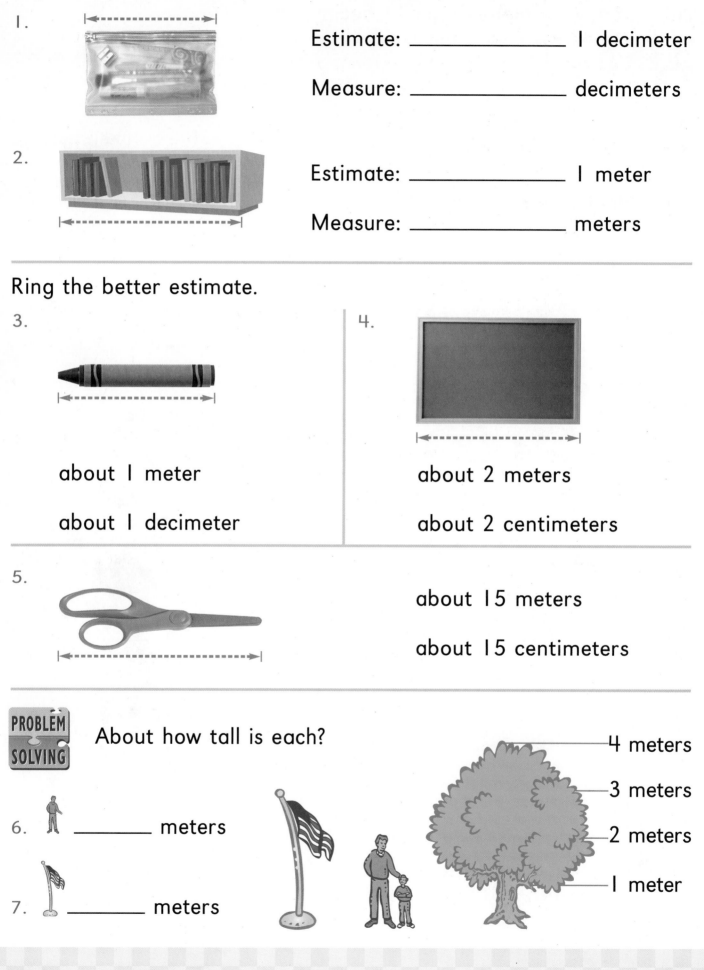

Estimate: _____ 1 decimeter

Measure: _____ decimeters

2.

Estimate: _____ 1 meter

Measure: _____ meters

Ring the better estimate.

3.

about 1 meter

about 1 decimeter

4.

about 2 meters

about 2 centimeters

5.

about 15 meters

about 15 centimeters

PROBLEM SOLVING About how tall is each?

6. _____ meters

7. _____ meters

4 meters

3 meters

2 meters

1 meter

Name _____

The distance around
a figure is the
<mark>perimeter.</mark>

Add the lengths
of the sides.

_3_ + _1_ + _3_ + _1_ = _8_

The perimeter is _8_ in.

____ in.

___3___ in.

_1_ in.

____ in.

Use your inch ruler.
Measure around each figure. Then add.

1.

____ + ____ + ____ = ____

____ in. around

2.

+
_____

____ in. around

3.

____ + ____ + ____ + ____ = ____ in. around

4.

____ + ____ + ____ + ____ + ____ = ____

____ in. around

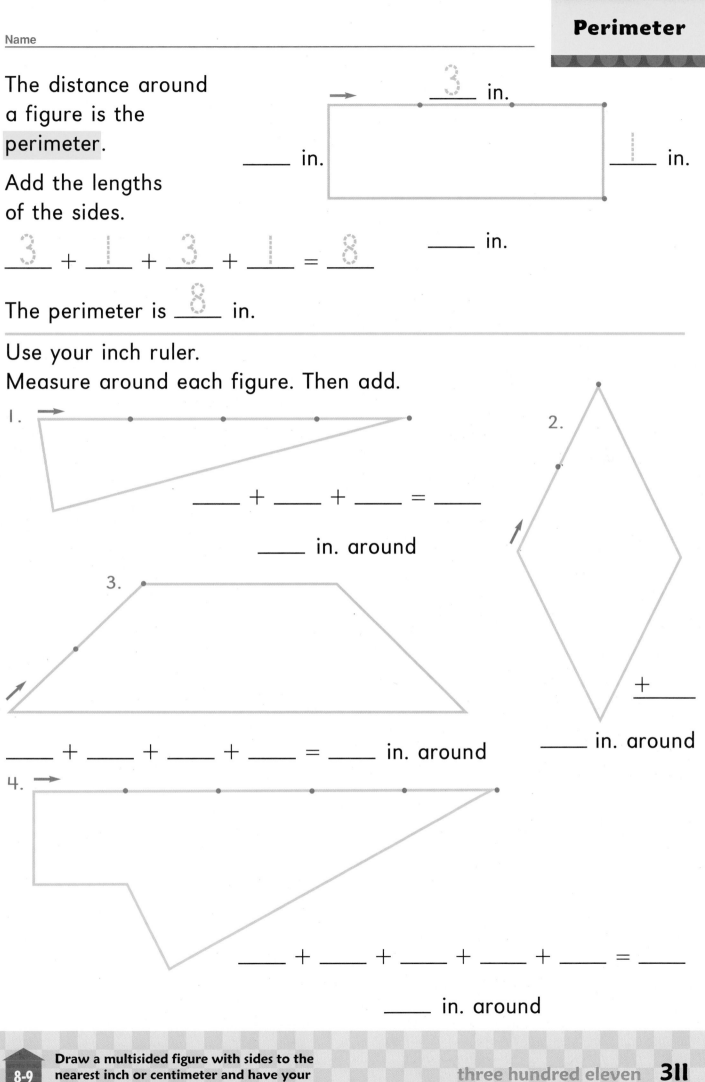

Use your centimeter ruler.
Find the perimeter.

cm means centimeter.

1.
_7_ cm

_7_ cm

_2_ cm

_7_ + _2_ + _7_ = _16_

The perimeter is _16_ cm.

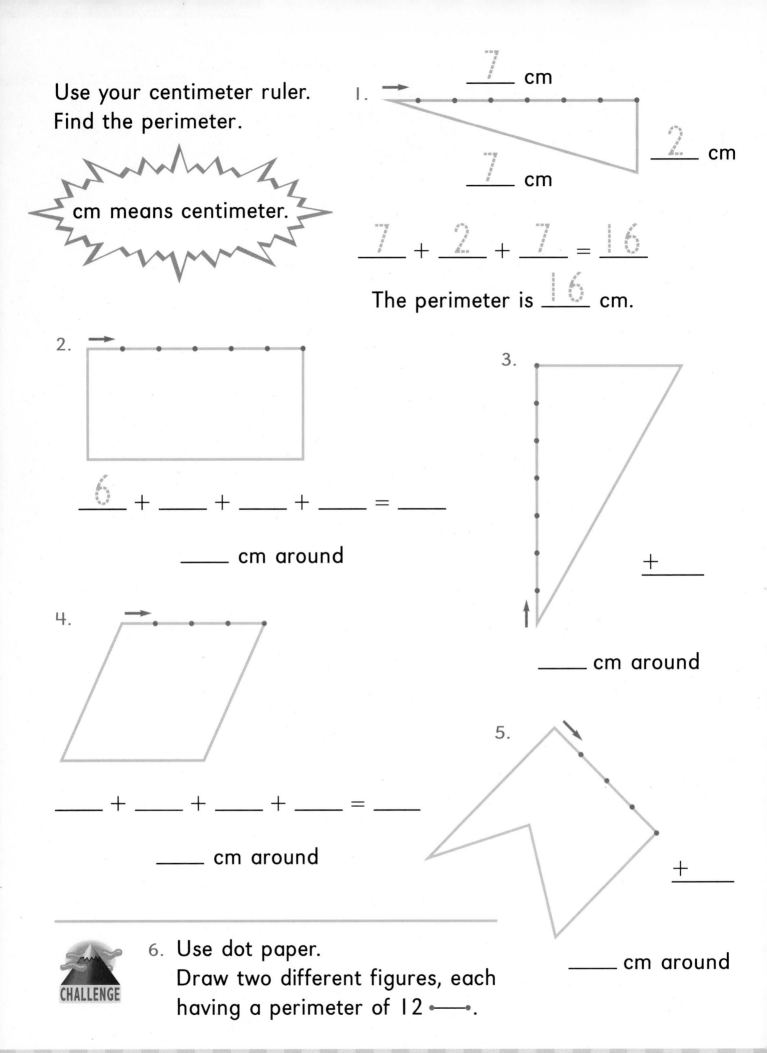

2.
_6_ + ___ + ___ + ___ = ___

_____ cm around

3.
+
_____

_____ cm around

4.
___ + ___ + ___ + ___ = ___

_____ cm around

5.
+
_____

_____ cm around

6. Use dot paper.
Draw two different figures, each
having a perimeter of 12 •——•.

CHALLENGE

You can cover figures
with square units.

How many square units
will cover this rectangle?

I square unit

1st Estimate how many
square units.

about _____

2nd Then measure.

_____

Area is the number
of square units it takes
to cover a figure.

First estimate how many square units will cover each.
Then measure.

1.

about _____

_____

2.

about _____

_____

3.

about _____

_____

CRITICAL THINKING Describe the pattern in 1–3. Predict the area
of the next square. Then draw it in your Math Journal.

8-10 Ask your child how she/he would use a 1-inch
square to find the area of a placemat.

three hundred thirteen **313**

Find the area in square units. Draw a
different figure that has the same area.

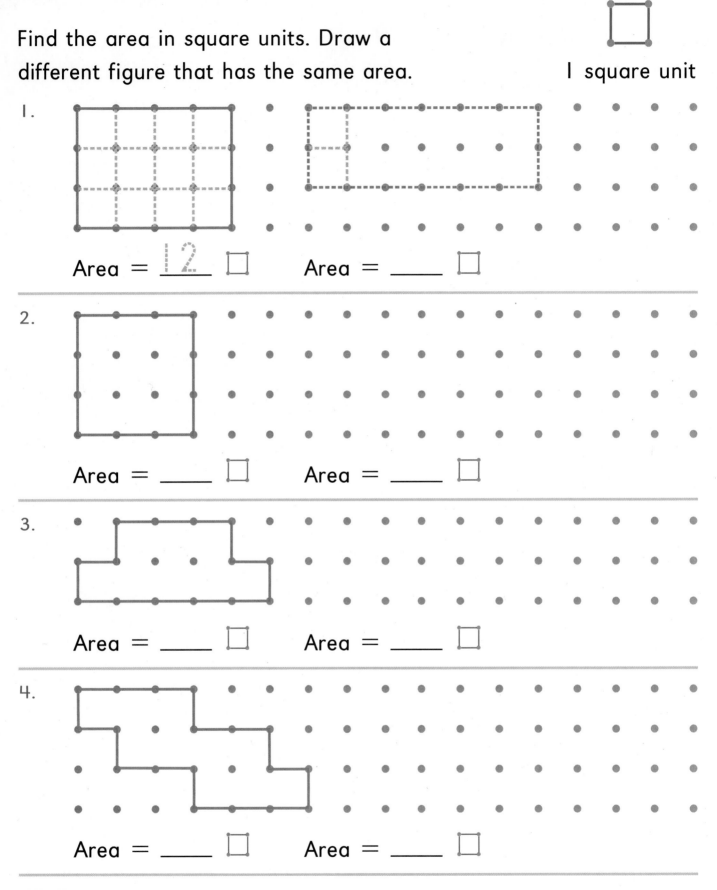

1 square unit

1.

Area = __12__ □   Area = _____ □

2.

Area = _____ □   Area = _____ □

3.

Area = _____ □   Area = _____ □

4.

Area = _____ □   Area = _____ □

FINDING
TOGETHER

5. On dot paper, draw three different rectangles,
   each having an area of 18 □.

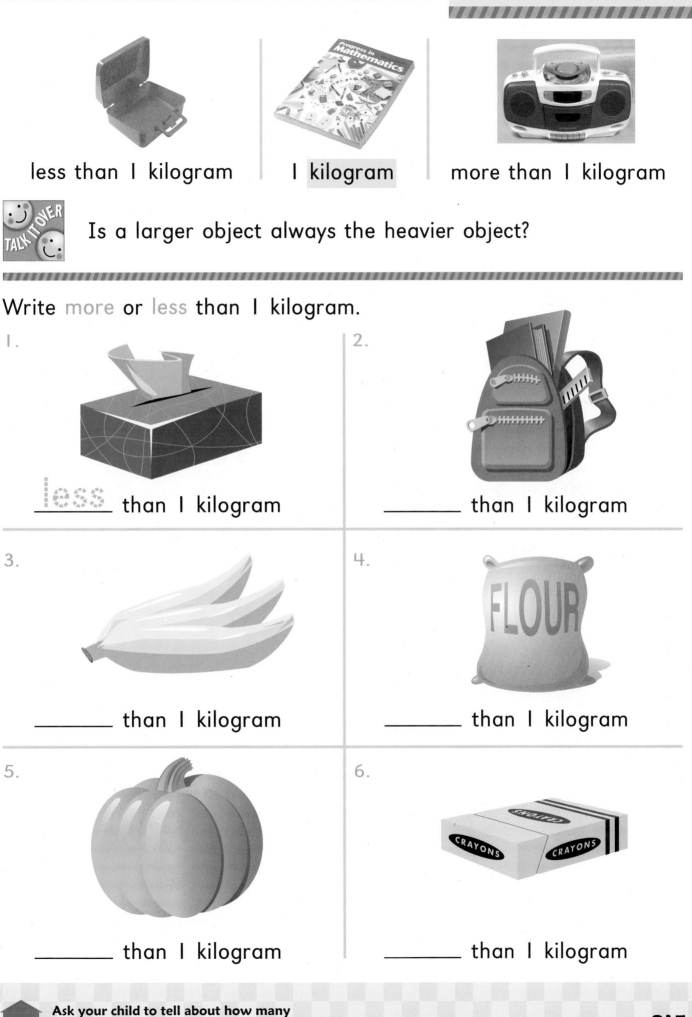

less than I kilogram | I kilogram | more than I kilogram

TALK IT OVER Is a larger object always the heavier object?

Write more or less than I kilogram.

1. _____less_____ than I kilogram

2. _____ than I kilogram

3. _____ than I kilogram

4. _____ than I kilogram

5. _____ than I kilogram

6. _____ than I kilogram

8-11 Ask your child to tell about how many kilograms an object is and use a scale to verify the estimate.

three hundred fifteen **315**

A bunch of bananas
is about 1 kilogram.

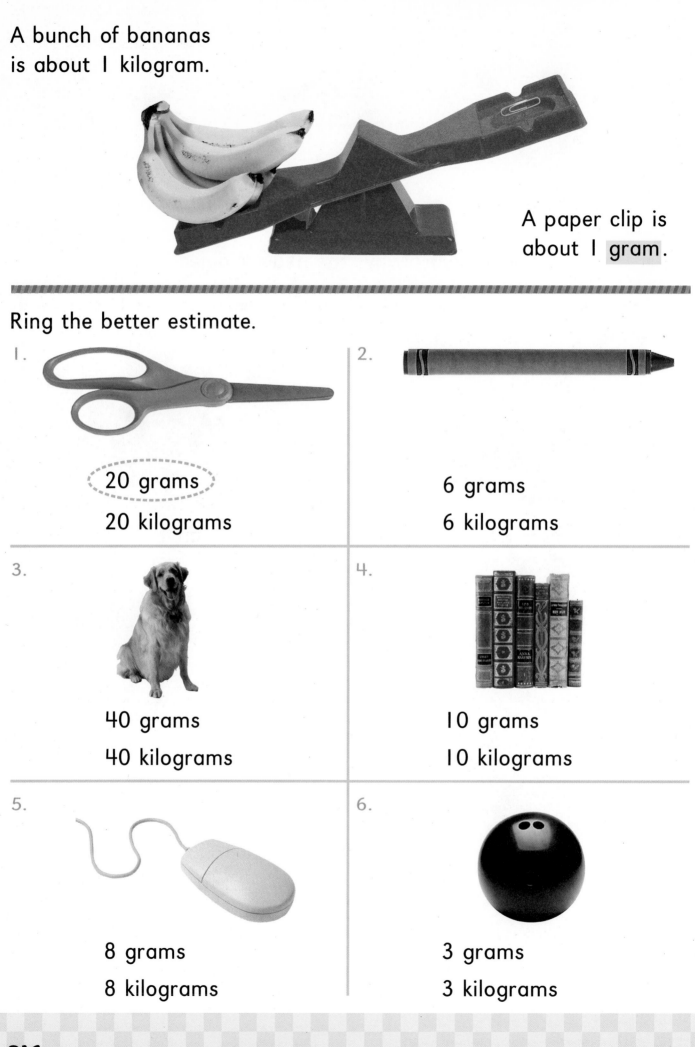

A paper clip is
about 1 gram.

Ring the better estimate.

1.

(20 grams)

20 kilograms

2.

6 grams

6 kilograms

3.

40 grams

40 kilograms

4.

10 grams

10 kilograms

5.

8 grams

8 kilograms

6.

3 grams

3 kilograms

less than 1 liter | 1 liter | more than 1 liter

Write more or less than 1 liter.

1. ___more___ than 1 liter

2. _____ than 1 liter

3. _____ than 1 liter

4. _____ than 1 liter

5. _____ than 1 liter

6. _____ than 1 liter

MATH JOURNAL

7. Draw a container that holds 1 liter, a container that holds more than 1 liter, and a container that holds less than 1 liter.

8-12 Prompt your child to estimate the capacity of several containers in liters and then use a liter container to check.

three hundred seventeen **317**

Ring the better estimate.

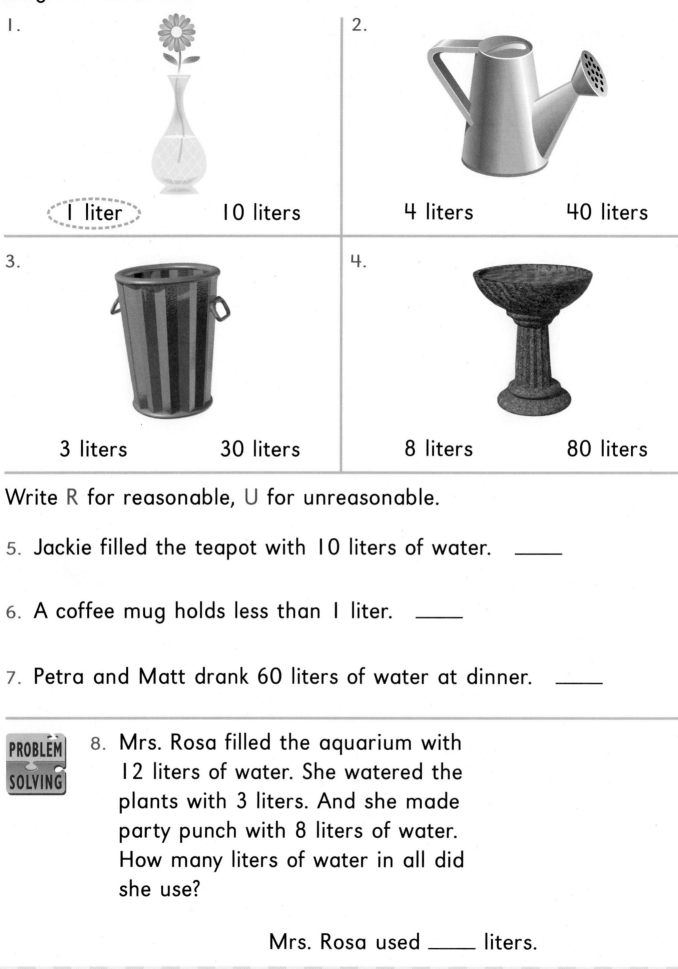

1.

1 liter   10 liters

2.

4 liters   40 liters

3.

3 liters   30 liters

4.

8 liters   80 liters

Write R for reasonable, U for unreasonable.

5. Jackie filled the teapot with 10 liters of water. _____

6. A coffee mug holds less than 1 liter. _____

7. Petra and Matt drank 60 liters of water at dinner. _____

PROBLEM SOLVING

8. Mrs. Rosa filled the aquarium with 12 liters of water. She watered the plants with 3 liters. And she made party punch with 8 liters of water. How many liters of water in all did she use?

Mrs. Rosa used _____ liters.

Name

This thermometer shows temperature in degrees Fahrenheit (°F).

55° is halfway between 50° and 60°.

hot
warm
cold

Write the temperature.

1.

play a game of softball

70°F

2.

spend a day at the beach

_____°F

3.

build a snowman

_____°F

Color the temperature.

4.

50°F

5.

75°F

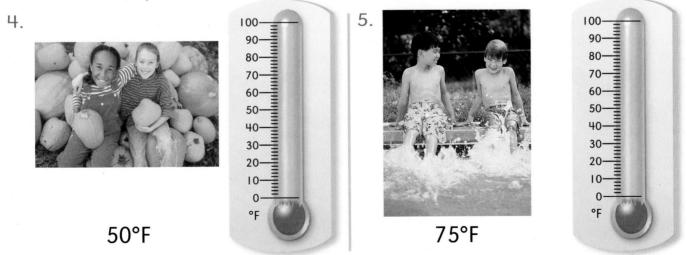

 8-13 Provide your child with the outline of a thermometer and ask him/her to shade to show today's temperature.

three hundred nineteen **319**

Ring the temperature. Color the thermometer.

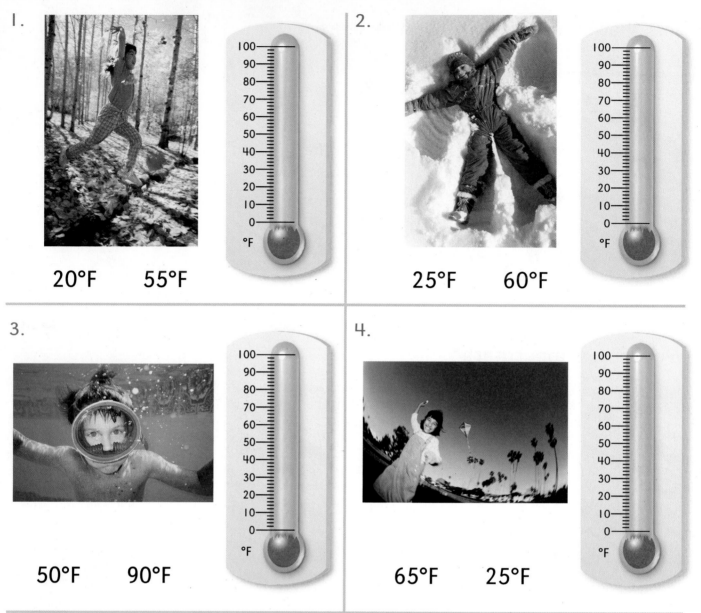

1.

20°F   55°F

2.

25°F   60°F

3.

50°F   90°F

4.

65°F   25°F

CHALLENGE

People in some countries measure temperature in degrees Celsius (°C).

Write the temperature.

5. sledding   _____°C

6. swimming   _____°C

7. picking pumpkins   _____°C

I measure how **tall** the carton is in inches.

I measure how much the carton **holds** in quarts.

An inch and a quart are **units of measure**.

Ring the unit of measure you would use.

1. How long is it?

(meter)   kilogram

2. How much will it hold?

kilogram   liter

3. How wide is it?

centimeter   liter

4. How much will it hold?

gallon   yard

5. How tall is it?

foot   pound

6. How heavy is it?

gallon   pound

7. How long is it?

meter   kilogram

8. How heavy is it?

kilogram   liter

9. How wide is it?

centimeter   liter

These are measuring tools you have used.

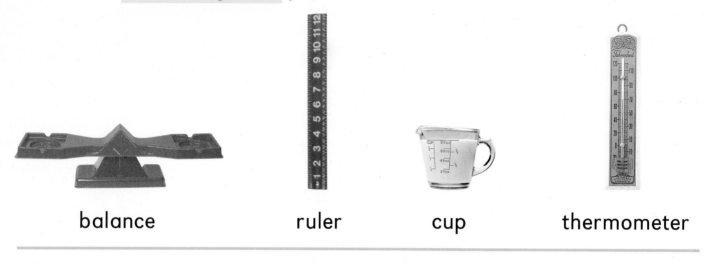

balance   ruler   cup   thermometer

Write the name of the tool you would use to measure.

1. the amount of water to fill a pot _____

2. the distance from your foot to your hip _____

3. the temperature on a very sunny day _____

4. the amount of milk in a pitcher _____

5. the weight of a bag of apples _____

6. the length of a vine _____

FINDING TOGETHER List one object you can measure in each unit. Then draw a picture of each in your Math Journal.

7. in kilograms	8. in meters	9. in liters
10. in inches	11. in pounds	12. in quarts

LOGO: Relate Area and Perimeter

A **LOGO** turtle
can draw shapes.

Square Corners

FD 30	FD 30
RT 90	LT 90
FD 30	FD 30

Follow the LOGO commands.
Start at 🐢. Connect the dots.
Make ☐ to find the area in square units.

☐ = I square unit

1. FD 40
   RT 90
   FD 20
   RT 90
   FD 40
   RT 90
   FD 20

   Area ___8___ ☐

2. FD 30
   LT 90
   FD 30
   LT 90
   FD 30
   LT 90
   FD 30

   Area _____ ☐

3. FD 10
   LT 90
   FD 40
   LT 90
   FD 40
   LT 90
   FD 40
   LT 90
   FD 30

   Area _____ ☐

4. RT 90
   FD 20
   LT 90
   FD 30
   LT 90
   FD 20
   LT 90
   FD 30

   Area _____ ☐

8-15 This lesson teaches how to follow **LOGO** commands to draw figures and how to find their area and perimeter.

three hundred twenty-three **323**

Draw a rectangle that has a perimeter of 10 ⟶. What is its area?

1.

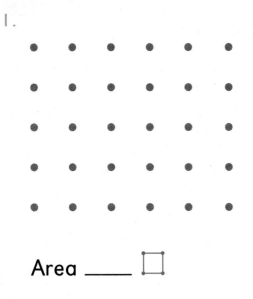

Area _____ ☐

Draw another rectangle with a perimeter of 10 ⟶. What is its area?

2.

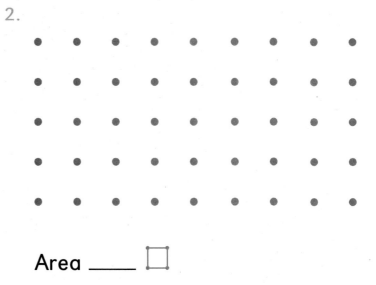

Area _____ ☐

Draw a rectangle that has an area of 8 ☐. What is its perimeter?

3.

Perimeter _____ ⟶

Draw another rectangle that has an area of 8 ☐. What is its perimeter?

4.

Perimeter _____ ⟶

FINDING TOGETHER

5. For each figure above, enter the LOGO commands into a computer. Compare the figures you drew with those on the screens.

Name _____

Read → Think → Write → Check

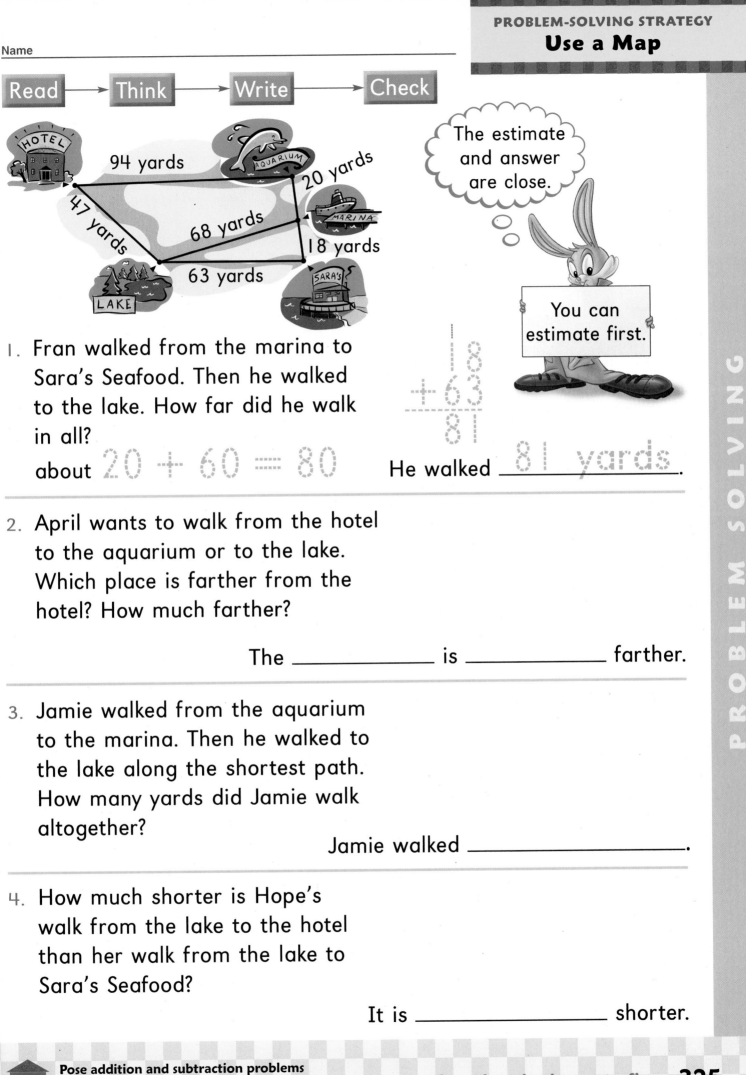

The estimate and answer are close.

You can estimate first.

1. Fran walked from the marina to Sara's Seafood. Then he walked to the lake. How far did he walk in all?

   about 20 + 60 = 80

   He walked ___81___ yards.

2. April wants to walk from the hotel to the aquarium or to the lake. Which place is farther from the hotel? How much farther?

   The _____ is _____ farther.

3. Jamie walked from the aquarium to the marina. Then he walked to the lake along the shortest path. How many yards did Jamie walk altogether?

   Jamie walked _____.

4. How much shorter is Hope's walk from the lake to the hotel than her walk from the lake to Sara's Seafood?

   It is _____ shorter.

8-16 Pose addition and subtraction problems for your child to solve using the map on this page.

three hundred twenty-five **325**

PROBLEM SOLVING

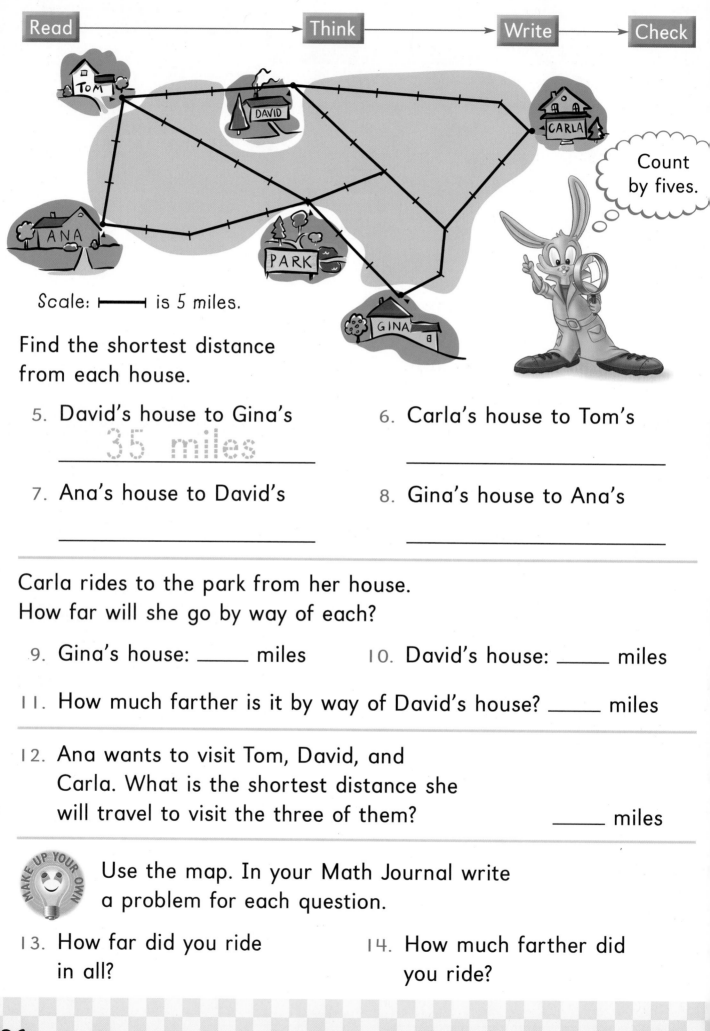

Count by fives.

Scale: ⊢——⊣ is 5 miles.

Find the shortest distance from each house.

5. David's house to Gina's

   35 miles

6. Carla's house to Tom's

   _____

7. Ana's house to David's

   _____

8. Gina's house to Ana's

   _____

Carla rides to the park from her house.
How far will she go by way of each?

9. Gina's house: _____ miles

10. David's house: _____ miles

11. How much farther is it by way of David's house? _____ miles

---

12. Ana wants to visit Tom, David, and Carla. What is the shortest distance she will travel to visit the three of them?   _____ miles

---

Use the map. In your Math Journal write a problem for each question.

13. How far did you ride in all?

14. How much farther did you ride?

PROBLEM SOLVING

Name _____

Read ⟶ Think ⟶ Write ⟶ Check

Can you find the missing information I took from each?

1. Mrs. Brown had 34 inches of ribbon.
   She bought some more ribbon.
   How many inches did she have then?

   <u>17 inches</u>
   missing information

   $$\begin{array}{r} 34 \\ +17 \\ \hline 51 \end{array}$$

   She had <u>51 inches</u>

   58 centimeters
   17 inches
   42 pounds
   26 gallons

2. Greg's vine is 76 centimeters long.
   Ali's vine is shorter. How many
   centimeters longer is Greg's vine
   than Ali's?

   _____
   missing information

   It is _____ longer.

3. Mr. Shaw bought some paint. He
   used some of it. Then 19 gallons
   were left. How many gallons of
   paint did Mr. Shaw use?

   _____
   missing information

   He used _____.

4. Thuy's class recycled 8 pounds of
   plastic, 28 pounds of glass, and
   some paper. How many pounds
   did her class recycle altogether?

   _____
   missing information

   _____ altogether

8-17 Tell your child to model problems like these for addition and subtraction.

three hundred twenty-seven **327**

PROBLEM SOLVING

Use a strategy you have learned.

STRATEGY FILE

Choose the Operation
Extra Information
Two-Step Problem
Hidden Information

5. At the zoo, the first cart sold 2 dozen liters of water, the second sold 33 liters, and the third sold 41 liters. How many liters of water did the 3 carts sell?

They sold _____.

6. The giraffe is 18 feet tall. The elephant is 11 feet tall. The giraffe has 45 spots. How much shorter is the elephant than the giraffe?

The elephant is _____ shorter.

7. The zookeeper spent 18 minutes feeding bears, and double that time feeding lions and tigers. Did it take more or less than 1 hour to feed these animals?

It took _____ 1 hour.

8. One liter of juice cost 45¢. Josh gave the cashier 95¢ for 2 liters. What was Josh's change?

Josh's change was _____.

CHALLENGE

9. The perimeter of the tiger's playground is 88 meters. How much less is this than the perimeter of the monkey's playground?

25 meters 32 meters
Monkey
Playground
38 meters

This is _____ less.

Start at the mark. Draw a line for each measure.

1. 4 $\frac{1}{2}$ in. |

2. 11 cm |

Choose the reasonable measure. Ring.

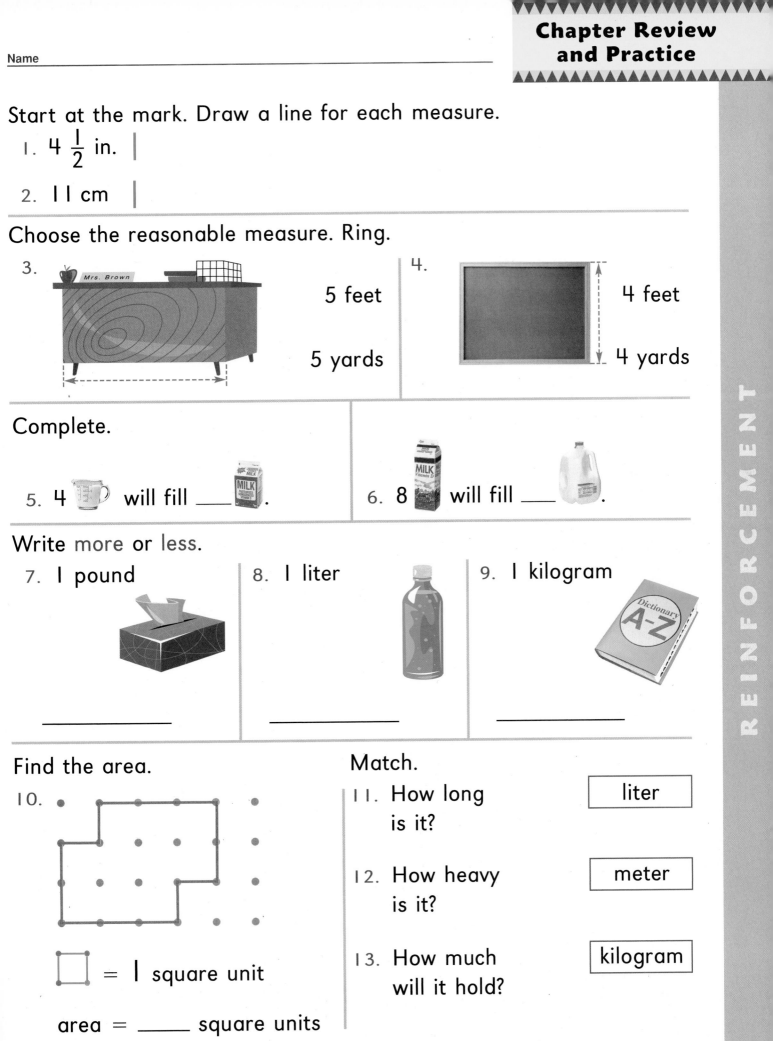

3. 5 feet

   5 yards

4. 4 feet

   4 yards

Complete.

5. 4 ☕ will fill _____ 🥛.

6. 8 🥛 will fill _____ 🍶.

Write more or less.

7. 1 pound

_____

8. 1 liter

_____

9. 1 kilogram

_____

Find the area.

10.

⬜ = 1 square unit

area = _____ square units

Match.

11. How long is it? | liter |

12. How heavy is it? | meter |

13. How much will it hold? | kilogram |

🏠 8 This page reviews the mathematical content presented in Chapter 8.

three hundred twenty-nine **329**

Name _____

You can measure around a **curved** object or path.

Use a string to estimate the length. Then place the string along an inch ruler to measure.

Estimate. Then measure.

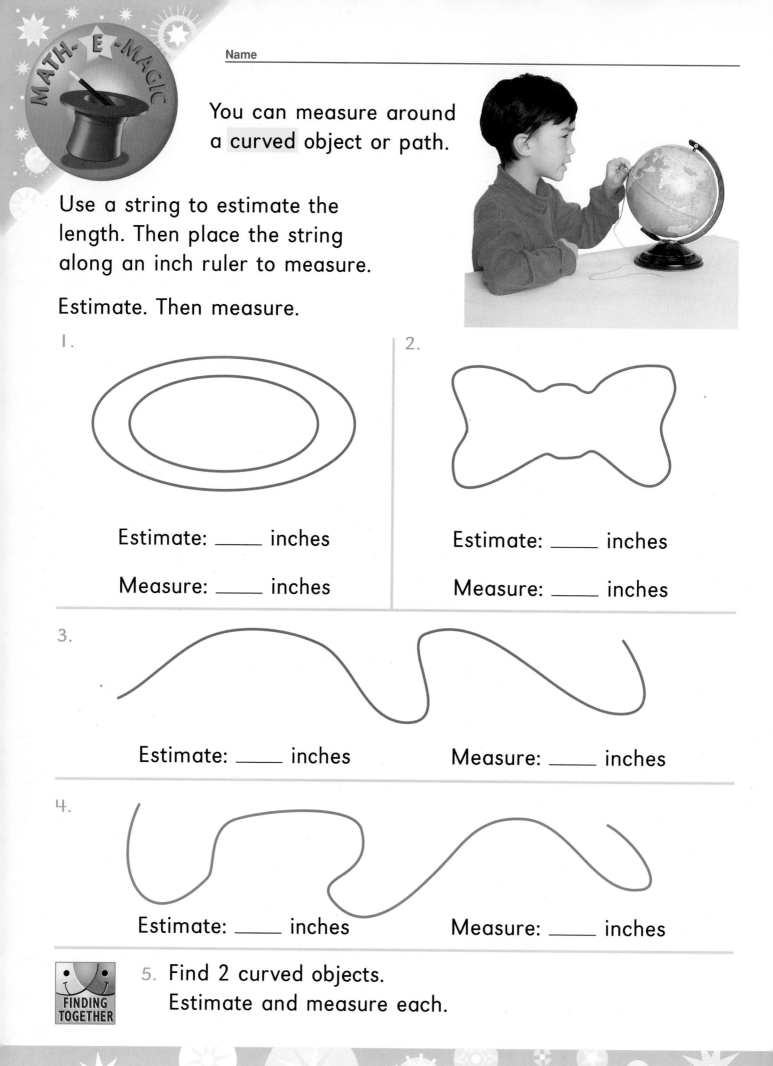

1.

Estimate: _____ inches

Measure: _____ inches

2.

Estimate: _____ inches

Measure: _____ inches

3.

Estimate: _____ inches   Measure: _____ inches

4.

Estimate: _____ inches   Measure: _____ inches

FINDING TOGETHER

5. Find 2 curved objects. Estimate and measure each.

This page reviews your child's understanding of measuring curved objects and paths.

8

Name _____

1. Use 2 different units to measure from your knee to your toe.

   Predict: _____ paper clips

   _____ inches

   Measure: _____ paper clips

   _____ inches

Did you use more paper clips or inches? Why?

2. Complete the tables to show equal amounts.

1 pint	2 pints	3 pints	4 pints
2 cups	cups	cups	cups

1 gallon	2 gallons	3 gallons	4 gallons
4 quarts	quarts	quarts	quarts

PORTFOLIO

Choose 1 of these projects.
Use a separate sheet of paper.

3. Make a poster to show how people in three different careers might use measuring tools.

   For example: nurse—ruler for height, thermometer for temperature.

4. Make a rectangle on dot paper. Each side must be longer than 4 ⟷.

   Measure each side to the nearest in. or cm. Record the perimeter. Find the area in ☐.

8

This page provides a variety of informal assessment opportunities in order to measure your child's understanding of Chapter 8.

three hundred thirty-one **331**

ASSESSMENT

Start at the mark. Draw a line for each measure.

1. $3\frac{1}{2}$ in. |

2. 9 cm |

Choose the reasonable measure. Ring.

3. 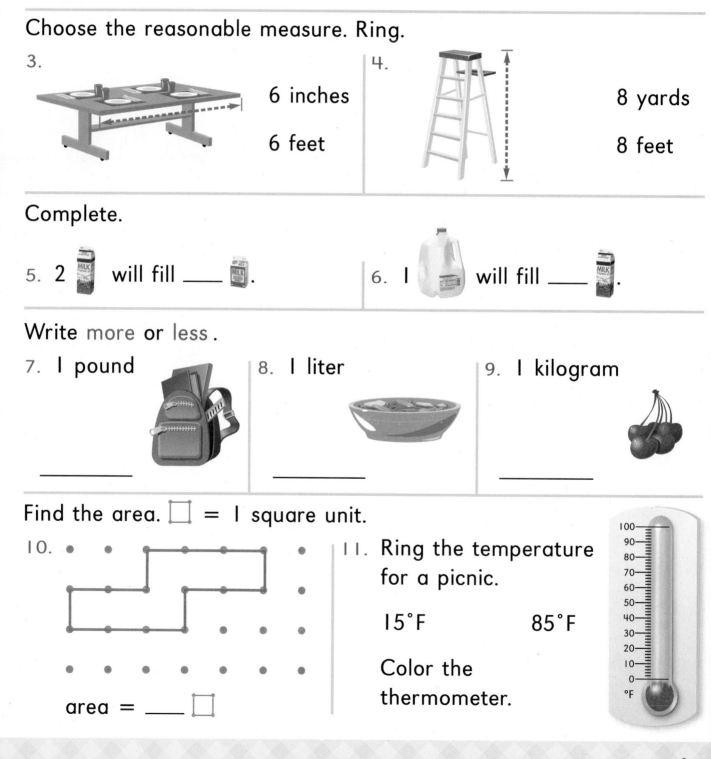 6 inches

6 feet

4. 8 yards

8 feet

Complete.

5. 2 [MILK] will fill ____ [MILK].

6. 1 [jug] will fill ____ [MILK].

Write more or less.

7. 1 pound

8. 1 liter

9. 1 kilogram

_____   _____   _____

Find the area. ☐ = 1 square unit.

10.

area = ____ ☐

11. Ring the temperature for a picnic.

15°F   85°F

Color the thermometer.

100
90
80
70
60
50
40
30
20
10
0
°F

This page is a formal assessment of your child's understanding of the content presented in Chapter 8.

8

Place Value to 1000 and Addition

9

CRITICAL THINKING

The total number of teeth the toothy Crocodile has is the sum of a double. He has 32 teeth in all. How many upper teeth does he have?

For more information about Chapter 9, visit the Family Information Center at **www.sadlier-oxford.com**

Internet

Dear Family,

Today your child began Chapter 9. As she/he studies place value and addition of 3-digit numbers, you may want to read the poem below, which was read in class. Have your child talk about some of the math ideas pictured on page 333.

Look for the 📖 at the bottom of each skills lesson. The suggestion on the page gives you an opportunity to improve your child's understanding of math and to reinforce her/his math language. You may want to have countables and dollars, dimes, and pennies available for your child to use throughout this chapter.

Home Reading Connection

Excerpt from **Toothy Crocodile**

There was a toothy Crocodile
Who wallowed in the muck,
Counting up how many
Of his many teeth were buck.
One day he'd count to twenty-nine,
The next day fifty-two,
For counting Crocodile teeth
Is difficult to do.

*   *   *

So weeks went by and thus
The counting Crocodile grew thinner,
Because the Alligators ate
The Crocodile's dinner,
While he was much too busy
With a cavern to explore—

   "Twenty-seven?"
   "Sixty-six?"
   "One hundred thirty-four?!"
   *   *   *

J. Patrick Lewis

Home Activity

The Hundreds Vine

Make a hanging-vine number line. Use yarn or string for the vine. From construction paper, cut out large leaves to attach to the vine. On self-stick sheets, write 3-digit numbers (100–900) and place each number on one of the leaves, scrambling their order. Have your child untangle the vine by placing the numbers in the correct order. As your child progresses through this chapter, change the numbers and the skill.

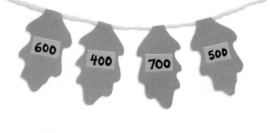

Name

Visitors to the rain forest saw many animals.

1st
Estimate the number of animals seen.

2nd
Make a pictograph to show the data.

Rain Forest Tour

Animal	Number Seen	Closer to
Monkeys	33	(30) or 40
Parrots	89	80 or 90
Crocodiles	21	20 or 30
Snakes	18	10 or 20
Butterflies	97	90 or 100

Animals Seen

Animal	Estimate
Monkeys	◯ ◯ ◯
Crocodiles	
Parrots	
Snakes	
Butterflies	

Key: Each 👁 stands for 10 animals.

1. About how many more monkeys than snakes did the visitors see? _____

2. About how many fewer crocodiles than parrots? _____

3. Were more or less than 2 hundred animals seen in all? _____

FINDING TOGETHER

4. On their next trip the visitors saw double the number of each kind of animal. Make a bar graph to show the data.

9 PORTFOLIO You can put this in your Math Portfolio.

The leaf-cutting ants are gathering leaves in the rain forest.

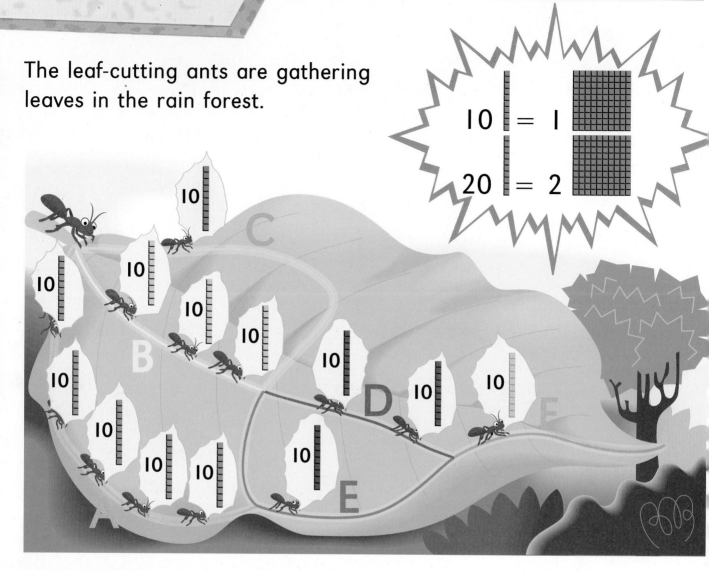

$10 = 1$

$20 = 2$

Complete the chart to show the number of leaves in all gathered on each path.

Path	Tens	Leaves in all
CEF	30 tens	3 hundreds = 300
CDF		

**You can put this in your
Math Portfolio.** PORTFOLIO

Name _____

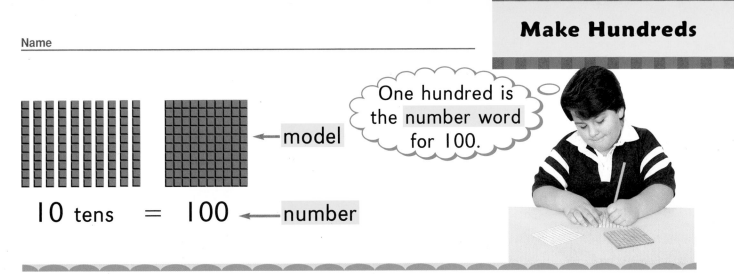

One hundred is the number word for 100.

10 tens = 100 ← number

← model

Write the place value and the number word.

1. __2__ hundreds __0__ tens __0__ ones

hundreds	tens	ones
2	0	0

__two hundred__

2. ____ hundreds ____ tens ____ ones

hundreds	tens	ones

____ hundred

3. ____ hundreds ____ tens ____ ones

hundreds	tens	ones

____ hundred

4. ____ hundreds ____ tens ____ ones

hundreds	tens	ones

____ hundred

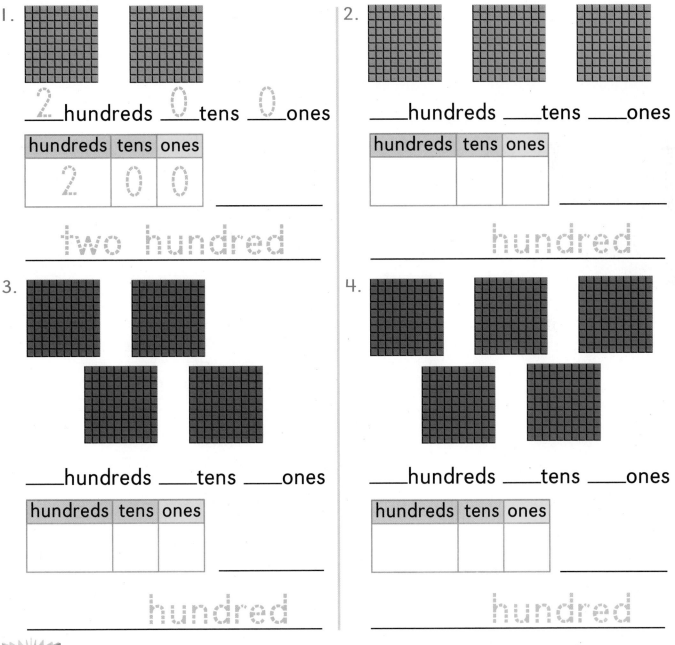

MATH JOURNAL

5. The numbers in 1–4 make a pattern. Draw and write the next four numbers in the pattern.

9-1 Provide 9 squares for your child to use to count from 100 to 900 and tell the place value of each.

three hundred thirty-seven **337**

Write the number.

1.
hundreds	tens	ones
6	0	0

_600_

2.
hundreds	tens	ones
7	0	0

_____

3.
hundreds	tens	ones
5	0	0

_____

4.
hundreds	tens	ones
3	0	0

_____

Write the number and the number word.

5. 4 hundreds 0 tens 0 ones

_400_ _four hundred_

6. 1 hundred 0 tens 0 ones

_____ _____

7. 8 hundreds 0 tens 0 ones

_____ _____

8. 9 hundreds 0 tens 0 ones

_____ _____

Color to match.

9. three hundred

10. two hundred

11. six hundred

12. seven hundred

13. five hundred

| 700 |
| 500 |
| 200 |
| 300 |
| 600 |

| 6 hundreds 0 tens 0 ones |
| 7 hundreds 0 tens 0 ones |
| 5 hundreds 0 tens 0 ones |
| 2 hundreds 0 tens 0 ones |
| 3 hundreds 0 tens 0 ones |

CRITICAL THINKING Write the missing numbers on the number line.

14.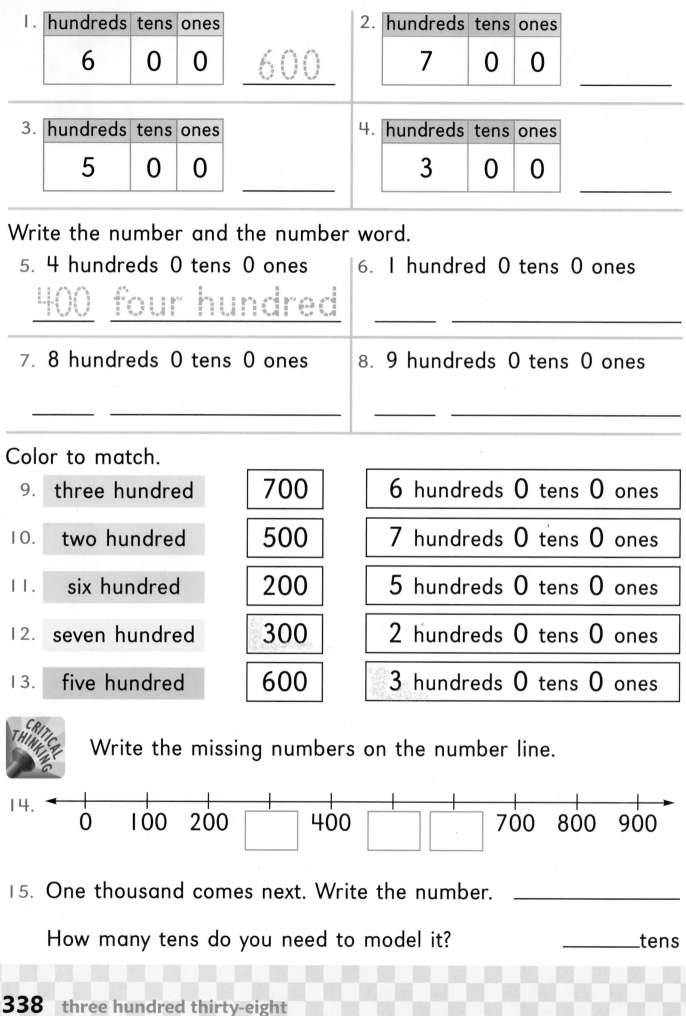

0   100   200   ☐   400   ☐   ☐   700   800   900

15. One thousand comes next. Write the number. _____

How many tens do you need to model it? _____ tens

Name _____

I can show a 3-digit number in different ways.

4 hundreds 2 tens 3 ones

place-value chart

hundreds	tens	ones
4	2	3

423

four hundred twenty-three

Complete the place-value chart, number, and word.

1. 5 hundreds 1 ten 6 ones

hundreds	tens	ones
5	1	6

516

five hundred sixteen

2. 2 hundreds 8 tens 4 ones

hundreds	tens	ones

_____

_____ eighty-four

3. 6 hundreds 5 tens 1 one

hundreds	tens	ones

_____

_____ fifty-one

4. 1 hundred 3 tens 9 ones

hundreds	tens	ones

_____

_____ thirty-nine

5. 8 hundreds 6 tens 0 ones

hundreds	tens	ones

_____

_____ sixty

6. 9 hundreds 0 tens 7 ones

hundreds	tens	ones

_____

_____ seven

TALK IT OVER How would you model 1—6?

9-2 Provide three number cards such as 7, 4, and 1 for your child to make different 3-digit numbers and to tell you the value of each digit.

three hundred thirty-nine **339**

Write the number.

1.
hundreds	tens	ones
5	0	4

504

2.
hundreds	tens	ones
7	1	5

_____

3.
hundreds	tens	ones
3	7	2

_____

4.
hundreds	tens	ones
4	9	0

_____

5.
hundreds	tens	ones
6	4	8

_____

6.
hundreds	tens	ones
8	0	8

_____

SHARE YOUR THINKING How do the 4s in 504, 490, and 648 differ?
Which 8 in 808 has a greater value? Why?

Write the number.

7. 3 hundreds 4 tens 7 ones

347

8. 8 hundreds 1 ten 6 ones

_____

9. 9 hundreds 0 tens 5 ones

_____

Write the number word.

10. 6 hundreds 9 tens 1 one

six hundred ninety-one

11. 1 hundred 8 tens 6 ones

_____

12. 5 hundreds 6 tens 0 ones

_____

PROBLEM SOLVING

13. Marta, Jenna, and Rory each model a different 3-digit number using these cards. Write each one's number.

3 4 2

Marta's has the most hundreds and fewest ones.

_____

Jenna's has an odd number as the tens digit.

_____

Rory's model has 2 more ones than tens.

_____

Name _____

June adds hundreds, tens, and ones to show a 3-digit number.

325 =

3 ▦ 2▮ 5▮

300 + 20 + 5

	hundreds	tens	ones
325	3	2	5

three hundred twenty-five

Complete each place-value chart. Add the place values.

1. 873

hundreds	tens	ones
8	7	3

800 + 70 + 3 _____

2. 456

hundreds	tens	ones

_____

3. 198

hundreds	tens	ones

_____

4. 281

hundreds	tens	ones

_____

5. 716

hundreds	tens	ones

_____

6. 549

hundreds	tens	ones

_____

7. 604

hundreds	tens	ones

_____

8. 930

hundreds	tens	ones

_____

9-3 Provide number cards 0 through 9 for your child to make 3-digit numbers and to tell the expanded form.

three hundred forty-one **341**

Complete the place value.

1. 402 =

   __4__ hundreds __0__ tens __2__ ones

2. 830 =

   ____ hundreds ____ tens ____ ones

3. 981 =

   ____ hundreds ____ tens ____ one

4. 627 =

   ____ hundreds ____ tens ____ ones

Add the place values.

5. 596 =

   500 + 90 + 6

6. 309 =

   _____

7. 744 =

   _____

8. 201 =

   _____

Write the value of 5 in each number.

9. 571 __500__   751 _____   517 _____   175 _____

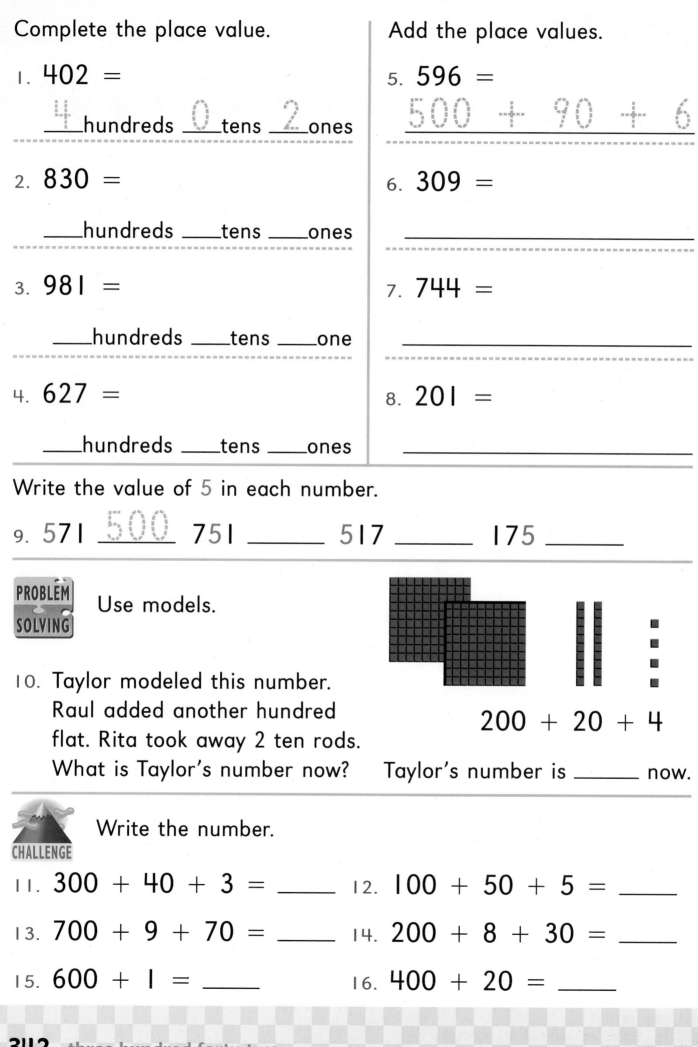

PROBLEM SOLVING   Use models.

10. Taylor modeled this number. Raul added another hundred flat. Rita took away 2 ten rods. What is Taylor's number now?

200 + 20 + 4

Taylor's number is _____ now.

CHALLENGE   Write the number.

11. 300 + 40 + 3 = ____   12. 100 + 50 + 5 = ____

13. 700 + 9 + 70 = ____   14. 200 + 8 + 30 = ____

15. 600 + 1 = ____   16. 400 + 20 = ____

You can count on by 1, 10, and 100 using hundred charts.

 1st   Use a blank hundred chart. Start with 101, 201, or 301.

2nd   Fill in the numbers as you count on by 1s.

3rd   Color the numbers that have 1 one. Which column or row did you color?   1st column

4th   Color the numbers that have 6 tens. Where are most of these numbers? _____

   Say each number in one row. Are you counting by 1s or by 10s? Say each number in one column. Are you counting by 1s or by 10s?

Complete the hundred chart.

431	432	433	434	435	436			439	
441	442				446				450
		453		455					
			464						470
		473		475					480
	482				486				490
491						497			500

Count by 100s. Write the missing numbers.

1. 203, _303_, 403, _____, 603, _____, 803, _____

2. 61, 161, 261, _____, _____, _____, 661, _____

3. 110, 210, _____, _____, 510, _____, 710, _____

Count by 10s. Write the missing numbers.

4. 90, 100, 110, _120_, _____, 140, 150, _____

5. 482, 492, _____, 512, _____ 532, _____

6. 746, 756, _____, _____, 786, _____ , _____

Write the number 1 more than each.

7. 999 _1000_ 8. 718 _____ 9. 689 _____ 10. 574 _____

Write the number 10 more than each.

11. 407 _417_ 12. 718 _____ 13. 689 _____ 14. 574 _____

PROBLEM SOLVING

15. Jay's number is 2 tens more than 700. Dee's number is 2 hundreds more than Jay's. Brad's is 2 ones more than Dee's. Write their numbers.

Jay _____

Dee _____

Brad _____

CRITICAL THINKING

Write the missing numbers. Color the count by 2s (yellow) and the count by 5s (blue).

16. < 970 > 975 < > < 990 > < 1000 >

17. 700) 702) 706)) 712

Name _____

Add: 123 + 112 = ___?___

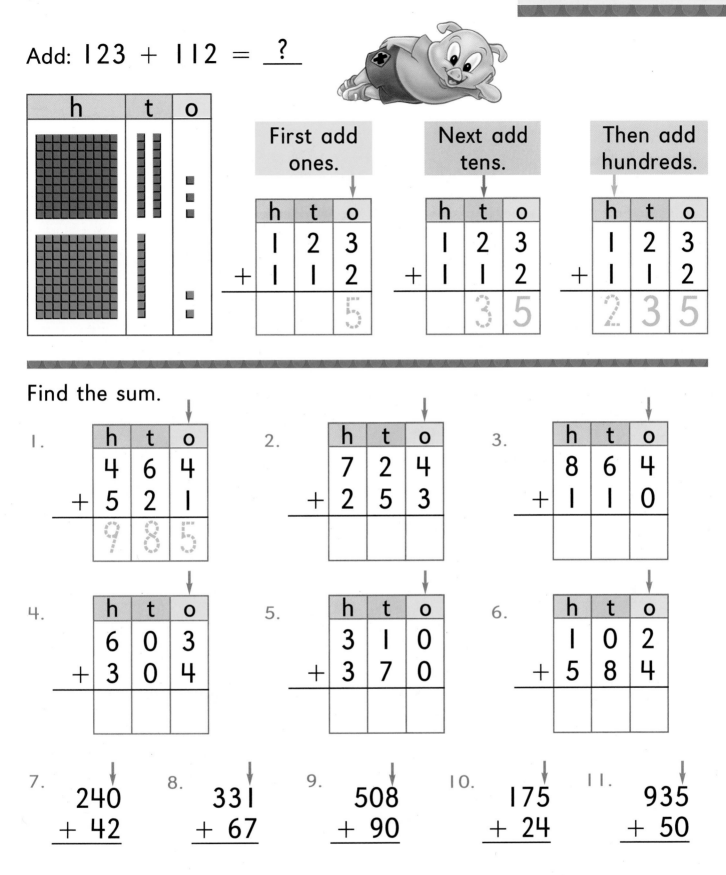

h	t	o

First add ones.

h	t	o
1	2	3
+ 1	1	2
		5

Next add tens.

h	t	o
1	2	3
+ 1	1	2
	3	5

Then add hundreds.

h	t	o
1	2	3
+ 1	1	2
2	3	5

Find the sum.

1.

h	t	o
4	6	4
+ 5	2	1
9	8	5

2.

h	t	o
7	2	4
+ 2	5	3

3.

h	t	o
8	6	4
+ 1	1	0

4.

h	t	o
6	0	3
+ 3	0	4

5.

h	t	o
3	1	0
+ 3	7	0

6.

h	t	o
1	0	2
+ 5	8	4

7.
```
  240
+  42
```

8.
```
  331
+  67
```

9.
```
  508
+  90
```

10.
```
  175
+  24
```

11.
```
  935
+  50
```

SHARE YOUR THINKING Which sum is greatest: 215 + 200, 215 + 20, or 215 + 2? Why?

Sometimes you can add mentally by counting on.

$202 + 300 = 502$

202, 302, 402, 502

$564 + 20 = 584$

564, 574, 584

Add. ✔ when you use mental math.

1.
100	400	213	527	674
+200	+200	+764	+241	+225

300 ✔

2.
750	610	425	335	624
+ 40	+ 80	+163	+211	+124

3.
600	700	203	348	131
+ 35	+ 47	+ 53	+31	+465

PROBLEM SOLVING

4. Leroy read two books about crocodiles. One book had 120 pages and the other one had 76 pages. How many pages did Leroy read in all?

_____ pages

5. Tami has 123 stickers of rain forest birds and 254 stickers of rain forest insects. How many stickers does she have in all?

_____ stickers

MENTAL MATH

6. Add 111 to
7. Make up your own.

542	303	123	876	777

Name _____

Sometimes when you add you have enough tens to make 1 hundred.

> 10 tens = 1 hundred
> 11 tens = 1 hundred 1 ten

Add 6 tens to 6 tens.
Can you make 1 hundred?

| 1st | Model each addend. Join tens. |
| 2nd | Regroup if you have enough tens. |

6 tens + 6 tens = __12__ tens

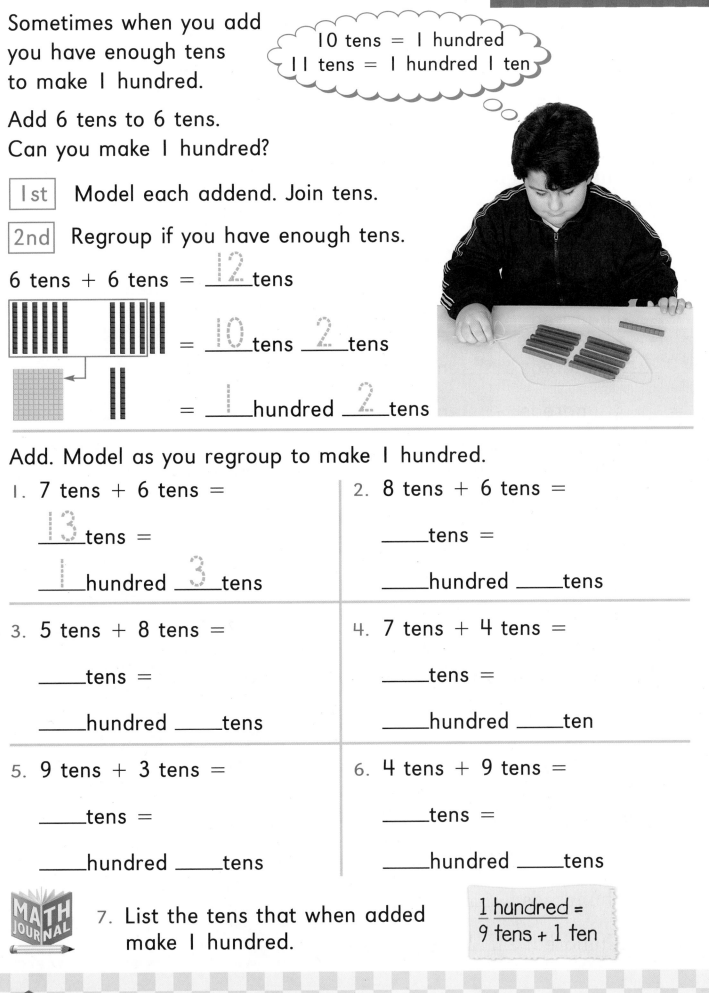

= __10__ tens __2__ tens

= __1__ hundred __2__ tens

Add. Model as you regroup to make 1 hundred.

1. 7 tens + 6 tens =

   __13__ tens =

   __1__ hundred __3__ tens

2. 8 tens + 6 tens =

   _____ tens =

   _____ hundred _____ tens

3. 5 tens + 8 tens =

   _____ tens =

   _____ hundred _____ tens

4. 7 tens + 4 tens =

   _____ tens =

   _____ hundred _____ ten

5. 9 tens + 3 tens =

   _____ tens =

   _____ hundred _____ tens

6. 4 tens + 9 tens =

   _____ tens =

   _____ hundred _____ tens

MATH JOURNAL

7. List the tens that when added make 1 hundred.

> 1 hundred =
> 9 tens + 1 ten

9-6 Provide number cards that say *1 ten* through *9 tens* for your child to explain how to add and regroup tens as hundreds.

three hundred forty-seven **347**

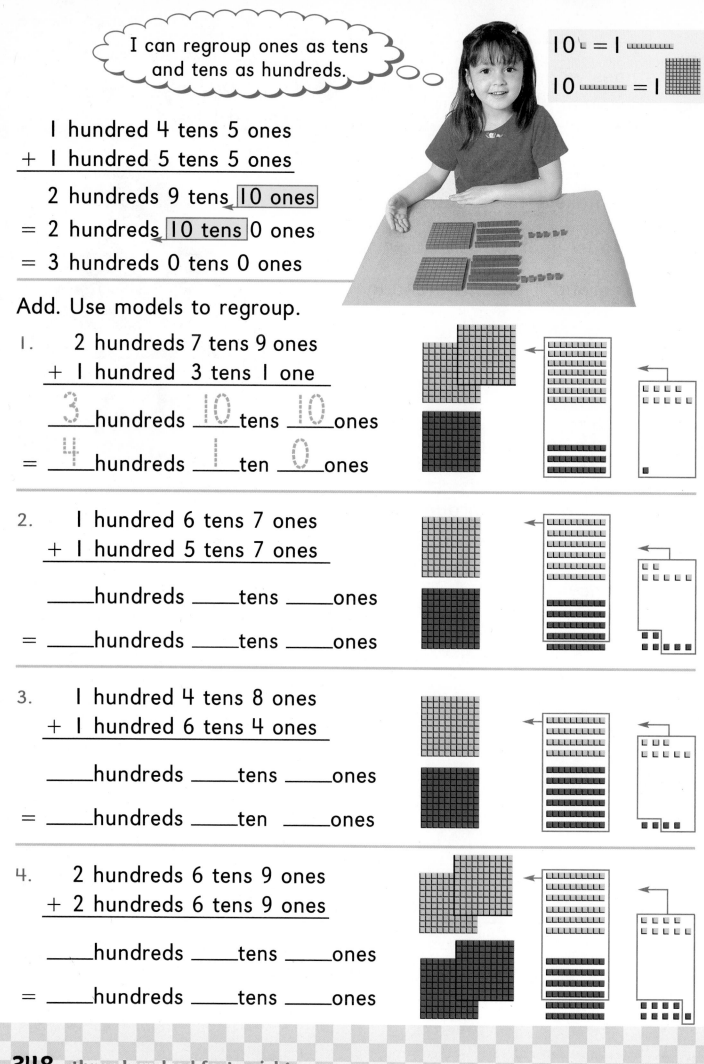

I can regroup ones as tens and tens as hundreds.

$10 \llcorner = 1$ ⸺

10 ⸺ $= 1$ ▦

   1 hundred 4 tens 5 ones
+ 1 hundred 5 tens 5 ones

   2 hundreds 9 tens 10 ones

= 2 hundreds 10 tens 0 ones

= 3 hundreds 0 tens 0 ones

Add. Use models to regroup.

1.   2 hundreds 7 tens 9 ones
   + 1 hundred 3 tens 1 one

   __3__ hundreds __10__ tens __10__ ones

= __4__ hundreds __1__ ten __0__ ones

2.   1 hundred 6 tens 7 ones
   + 1 hundred 5 tens 7 ones

   _____ hundreds _____ tens _____ ones

= _____ hundreds _____ tens _____ ones

3.   1 hundred 4 tens 8 ones
   + 1 hundred 6 tens 4 ones

   _____ hundreds _____ tens _____ ones

= _____ hundreds _____ ten _____ ones

4.   2 hundreds 6 tens 9 ones
   + 2 hundreds 6 tens 9 ones

   _____ hundreds _____ tens _____ ones

= _____ hundreds _____ tens _____ ones

Name _____

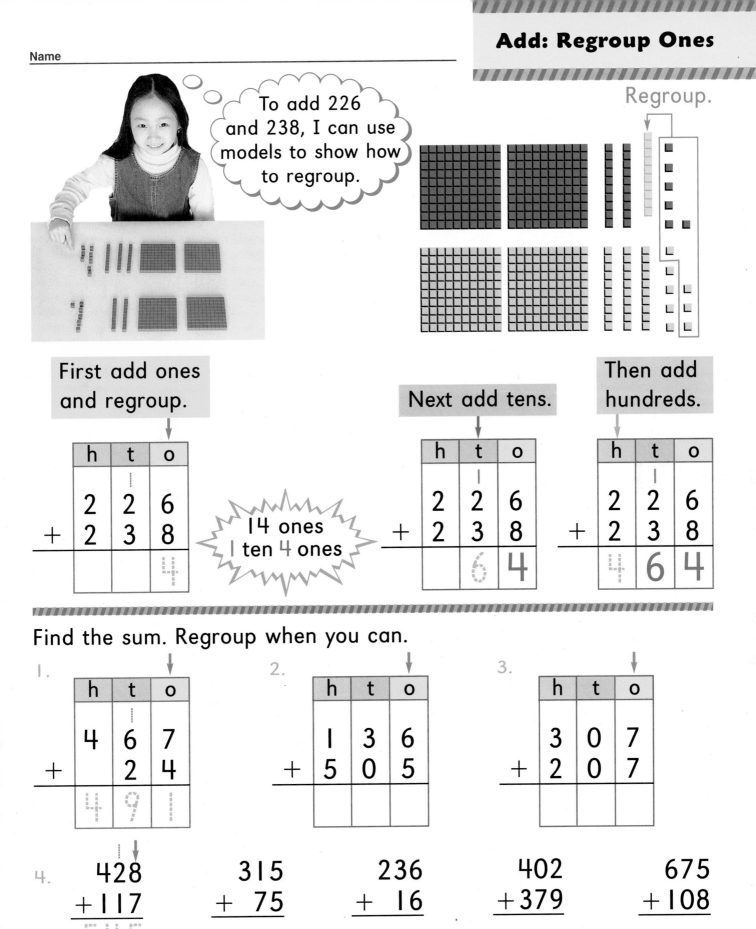

To add 226 and 238, I can use models to show how to regroup.

Regroup.

First add ones and regroup.

h	t	o
2	2	6
+ 2	3	8
		4

14 ones
1 ten 4 ones

Next add tens.

h	t	o
2	2	6
+ 2	3	8
	6	4

Then add hundreds.

h	t	o
2	2	6
+ 2	3	8
4	6	4

Find the sum. Regroup when you can.

1.

h	t	o
4	6	7
+	2	4
4	9	1

2.

h	t	o
1	3	6
+ 5	0	5

3.

h	t	o
3	0	7
+ 2	0	7

4.
```
  428        315        236        402        675
+ 117      +  75      +  16      + 379      + 108
─────
  545
```

Do you need to regroup when the sum of the ones is greater than or less than 9?

9-7 Ask your child to add 132 + 118 and 132 + 18 and to explain regrouping.

three hundred forty-nine **349**

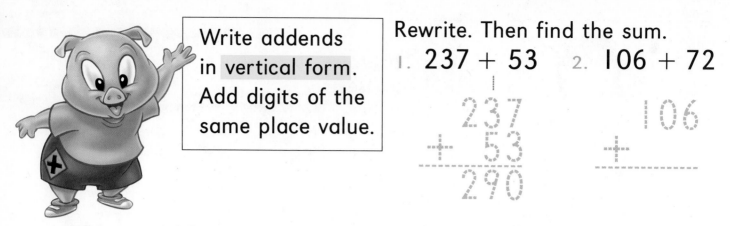

Write addends in vertical form. Add digits of the same place value.

Rewrite. Then find the sum.

1. 237 + 53

$$
\begin{array}{r}
237 \\
+\ 53 \\
\hline
290
\end{array}
$$

2. 106 + 72

$$
\begin{array}{r}
106 \\
+\ \\
\hline
\end{array}
$$

3. 521 + 109

4. 636 + 27

5. 314 + 314

6. 408 + 48

7. 455 + 27

8. 217 + 217

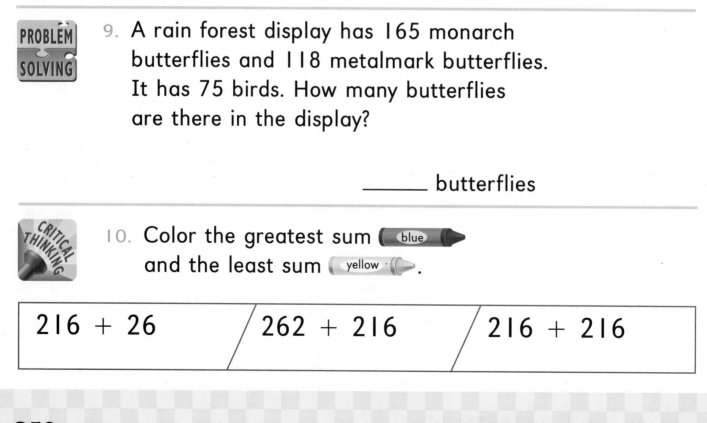

PROBLEM SOLVING

9. A rain forest display has 165 monarch butterflies and 118 metalmark butterflies. It has 75 birds. How many butterflies are there in the display?

_____ butterflies

CRITICAL THINKING

10. Color the greatest sum (blue) and the least sum (yellow).

| 216 + 26 | 262 + 216 | 216 + 216 |

Name

Add: 166 + 242 = ?

Regroup 10 tens as 1 hundred 0 tens.

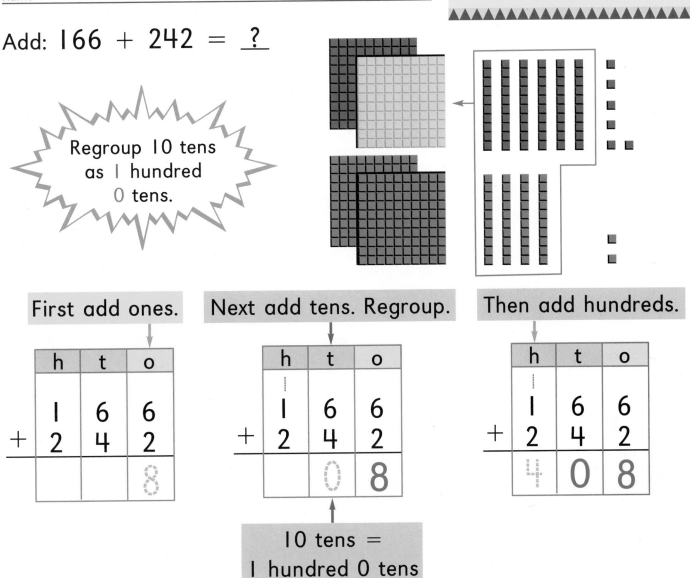

First add ones.		
h	t	o
1	6	6
+ 2	4	2
		8

Next add tens. Regroup.		
h	t	o
1	6	6
+ 2	4	2
	0	8

10 tens = 1 hundred 0 tens

Then add hundreds.		
h	t	o
1	6	6
+ 2	4	2
4	0	8

Find the sum. Regroup tens as hundreds when needed.

1.

h	t	o
3	7	4
+	9	4
4	6	8

2.

h	t	o
4	5	6
+ 2	9	1

3.

h	t	o
6	4	5
+ 1	7	4

4.
```
  492        345        482        236        794
+ 127      +  73      + 363      +  82      + 195
```

9-8 Ask your child to add 132 + 128 and
123 + 182 and to tell how regrouping tens is
like regrouping ones and how it is different.

three hundred fifty-one **351**

Add. Regroup if needed.

1.
$$\begin{array}{r} 244 \\ +385 \\ \hline 629 \end{array}$$
$$\begin{array}{r} 421 \\ +187 \\ \hline \end{array}$$
$$\begin{array}{r} 365 \\ + 93 \\ \hline \end{array}$$
$$\begin{array}{r} 192 \\ +363 \\ \hline \end{array}$$
$$\begin{array}{r} 232 \\ + 82 \\ \hline \end{array}$$

2.
$$\begin{array}{r} 544 \\ +192 \\ \hline \end{array}$$
$$\begin{array}{r} 742 \\ + 76 \\ \hline \end{array}$$
$$\begin{array}{r} 590 \\ +389 \\ \hline \end{array}$$
$$\begin{array}{r} 581 \\ +226 \\ \hline \end{array}$$
$$\begin{array}{r} 896 \\ + 30 \\ \hline \end{array}$$

3.
$$\begin{array}{r} 219 \\ +711 \\ \hline \end{array}$$
$$\begin{array}{r} 507 \\ +286 \\ \hline \end{array}$$
$$\begin{array}{r} 638 \\ + 55 \\ \hline \end{array}$$
$$\begin{array}{r} 604 \\ +208 \\ \hline \end{array}$$
$$\begin{array}{r} 729 \\ + 55 \\ \hline \end{array}$$

 TALK IT OVER How is 3 different from 1 and 2?

 PROBLEM SOLVING

4. Sela counted 175 catfish. Rico counted 140 catfish. What is the total number of fish Sela and Rico counted?

_____ fish

5. Last year 89 inches of rain fell in the forest. A dozen more inches fell this year than last year. How many inches of rain fell in all in 2 years?

_____ inches

MATH JOURNAL

6. Make 3 different sums that have 5 hundreds. Use each card only once in each addition.

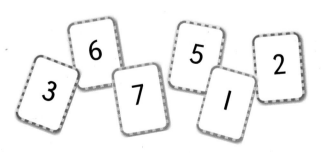

Name

What is the sum of 145 and 259?

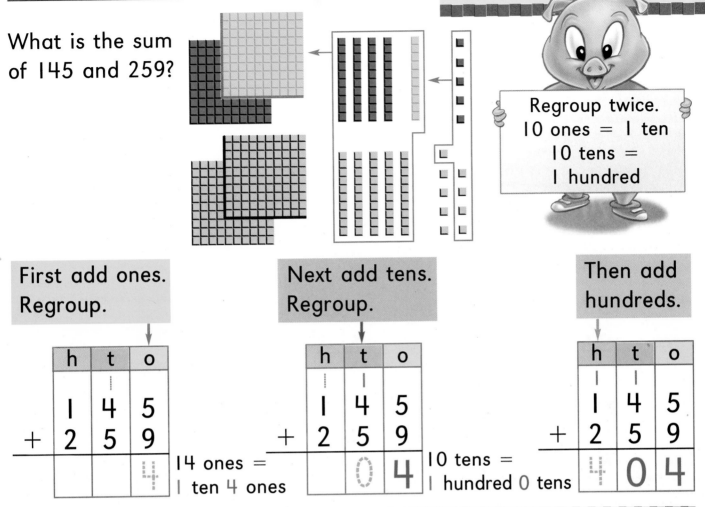

Regroup twice.
10 ones = 1 ten
10 tens = 1 hundred

First add ones. Regroup.

h	t	o
	¦	
1	4	5
+ 2	5	9
		4

14 ones = 1 ten 4 ones

Next add tens. Regroup.

h	t	o
¦	1	
1	4	5
+ 2	5	9
	0	4

10 tens = 1 hundred 0 tens

Then add hundreds.

h	t	o
1	1	
1	4	5
+ 2	5	9
4	0	4

Add. Regroup twice if needed.

1.

h	t	o
¦	¦	
3	9	7
+ 4	6	5
8	6	2

2.

h	t	o
6	9	8
+ 2	0	9

3.

h	t	o
3	4	8
+	6	7

4.
```
  563        286        395        467        193
+ 288      + 147      + 315      + 194      + 157
```

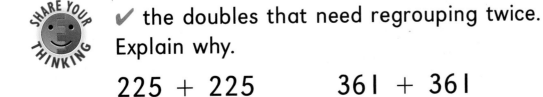
SHARE YOUR THINKING

✔ the doubles that need regrouping twice.
Explain why.

225 + 225   361 + 361   178 + 178

Ask your child to add 267 + 267 and to explain the regrouping.

9-9

three hundred fifty-three **353**

Find the sum. Regroup twice if needed.

1. 286   178   497   368   863
   + 87   + 32   + 47   + 99   + 78
    373

2. 439   596   474   385   173
   +374   +275   +357   +569   +429

3. 439   176   324   179   324
   +462   +473   +156   +289   +184

4. 528   729   345   615   555
   +387   +195   +577   +286   + 55

PROBLEM SOLVING

5. The hikers saw 368 butterflies yesterday.
   Today they saw 226. How many butterflies
   in all did the hikers see on both days?

   _____ butterflies

CRITICAL THINKING

6. Compare the sum of the numbers
   with 6 tens to the sum of the numbers
   with 6 ones. Write <, =, >.

647   416   667

268   506

_____ ◯ _____

Name _____

Are these measurements reasonable?

Color [blue] for Yes and

[red] for No.

John puts 4 cups of soup into a 1 quart pot.

weighs less than 1 kilogram.

will hold 1 liter.

Spencer's brother drank 1 cup of milk.

Paula's pencil is 1 meter long.

Cal's hat weighs 5 kilograms.

weigh less than 1 pound.

Ed has a ladder that is 8 feet tall.

The height of the flagpole is 10 cm.

Irving's foot is more than 1 inch long.

I used 1 liter of water to wash the car.

Kelly put 2 quarts of milk into a 1-pint bottle.

= 3 half inches.

Sandy's bike weighs more than 1 pound.

9 This page reviews the mathematical content presented in Chapter 8.

three hundred fifty-five **355**

Marta spent $1.55 yesterday and $1.24 today. How much did she spend altogether?

Yesterday	Today

First add. Begin with pennies.

$$\begin{array}{r} \$1.55 \\ +1.24 \\ \hline 279 \end{array}$$

Then write the . and $.

$$\begin{array}{r} \$1.55 \\ +1.24 \\ \hline \$2.79 \end{array}$$

She spent **$2.79** altogether.

Find the sum. Model to check.

1.
$$\begin{array}{r} \$3.23 \\ +1.24 \\ \hline \$4.47 \end{array}$$
$$\begin{array}{r} \$4.61 \\ +3.17 \\ \hline \end{array}$$
$$\begin{array}{r} \$1.30 \\ +3.15 \\ \hline \end{array}$$
$$\begin{array}{r} \$4.03 \\ +2.13 \\ \hline \end{array}$$

2.
$$\begin{array}{r} \$5.21 \\ +2.02 \\ \hline \end{array}$$
$$\begin{array}{r} \$7.20 \\ +0.46 \\ \hline \end{array}$$
$$\begin{array}{r} \$8.22 \\ +0.63 \\ \hline \end{array}$$
$$\begin{array}{r} \$2.31 \\ +6.47 \\ \hline \end{array}$$

3.
$$\begin{array}{r} \$1.72 \\ +2.27 \\ \hline \end{array}$$
$$\begin{array}{r} \$1.45 \\ +3.54 \\ \hline \end{array}$$
$$\begin{array}{r} \$2.36 \\ +2.63 \\ \hline \end{array}$$
$$\begin{array}{r} \$4.80 \\ +1.09 \\ \hline \end{array}$$

 PROBLEM SOLVING How much will Marta pay altogether?

4.

$4.25

$5.34

___ + ___

5.
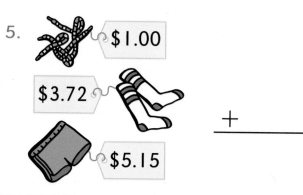
$1.00

$3.72

$5.15

___ + ___

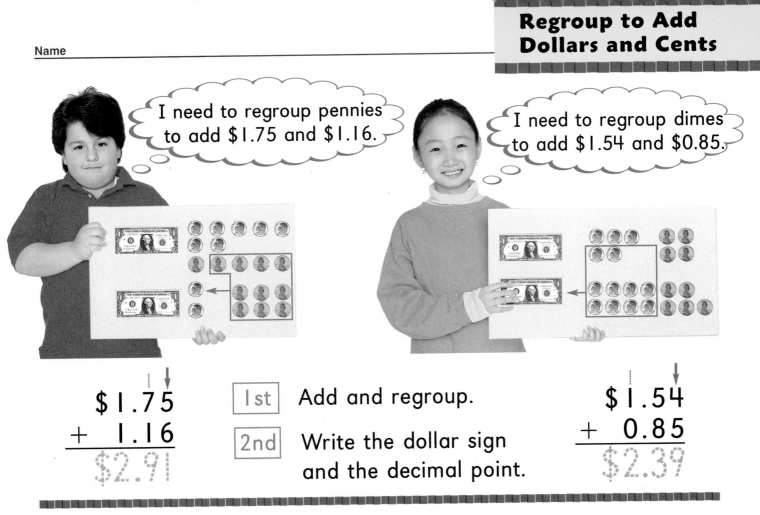

I need to regroup pennies to add $1.75 and $1.16.

I need to regroup dimes to add $1.54 and $0.85.

$1.75
+ 1.16
$2.91

| 1st | Add and regroup. |
| 2nd | Write the dollar sign and the decimal point. |

$1.54
+ 0.85
$2.39

Find the sum. Regroup pennies to dimes.

1.

$.	d	p
$ 2.	6	8
+ 3.	0	5
$5.	7	3

2.

$.	d	p
$ 4.	1	6
+ 0.	2	9
$		

3.

$.	d	p
$ 3.	1	5
+ 1.	2	7

Find the sum. Regroup dimes to dollars.

4.

$.	d	p
$ 1.	6	2
+ 0.	6	3
$2.	2	5

5.

$.	d	p
$ 1.	9	0
+ 1.	4	6

6.

$.	d	p
$ 3.	6	1
+ 3.	8	8

TALK IT OVER How is adding $1.99 and $1.50 like adding 199 and 150? How is it different?

9-11 Ask your child to add $1.75 + $1.16 and to explain the regrouping.

three hundred fifty-seven **357**

Ring addends you can regroup. Add.

1. $5.07⃝ $4.63 $1.29 $3.07 $6.06
   + 4.34   + 2.18   + 0.21   + 4.08   + 0.84
    $9.41

2. $6.51 $3.13 $8.70 $2.80 $3.24
   + 1.78   + 3.94   + 0.65   + 4.80   + 1.84

3. $4.71 $2.28 $4.93 $1.42 $7.34
   + 2.81   + 1.24   + 0.73   + 3.48   + 0.92

PROBLEM SOLVING

4. paid $6.84 for sunglasses. He paid $2.12 for lotion. How much did Leo spend for both?

_____

5. saved $3.48. Her aunt gave her $4.19. How much money did Gia have then?

_____

MENTAL MATH

6.

Rule: + $.09	$1.19	$1.18	$1.17	$1.16	$1.15
	$1.28				

7.

Rule: + $.90	$1.10	$1.20	$1.30	$1.40	$1.50

Name _____

Add: **$1.65 + $1.45 = ?**

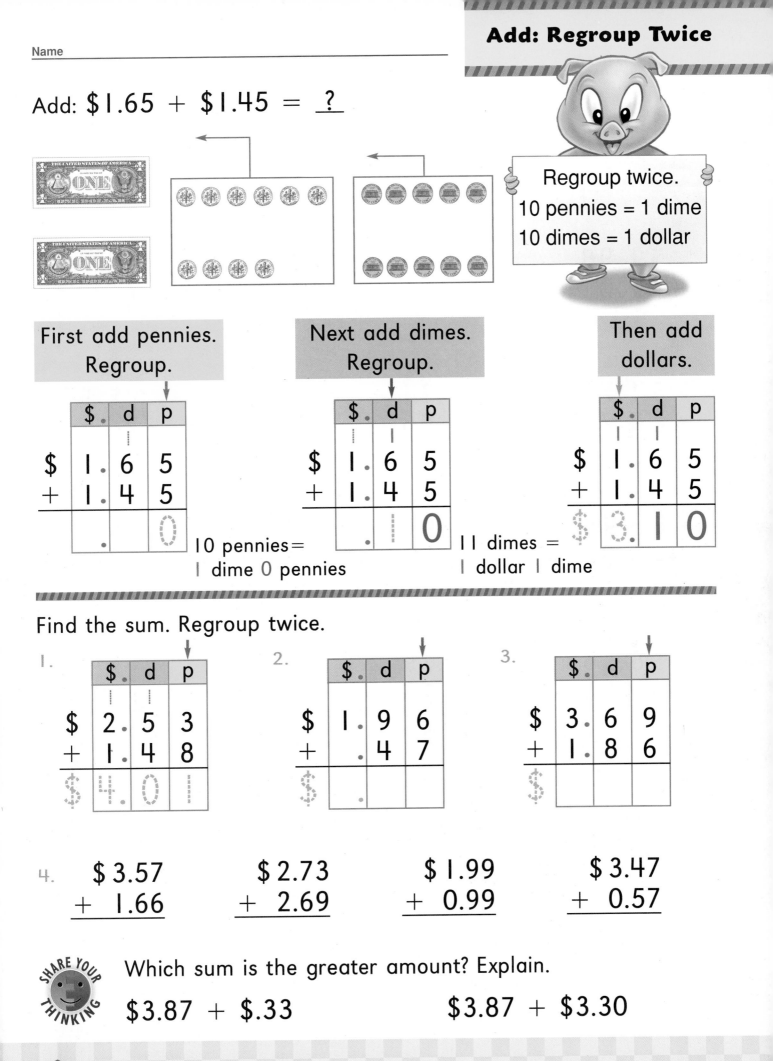

Regroup twice.
10 pennies = 1 dime
10 dimes = 1 dollar

First add pennies. Regroup.

$.	d	p
$ 1.	6	5
+ 1.	4	5
		0

10 pennies =
1 dime 0 pennies

Next add dimes. Regroup.

$.	d	p
$ 1.	6	5
+ 1.	4	5
		0

11 dimes =
1 dollar 1 dime

Then add dollars.

$.	d	p
$ 1.	6	5
+ 1.	4	5
$ 3.	1	0

Find the sum. Regroup twice.

1.

$.	d	p
$ 2.	5	3
+ 1.	4	8
$ 4.	0	1

2.

$.	d	p
$ 1.	9	6
+ .	4	7

3.

$.	d	p
$ 3.	6	9
+ 1.	8	6

4.

$ 3.57
+ 1.66

$ 2.73
+ 2.69

$ 1.99
+ 0.99

$ 3.47
+ 0.57

SHARE YOUR THINKING

Which sum is the greater amount? Explain.

$3.87 + $.33 $3.87 + $3.30

9-12 Ask your child to add $2.95 + $1.45 and to explain the regrouping.

three hundred fifty-nine **359**

Find the sum.

1. $6.82 $5.39 $4.16 $8.07
   + 0.28   + 3.91   + 4.87   + 0.98
   $7.10

2. $1.98 $2.25 $7.61 $4.64
   + 2.09   + 2.96   + 0.99   + 4.39

3. $1.76 $3.24 $1.89 $3.24
   + 4.73   + 1.56   + 2.78   + 1.84

4. $5.48 $7.99 $3.45 $5.53
   + 3.88   + 1.25   + 5.77   + 0.57

5. $8.56 $3.79 $2.17 $4.97
   + 0.86   + 0.23   + 0.84   + 0.69

6. The Rain Forest Club saved $4.78 this month and $3.56 last month. How much money did the club save in both months?

7. The book about jaguars costs $5.96. The book about butterflies costs $2.05. How much do both books cost?

_____   _____

Name _____

1. Read — How many numbers between 100 and 200 have 3 tens?

Think — 3 tens = _30_

Write — 130, 131, 132, 133, 134, 135, 136, 137, 138, 139

Check — _10_ numbers

2. Read — How many 3-digit numbers have 5 tens and 5 ones?

Think — 5 tens 5 ones = _55_

Write — 155,

Check — _____ numbers

3. Read — How many numbers between 345 and 445 have 4 tens?

Think — 4 tens = _____

Write — _____

Check — _____ numbers

4. Read — How many numbers between 200 and 800 have 2 tens 4 ones?

Think — 2 tens 4 ones = _____

Write — _____

Check — _____ numbers

9-13 Take turns making up and solving problems like the ones above with your child.

three hundred sixty-one **361**

5. How many numbers between 299 and 700 use the digits 3, 6, and 9 only once? (Use your number cards.)

369, 396, 639, 693
_____   4 numbers

6. How many 3-digit numbers greater than 200 use the digits 1, 2, and 5 only once?

_____   _____ numbers

7. How many 3-digit numbers less than 600 use the digits 2, 4, and 6 only once?

_____   _____ numbers

8. How many 3-digit numbers use the digits 2, 1, and 0 only once?

_____   _____ numbers

9. Ashley uses each of these 4 cards only once. How many 3-digit numbers less than 400 can she make?

2   4   6   8

_____   _____ numbers

10. Daryl has 3 cards for 2 and 3 cards for 1. How many 3-digit numbers can he make?

_____   _____ numbers

11. How many 3-digit numbers with 0, _____, and _____

are between _____ and _____?

_____ numbers

Name _____

Read → Find → Think → Write → Check

The 🧰 is across 6, up 1. 🧰 (6, 1)

Use this map to find the hidden money.

1. How much money can you find at each?

   (2, 1) $3.25   (3, 5) _____

   (1, 3) _____

2. Cody found money at (5, 6) and at (2,1). How much money did he find altogether?

   He found _____.

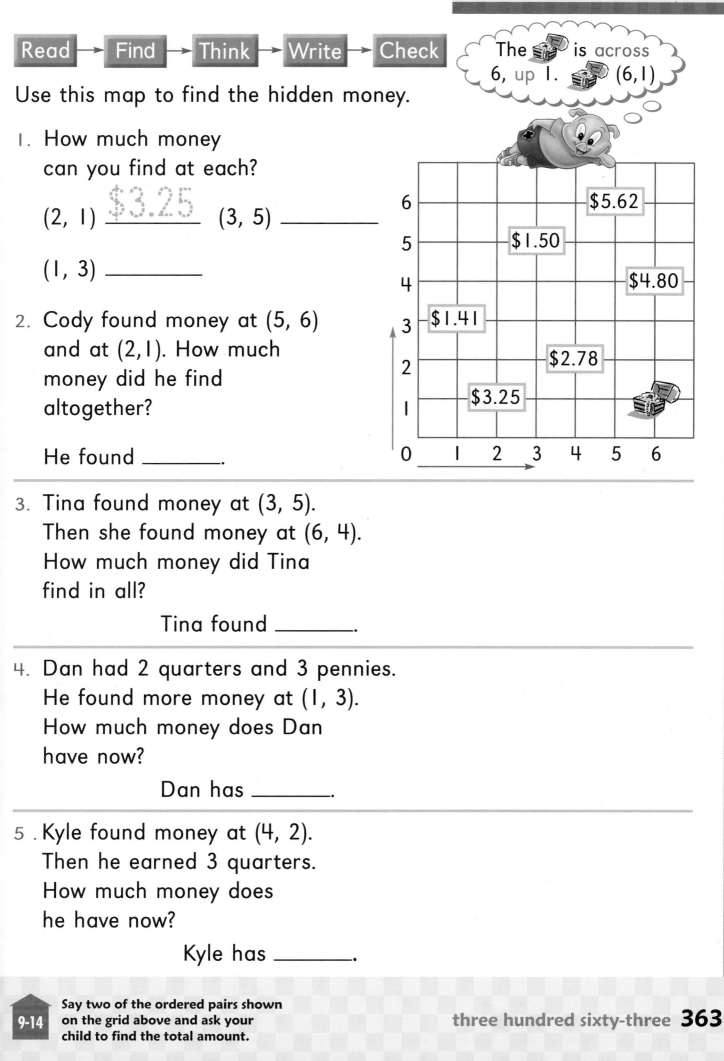

$5.62
$1.50
$4.80
$1.41
$2.78
$3.25

6
5
4
3
2
1
0   1   2   3   4   5   6

3. Tina found money at (3, 5). Then she found money at (6, 4). How much money did Tina find in all?

   Tina found _____.

4. Dan had 2 quarters and 3 pennies. He found more money at (1, 3). How much money does Dan have now?

   Dan has _____.

5. Kyle found money at (4, 2). Then he earned 3 quarters. How much money does he have now?

   Kyle has _____.

PROBLEM SOLVING

9-14 Say two of the ordered pairs shown on the grid above and ask your child to find the total amount.

three hundred sixty-three **363**

Use a strategy you have learned.

6. A canoe holds a total of 450 pounds. Color the combination of loads the canoe holds.

230 pounds 175 pounds 270 pounds 395 pounds 275 pounds

7. A scientist tagged 125 hornbills. 70 of them were green. She also tagged 85 toucans. How many birds did the scientist tag?

The scientist tagged _____ birds.

8. Briana paid $3.20 for a crocodile poster, $4.25 for a rain forest tee shirt, and $1.40 for parrot stickers. How much did she spend altogether?

She spent _____.

PROBLEM SOLVING

9. When were 23 pounds of paper recycled?

_____

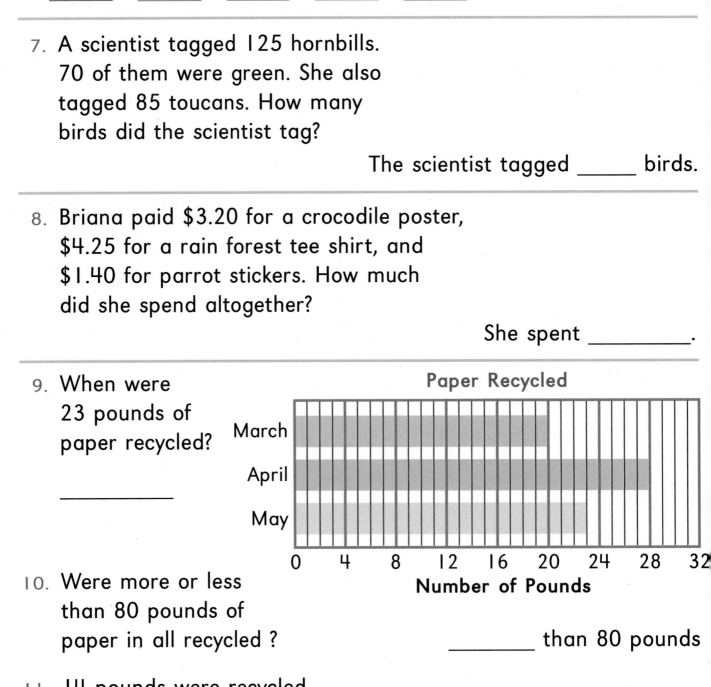

Paper Recycled

March
April
May

0   4   8   12   16   20   24   28   32
Number of Pounds

10. Were more or less than 80 pounds of paper in all recycled?

_____ than 80 pounds

11. 14 pounds were recycled in January. In which month was double this amount recycled?

_____

Name _____

Complete.

1. three hundred forty = _____hundreds _____tens _____ones

    _____ = _____ + _____ + _____

2. six hundred twelve = _____hundreds _____ten _____ones

    _____ = _____ + _____ + _____

3. Write the number that is 100 more: 652_____   191_____

   Write the number that is 10 more: 106 _____   895 _____

4. What is the value of 7 in 274 _____ and 716 _____ ?

Find the sum.

5.
```
  800        363        430        676        298
+ 167      +  28      +  72      +  53      +  45
  967
```

6.
```
  124        345        486        192        527
+  16      + 139      + 306      + 693      + 277
```

7.
```
$1.10      $2.22      $3.38      $4.38      $5.80
+ 0.26     + 1.23     + 3.07     + 1.92     + 1.45
```

PROBLEM SOLVING

8. Todd had 3 hundreds 4 tens 2 ones. He found 6 tens 8 ones. How many does he have in all?

   _____

9. The number is less than 300. It has the same number of tens and ones. The sum of its digits is 13.

   _____

This page reviews the mathematical content presented in Chapter 9.

9

three hundred sixty-five **365**

This is a magic square.
All of the sums are the same.

9	10	5	→ 9 + 10 + 5 = 24

9	10	5
4	8	12	→ 24
11	6	7	→ 24

24   24   24   24   24

Write the numbers to make a magic square.

1.

13	14	
8	12	
15	10	11

sum ___36___

2.

14	15	10
9		
16	11	

sum _____

3.

15	16	11
10		
17	12	

sum _____

4.

120		40
		180
160	60	80

sum _____

5.

	200	220
160	240	
260	280	

sum _____

6.

165		105
60		180
135	150	

sum _____

Name _____

1. Find the missing numbers. Tell how to model each pattern.

- 175, _____, 185, _____, 195, 200

- 220, 222, _____, 226, _____, _____

- 158, 168, 178, _____, _____, _____

2. Write an addend for each. Find the sum. Tell how to model the regrouping.

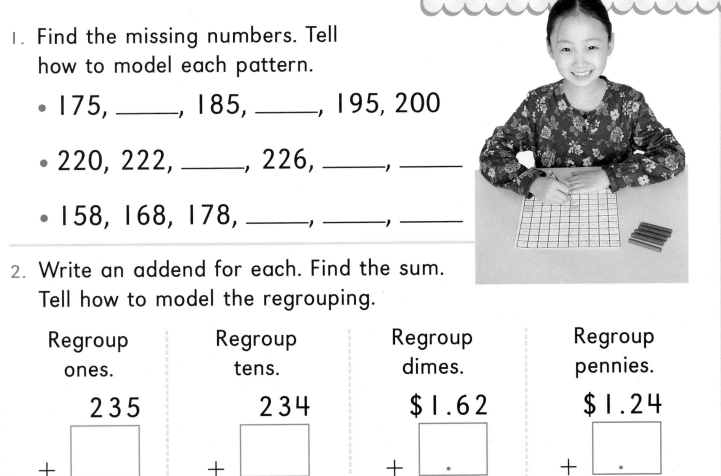

Regroup ones.	Regroup tens.	Regroup dimes.	Regroup pennies.
235	234	$1.62	$1.24
+ ☐	+ ☐	+ ☐ .	+ ☐ .

Choose 1 project.
Use a separate sheet of paper.

3. Show each counting pattern on a different number line.
Begin with 240.

240 ? ? ? ? ?

- Count by 100s.
- Count by 4s.
- Count by 3s.

4. Pick a 3-digit number less than 450. Write 10 sentences about your number. Here are some ideas.

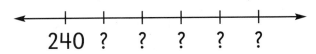

Two hundred forty-one
My number is between
The sum of its digits is
The double of this

9

This page provides a variety of informal assessment opportunities in order to measure your child's understanding of Chapter 9.

three hundred sixty-seven **367**

Write the number.

1. 3 hundreds 2 tens 8 ones

_____

2.
hundreds	tens	ones
4	0	7

_____

3. $800 + 20 + 7 =$ _____

4. $100 + 90 + 2 =$ _____

5. 100 more than 207 _____

6. 10 more than 401 _____

816 _____

195 _____

Complete. Add the place values.

7. $218 = 200 +$ _____

8. $530 =$ _____

Add. ✔ when you counted on by hundreds.

9. $276 + 30 =$ _____

10. $418 + 300 =$ _____

11. $608 + 2 =$ _____

12. $191 + 200 =$ _____

Find the sum.

13.
$$300 + 200$$ $$436 + 213$$ $$714 + 58$$ $$\$2.59 + 1.40$$ $$\$4.28 + 0.62$$

14.
$$673 + 154$$ $$375 + 82$$ $$287 + 263$$ $$\$5.60 + 1.59$$ $$\$6.19 + 1.92$$

15. Luis wrote numbers between 300 and 400 that have a 5 in the tens place. How many numbers did he write?

_____ numbers

This page is a formal assessment of your child's understanding of the content presented in Chapter 9.

Mark the ◯ for your answer.

Listening Section

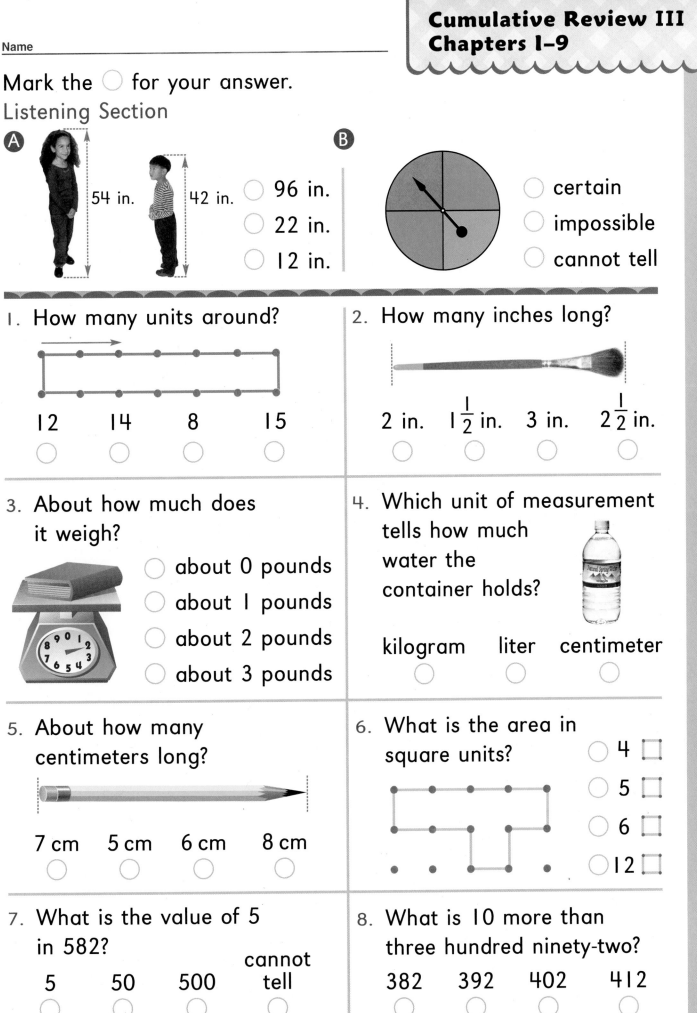

A

54 in.   42 in.
◯ 96 in.
◯ 22 in.
◯ 12 in.

B

◯ certain
◯ impossible
◯ cannot tell

---

1. How many units around?

12   14   8   15
◯   ◯   ◯   ◯

2. How many inches long?

2 in. $1\frac{1}{2}$ in.   3 in. $2\frac{1}{2}$ in.
◯   ◯   ◯   ◯

---

3. About how much does it weigh?

◯ about 0 pounds
◯ about 1 pounds
◯ about 2 pounds
◯ about 3 pounds

4. Which unit of measurement tells how much water the container holds?

kilogram   liter   centimeter
◯   ◯   ◯

---

5. About how many centimeters long?

7 cm   5 cm   6 cm   8 cm
◯   ◯   ◯   ◯

6. What is the area in square units?

◯ 4 ☐
◯ 5 ☐
◯ 6 ☐
◯ 12 ☐

---

7. What is the value of 5 in 582?

5   50   500   cannot tell
◯   ◯   ◯   ◯

8. What is 10 more than three hundred ninety-two?

382   392   402   412
◯   ◯   ◯   ◯

REINFORCEMENT

Mark the ○ for your answer:

9. What is the number?

238   283   832   234
○   ○   ○   ○

10. What is the temperature?

○ 15°F
○ 20°F
○ 25°F
○ 30°F

Find the missing addend.

11. $457 + \boxed{} = 557$

0   10   100   cannot tell
○   ○   ○   ○

12. What is the number for $300 + 20 + 2$?

302   31   320   322
○   ○   ○   ○

13. Count by 100. What is the missing number?

400, 500, _____, 700

800   600   300   900
○   ○   ○   ○

14.

What time will it be 2 hours from now?

○ 8:20
○ 9:20
○ 8:15
○ 9:15

15. Add.
$$\begin{array}{r} 823 \\ +174 \\ \hline \end{array}$$
○ 957
○ 997
○ 951
○ 900

16. Add.
$$\begin{array}{r} 394 \\ +112 \\ \hline \end{array}$$
○ 406
○ 496
○ 506
○ 512

17. Add.
$$\begin{array}{r} \$7.24 \\ +\ 0.56 \\ \hline \end{array}$$
○ $8.80
○ $8.70
○ $7.80
○ $7.70

18. Florence bought 250 goldfish and 135 tropical fish. How many fish did she buy altogether?

115   285   385   375
○   ○   ○   ○

19. Lanie drew 10 circles. She colored 2 circles red and four circles blue. What part of the set are not colored?

$\dfrac{1}{2}$ $\dfrac{2}{10}$ $\dfrac{4}{10}$ $\dfrac{6}{10}$
○   ○   ○   ○

Multiplication and Division

CRITICAL THINKING

If Jake puts his cards on 3 pages, how many cards will be on each page?

For more information about Chapter 10, visit the Family Information Center at **www.sadlier-oxford.com**

Internet

Dear Family,

Today your child began Chapter 10. As she/he studies multiplication and division, you may want to read the poem below which was read in class. Have your child talk about some of the math ideas pictured on page 371.

Look for the 🏠 at the bottom of each skills lesson. The suggestion on the page gives you an opportunity to improve your child's understanding of math and to reinforce her/his math language. You may want to have pennies and other countables available for your child to use throughout this chapter.

Home Reading Connection

Twos

Lots of things come in twos—
Ears and earmuffs, feet and shoes,
Ankles, shoulders, elbows, eyes,
Heels and shins and knees and thighs,
Galoshes, ice skates, mittens, socks,
Humps on camels, hands on clocks.
And heads on monsters also do—
Like that one...
Right in back of you!

Jeff Moss

Home Activity

By Twos

Try this activity with your child. Throughout this chapter have index cards or small paper plates and stickers available. Place five cards in a row and ask your child to put 2 stickers on each card. Then ask her/him to write the number sentence illustrated by each row of sticker cards. As your child progresses through this chapter, adjust the number of cards, stickers, and operation.

$$2 + 2 + 2 + 2 + 2 = 10$$

$$5 \text{ twos} = 10$$

$$5 \times 2 = 10$$

Name _____

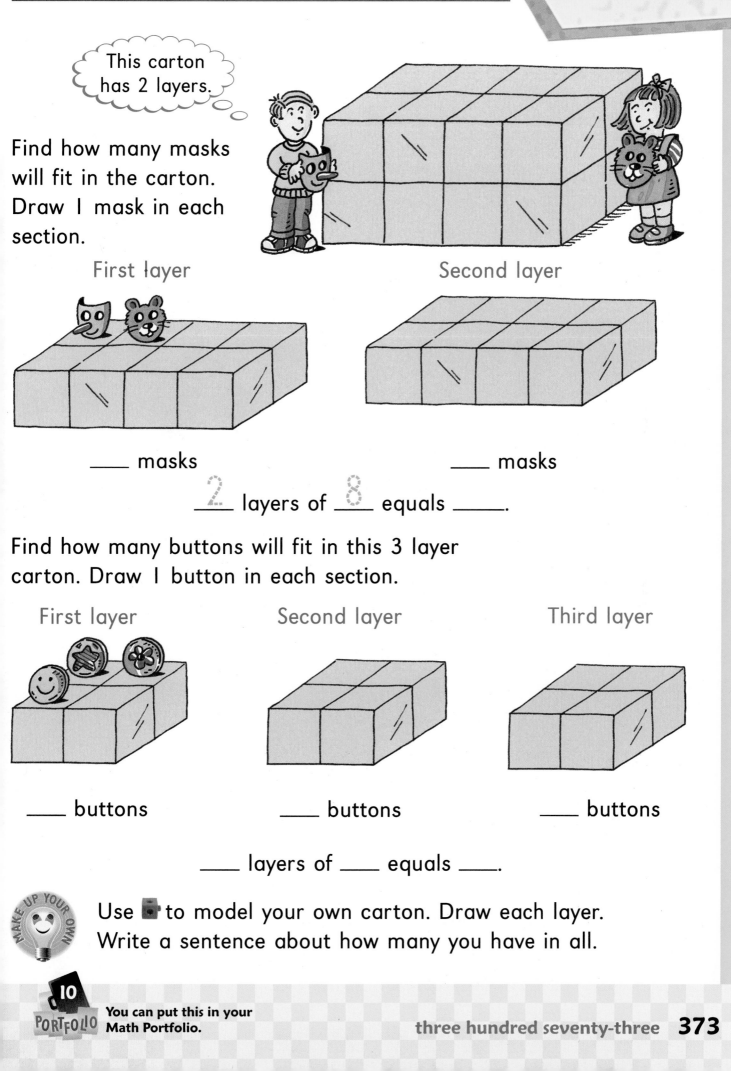

This carton has 2 layers.

Find how many masks will fit in the carton. Draw 1 mask in each section.

First layer

Second layer

____ masks

____ masks

_2_ layers of _8_ equals _____.

Find how many buttons will fit in this 3 layer carton. Draw 1 button in each section.

First layer

Second layer

Third layer

____ buttons

____ buttons

____ buttons

____ layers of ____ equals ____.

Use ▪ to model your own carton. Draw each layer. Write a sentence about how many you have in all.

CONNECTIONS

AUSTRALIA

CANADA

Emil has 8 decals to begin his decal collection. He put them in equal groups of 2.

CHINA

FINLAND

Make an organized list of how you can sort the decals.

1. Sort by kind of animal.

meat eater
dingo

INDIA

PERU

KENYA

JORDAN

2. Sort by shape.

0 sides
China

3. Draw an animal decal for your country. The decal should have five corners.

You can put this in your
Math Portfolio.

10
PORTFOLIO

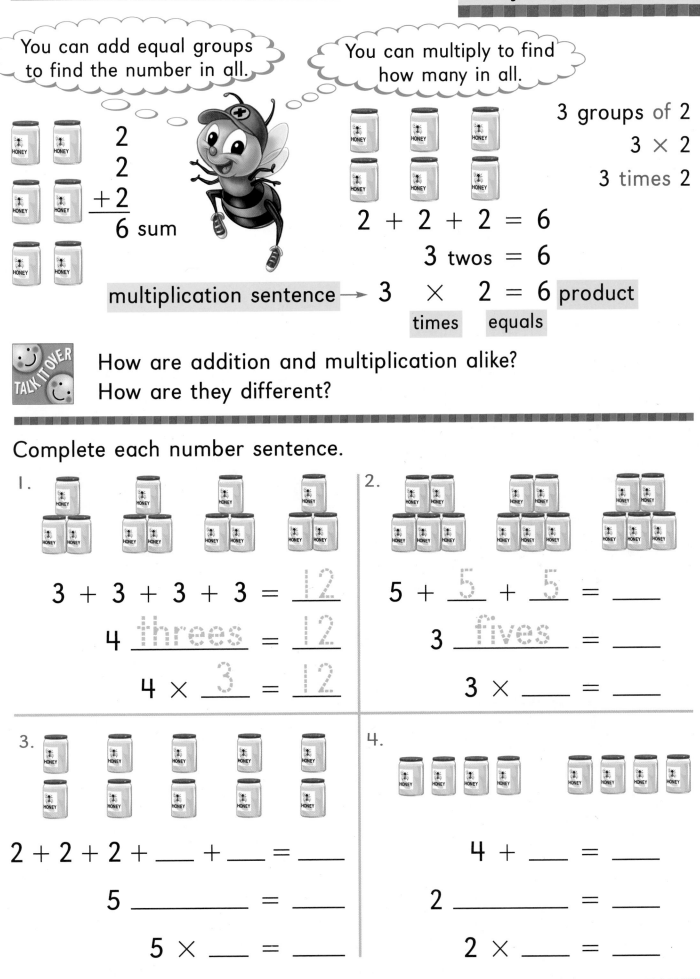

Name _____

You can add equal groups to find the number in all.

2
2
+2
6 sum

You can multiply to find how many in all.

2 + 2 + 2 = 6

3 twos = 6

3 groups of 2

3×2

3 times 2

multiplication sentence → 3 × 2 = 6 product

times   equals

TALK IT OVER How are addition and multiplication alike?
How are they different?

Complete each number sentence.

1.

$3 + 3 + 3 + 3 = \underline{12}$

$4 \ \underline{threes} = \underline{12}$

$4 \times \underline{3} = \underline{12}$

2.

$5 + \underline{5} + \underline{5} = \underline{}$

$3 \ \underline{fives} = \underline{}$

$3 \times \underline{} = \underline{}$

3.

$2 + 2 + 2 + \underline{} + \underline{} = \underline{}$

$5 \ \underline{} = \underline{}$

$5 \times \underline{} = \underline{}$

4.

$4 + \underline{} = \underline{}$

$2 \ \underline{} = \underline{}$

$2 \times \underline{} = \underline{}$

10-1 Ask your child to draw 4 groups of 5 and to write the addition and multiplication sentences for the drawing.

three hundred seventy-five **375**

Complete each number sentence.

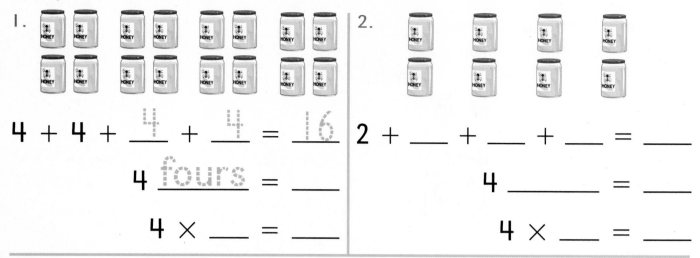

1. $4 + 4 + \underline{4} + \underline{4} = \underline{16}$

$4 \underline{\text{fours}} = \underline{\hspace{0.5cm}}$

$4 \times \underline{\hspace{0.5cm}} = \underline{\hspace{0.5cm}}$

2. $2 + \underline{\hspace{0.5cm}} + \underline{\hspace{0.5cm}} + \underline{\hspace{0.5cm}} = \underline{\hspace{0.5cm}}$

$4 \underline{\hspace{1cm}} = \underline{\hspace{0.5cm}}$

$4 \times \underline{\hspace{0.5cm}} = \underline{\hspace{0.5cm}}$

 PROBLEM SOLVING

3. Each pot must have 3 flowers.
Draw the missing flowers.
How many flowers are
needed for 6 pots?

$\underline{\hspace{1cm}}$ flowers are needed.

 FINDING TOGETHER

4. Use counters to model equal groups.
Complete the table.

Number of groups	Number in each equal group	Number in all
6	two	
3	four	
5	four	
2	ten	

I add to find the sum and multiply to find the product.

$2 + 2 + 2 + 2 =$ ___ sum

4 twos $=$ ___

$4 \times 2 =$ ___ product

Find the product.

1.

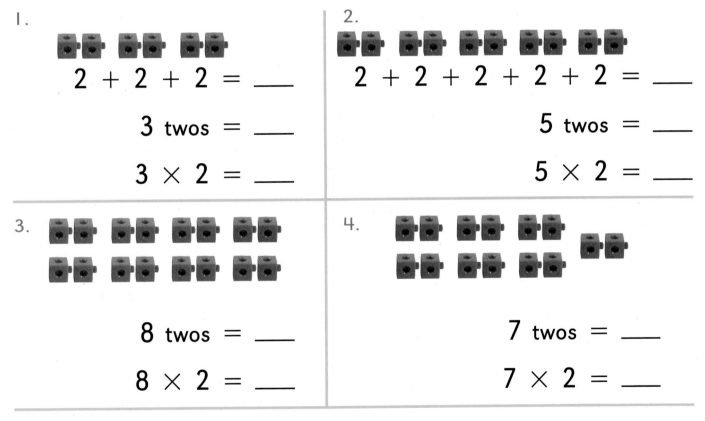

$2 + 2 + 2 =$ ___

3 twos $=$ ___

$3 \times 2 =$ ___

2.

$2 + 2 + 2 + 2 + 2 =$ ___

5 twos $=$ ___

$5 \times 2 =$ ___

3.

8 twos $=$ ___

$8 \times 2 =$ ___

4.

7 twos $=$ ___

$7 \times 2 =$ ___

Multiply. Use models to check.

5. $0 \times 2 =$ ___ $2 \times 2 =$ ___ $6 \times 2 =$ ___

6. $1 \times 2 =$ ___ $3 \times 2 =$ ___ $4 \times 2 =$ ___

TALK IT OVER When you make equal groups of 2, is the product odd or even?

10-2 Ask your child to draw 4 groups of twos and to write the addition and multiplication sentences for the drawing.

three hundred seventy-seven **377**

You can write multiplication in two ways. Complete.

1.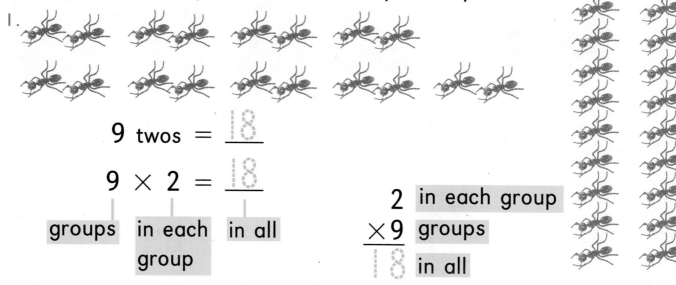

9 twos = 18

9 × 2 = 18

groups • in each group • in all

2   in each group
× 9   groups
18   in all

Find the product.

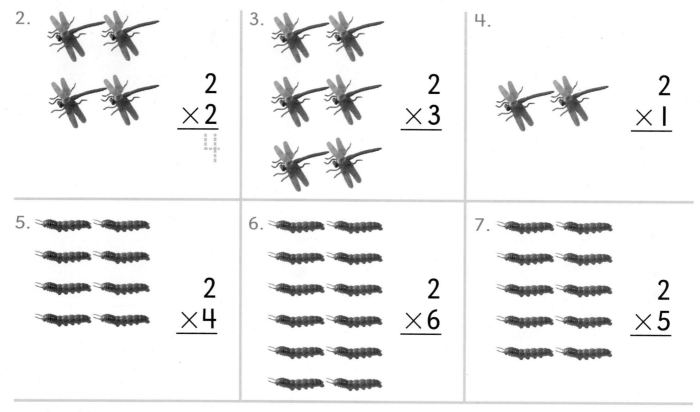

2.
2
× 2
4

3.
2
× 3

4.
2
× 1

5.
2
× 4

6.
2
× 6

7.
2
× 5

Multiply. Write the missing numbers.
Look for a pattern.

8.

										Number:
2	2	2	2	2	2	2	2	2	2	← in each group
× 0	1	2	3	4	5	6	7	8	9	← of groups
0	2	4								← in all

Name

You can multiply when the groups are equal.

Think:
3 + 3

Think:
3 + 3 + 3 + 3

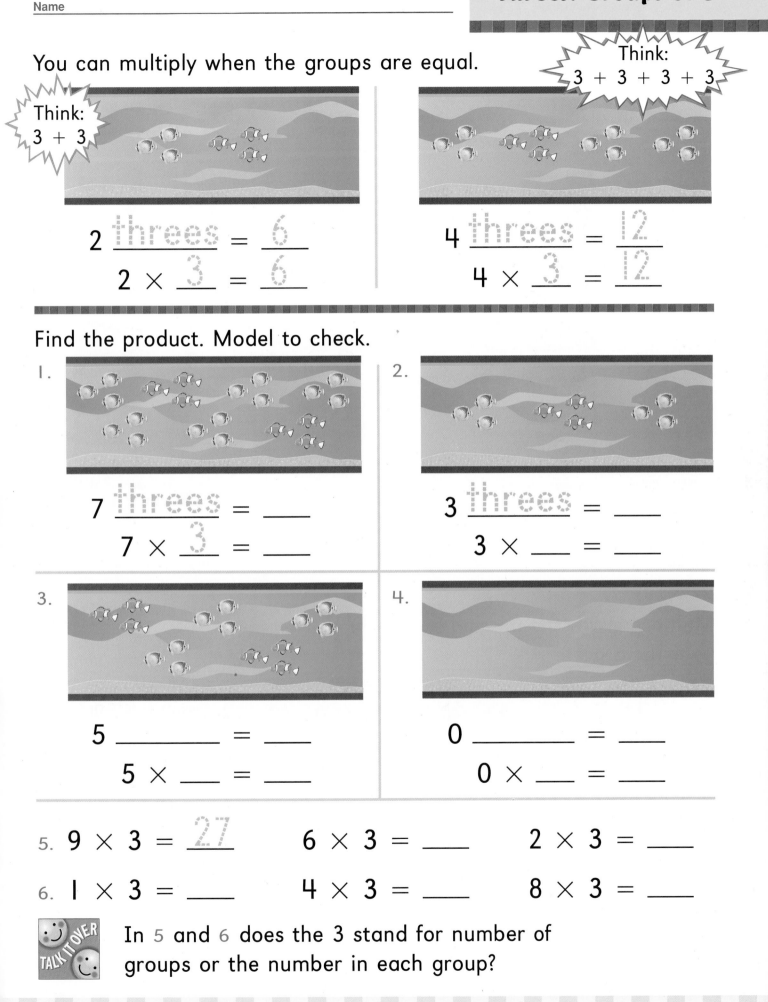

2 <u>threes</u> = <u>6</u>

2 × <u>3</u> = <u>6</u>

4 <u>threes</u> = <u>12</u>

4 × <u>3</u> = <u>12</u>

Find the product. Model to check.

1.

7 <u>threes</u> = ___

7 × <u>3</u> = ___

2.

3 <u>threes</u> = ___

3 × ___ = ___

3.

5 _____ = ___

5 × ___ = ___

4.

0 _____ = ___

0 × ___ = ___

5. 9 × 3 = <u>27</u>   6 × 3 = ___   2 × 3 = ___

6. 1 × 3 = ___   4 × 3 = ___   8 × 3 = ___

 In 5 and 6 does the 3 stand for number of
groups or the number in each group?

10-3 Ask your child to model the multiplication
sentences for exercise 5.

three hundred seventy-nine **379**

Draw or model to find each product.

1.   3 number in each group
     ×6 number of groups
     18 in all

2. 3
    ×8

3. 3
    ×7

4. 3
    ×9

5. 3   3   3   3
    ×4   ×1   ×5   ×2

Multiply. Write the missing numbers.
Look for a pattern.

6.

	3	3	3	3	3	3	3	3	3	3	← in each group
×	0	1	2	3	4	5	6	7	8	9	← of groups
	0	3	6								← in all

Number:

Are the products in 6 found by counting
by 2s, 3s, or 4s? Explain.

7. Maria, José, and Doug each have 3 marbles.
   How many marbles do they have in all? _____

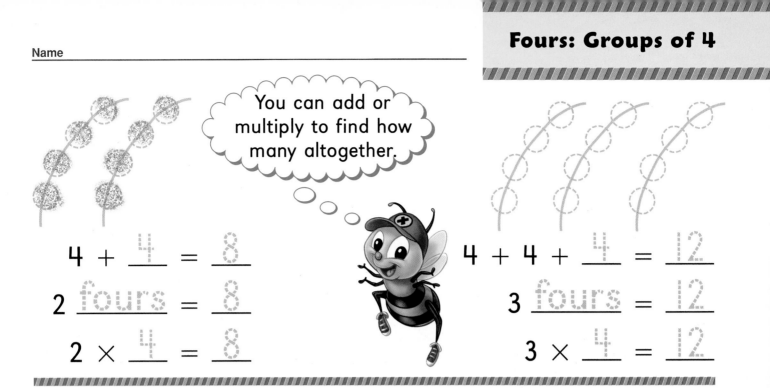

You can add or multiply to find how many altogether.

$4 + \underline{4} = \underline{8}$

$2 \text{ fours} = \underline{8}$

$2 \times \underline{4} = \underline{8}$

$4 + 4 + \underline{4} = \underline{12}$

$3 \text{ fours} = \underline{12}$

$3 \times \underline{4} = \underline{12}$

Draw 4 beads on each string. Complete.

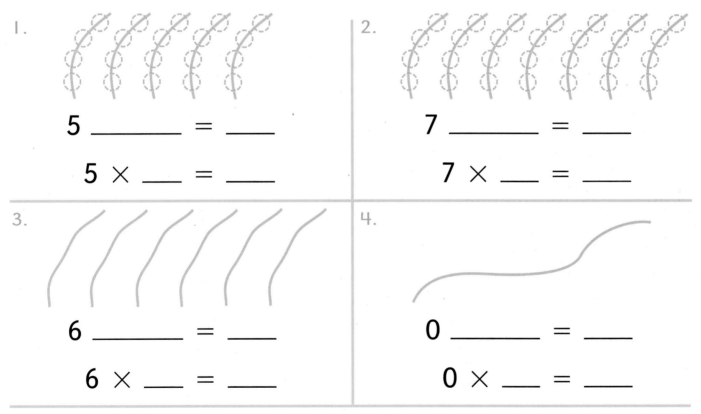

1.

$5 \underline{\hspace{2cm}} = \underline{\hspace{1cm}}$

$5 \times \underline{\hspace{1cm}} = \underline{\hspace{1cm}}$

2.

$7 \underline{\hspace{2cm}} = \underline{\hspace{1cm}}$

$7 \times \underline{\hspace{1cm}} = \underline{\hspace{1cm}}$

3.

$6 \underline{\hspace{2cm}} = \underline{\hspace{1cm}}$

$6 \times \underline{\hspace{1cm}} = \underline{\hspace{1cm}}$

4.

$0 \underline{\hspace{2cm}} = \underline{\hspace{1cm}}$

$0 \times \underline{\hspace{1cm}} = \underline{\hspace{1cm}}$

Multiply. Model each.

5. $8 \times 4 = \underline{\hspace{1cm}}$ $4 \times 4 = \underline{\hspace{1cm}}$ $9 \times 4 = \underline{\hspace{1cm}}$

Which addition sentence can you solve by multiplying? Explain.

$4 + 4 + 4 + 4 + 5 = 21$ $4 + 4 + 4 + 4 + 4 = 20$

10-4 Ask your child to model the multiplication sentences for exercise 5.

three hundred eighty-one **381**

Draw or model to find each product.

1.
$$\begin{array}{r} 4 \\ \times 3 \\ \hline 12 \end{array}$$

2.
$$\begin{array}{r} 4 \\ \times 5 \\ \hline \end{array}$$

3.
$$\begin{array}{r} 4 \\ \times 6 \\ \hline \end{array}$$

4.
$$\begin{array}{r} 4 \\ \times 4 \\ \hline \end{array}$$

5.
$$\begin{array}{r} 4 \\ \times 7 \\ \hline \end{array}$$

6.
$$\begin{array}{r} 4 \\ \times 8 \\ \hline \end{array}$$

7.
$$\begin{array}{r} 4 \\ \times 9 \\ \hline \end{array}$$

8.
$$\begin{array}{r} 4 \\ \times 2 \\ \hline \end{array}$$

9.
$$\begin{array}{r} 4 \\ \times 0 \\ \hline \end{array}$$

Multiply. Write the missing numbers.

10.

	4	4	4	4	4	4	4	4	4	4	← in each group
×	0	1	2	3	4	5	6	7	8	9	← of groups
	0	4	8								← in all

Find each product.

11. $1 \times 4 = $ ___ $1 \times 3 = $ ___ $1 \times 2 = $ ___ $1 \times 1 = $ ___

12. $0 \times 4 = $ ___ $0 \times 3 = $ ___ $0 \times 2 = $ ___ $0 \times 1 = $ ___

Write a rule for the facts in 11.

Write a rule for the facts in 12.

Name _____

I can model multiplication with grid paper.

3 equal rows
5 in each row

4 equal rows
5 in each row

3 _fives_ = _15_   4 _fives_ = ___

3 × _5_ = _15_   4 × _5_ = ___

TALK IT OVER   What addition sentence can you write for each of these examples?

Multiply. You can use grid paper for 3–6.

1.   5 equal rows
     5 in each row

     5 _fives_ = ___

     5 × ___ = ___

2.   2 equal rows
     5 in each row

     2 _____ = ___

     2 × ___ = ___

3.   7 equal rows
     5 in each row

     7 _____ = ___

     7 × ___ = ___

4.   9 equal rows
     5 in each row

     9 _____ = ___

     9 × ___ = ___

5.   6 equal rows
     5 in each row

     6 _____ = ___

     6 × ___ = ___

6.   8 equal rows
     5 in each row

     8 _____ = ___

     8 × ___ = ___

Draw or model to find each product.

1.
$$\begin{array}{r} 5 \\ \times 6 \\ \hline \end{array}$$

2.
$$\begin{array}{r} 5 \\ \times 3 \\ \hline \end{array}$$

3.
$$\begin{array}{r} 5 \\ \times 7 \\ \hline \end{array}$$

4.
$$\begin{array}{r} 5 \\ \times 4 \\ \hline \end{array}$$

5.
$$\begin{array}{r} 5 \\ \times 0 \\ \hline \end{array}$$

6.
$$\begin{array}{r} 5 \\ \times 1 \\ \hline \end{array}$$

7.
$$\begin{array}{r} 5 \\ \times 9 \\ \hline \end{array}$$

8.
$$\begin{array}{r} 5 \\ \times 8 \\ \hline \end{array}$$

9.
$$\begin{array}{r} 5 \\ \times 5 \\ \hline \end{array}$$

Multiply. Write the missing numbers.

10.

	5	5	5	5	5	5	5	5	5	5	← in each group
×	0	1	2	3	4	5	6	7	8	9	← of groups
	0	5	10								← in all

PROBLEM SOLVING Aristo puts 10 cubes in each can and 5 markers in each can.

11. How many cubes are in 3 cans?

_____ cubes

12. How many markers are in 3 cans?

_____ markers

Name _____

The product stays the same when you change the order.

2 equal rows

5 in each row

$2 \times 5 = 10$

5 equal rows

2 in each row

$5 \times 2 = 10$

Multiply. Change the order. Model to check.

1. $2 \times 4 = \underline{8}$
   $\underline{4} \times \underline{2} = \underline{8}$

2. $4 \times 3 = \underline{}$
   $\underline{3} \times \underline{4} = \underline{}$

3. $7 \times 2 = \underline{}$
   $\underline{2} \times \underline{7} = \underline{}$

4. $5 \times 3 = \underline{}$
   $\underline{} \times \underline{} = \underline{}$

5. $0 \times 4 = \underline{}$
   $\underline{} \times \underline{} = \underline{}$

6. $9 \times 5 = \underline{}$
   $\underline{} \times \underline{} = \underline{}$

7. $\begin{array}{r} 2 \\ \times 3 \\ \hline 6 \end{array}$ $\begin{array}{r} 3 \\ \times 2 \\ \hline 6 \end{array}$

8. $\begin{array}{r} 6 \\ \times 3 \\ \hline \end{array}$ $\begin{array}{r} 3 \\ \times 6 \\ \hline \end{array}$

9. $\begin{array}{r} 2 \\ \times 8 \\ \hline \end{array}$ $\begin{array}{r} 8 \\ \times 2 \\ \hline \end{array}$

10. $\begin{array}{r} 4 \\ \times 7 \\ \hline \end{array}$ $\begin{array}{r} \\ \times \\ \hline \end{array}$

11. $\begin{array}{r} 5 \\ \times 1 \\ \hline \end{array}$ $\begin{array}{r} \\ \times \\ \hline \end{array}$

12. $\begin{array}{r} 4 \\ \times 6 \\ \hline \end{array}$ $\begin{array}{r} \\ \times \\ \hline \end{array}$

13. $\begin{array}{r} 2 \\ \times 9 \\ \hline \end{array}$ $\begin{array}{r} \\ \times \\ \hline \end{array}$

14. $\begin{array}{r} 3 \\ \times 7 \\ \hline \end{array}$ $\begin{array}{r} \\ \times \\ \hline \end{array}$

15. $\begin{array}{r} 3 \\ \times 8 \\ \hline \end{array}$ $\begin{array}{r} \\ \times \\ \hline \end{array}$

SHARE YOUR THINKING

How does knowing that 9 rows of 4 equals 36 help you know what 4 rows of 9 equals?

 10-6 Ask your child to color rows on grid paper to show 4×8 and 8×4.

three hundred eighty-five **385**

Complete the multiplication table.

Number of Groups

1.

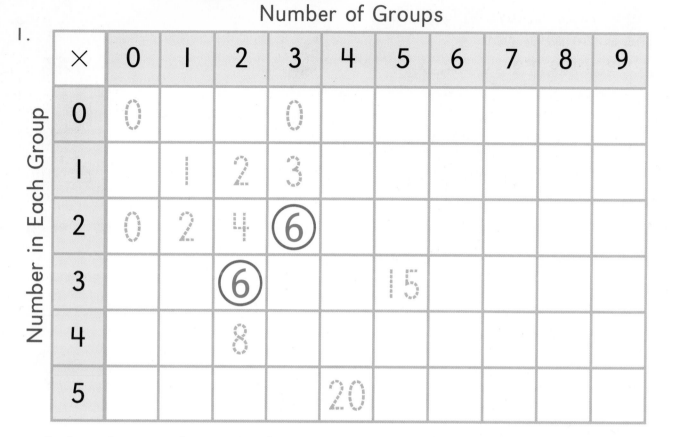

×	0	1	2	3	4	5	6	7	8	9
0	0			0						
1		1	2	3						
2	0	2	4	⑥						
3			⑥			15				
4			8							
5					20					

Number in Each Group

2. Color the products in the 1st row and the 1st column (blue).

3. Color the products in the 2nd row and the 2nd column (green).

Complete the sentences.

4. ____ times a number equals 0.

5. ____ times a number equals that number.

FINDING TOGETHER Write two multiplication sentences for each product.

6. 6 = _2_ × _3_   10 = _2_ × ___   20 = ___ × ___

    6 = _3_ × _2_   10 = ___ × _2_   20 = ___ × ___

7. 18 = ___ × ___   16 = ___ × ___   24 = ___ × ___

    18 = ___ × ___   16 = ___ × ___   24 = ___ × ___

MAKE UP YOUR OWN 8. Choose a product. Write two multiplication sentences in your Math Journal.

Name

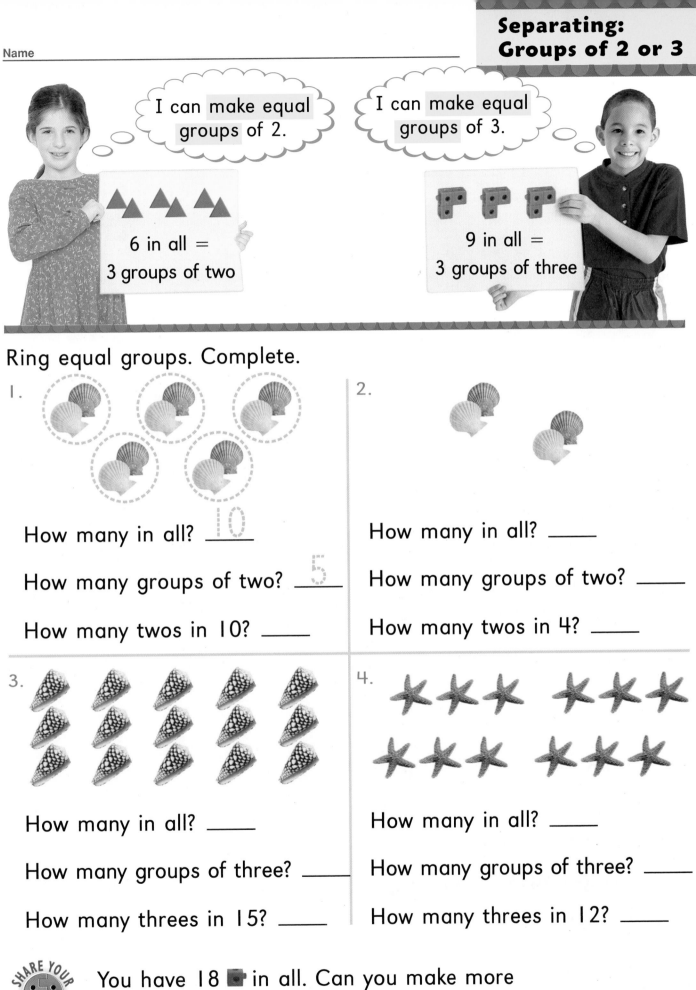

I can make equal groups of 2.

6 in all =
3 groups of two

I can make equal groups of 3.

9 in all =
3 groups of three

Ring equal groups. Complete.

1.

How many in all? ___10___

How many groups of two? ___5___

How many twos in 10? _____

2.

How many in all? _____

How many groups of two? _____

How many twos in 4? _____

3.

How many in all? _____

How many groups of three? _____

How many threes in 15? _____

4.

How many in all? _____

How many groups of three? _____

How many threes in 12? _____

SHARE YOUR THINKING You have 18 ▮ in all. Can you make more
groups of 2 or more groups of 3?

10-7 Put 16 pennies on a table and ask your child
to arrange them in equal groups of 2.

three hundred eighty-seven **387**

You can write a **division sentence** to find how many equal groups.

The answer in division is the **quotient**.

1. How many twos are in 14?

$$14 \div 2 = \underline{?}$$

$\underset{\text{divided by}}{} \quad \underset{\text{equals}}{}$

14 in groups of 2 equals ____. $\qquad$ $14 \div 2 = \underline{\quad}$

Model or draw. Complete each.

2. How many twos are in 12?

$$12 \div \underline{2} = \underline{\quad}$$

3. How many threes are in 6?

$$6 \div \underline{3} = \underline{\quad}$$

4. How many twos are in 16?

$$16 \div \underline{\quad} = \underline{\quad}$$

5. How many threes are in 21?

$$21 \div \underline{\quad} = \underline{\quad}$$

PROBLEM SOLVING Use a model.

6. There are 24 in all. 3 are in each stack. How many stacks are there?

____ stacks of 3

7. There are 8 in all. Each carton holds 2. How many cartons are there?

____ cartons of 2

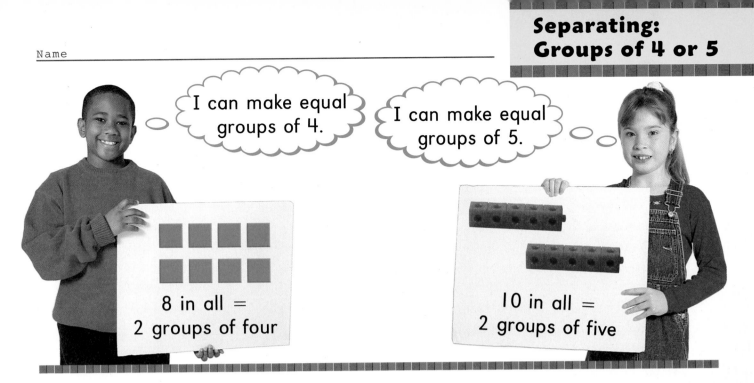

I can make equal groups of 4.

I can make equal groups of 5.

8 in all =
2 groups of four

10 in all =
2 groups of five

Ring equal groups. Complete.

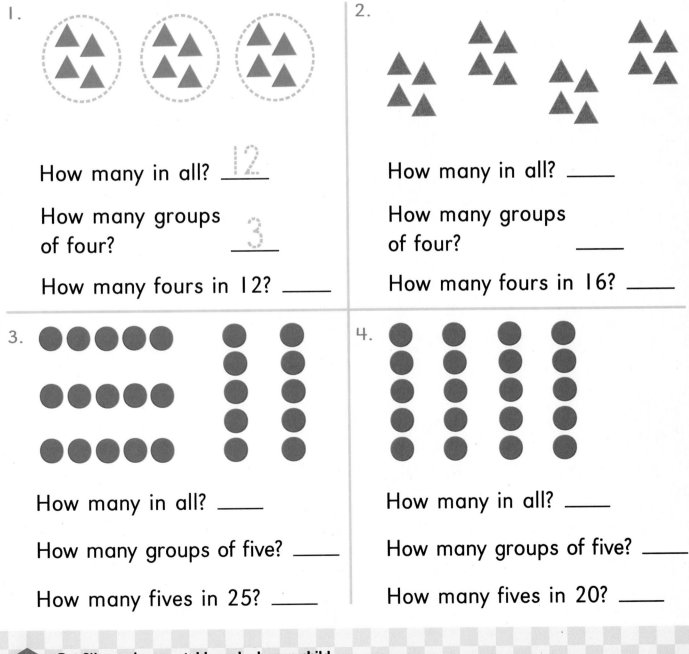

1.

How many in all? ___12___

How many groups
of four? ___3___

How many fours in 12? _____

2.

How many in all? _____

How many groups
of four? _____

How many fours in 16? _____

3.

How many in all? _____

How many groups of five? _____

How many fives in 25? _____

4.

How many in all? _____

How many groups of five? _____

How many fives in 20? _____

10-8   Put 24 pennies on a table and ask your child
to arrange them in equal groups of four.

three hundred eighty-nine   **389**

Draw or model each. Write a division sentence.

1. How many fives are in 35?

○○○○○ ○○○○○
○○○○○ ○○○○○
○○○○○ ○○○○○
   ○○○○○

35 ÷ 5 = _____

2. How many fours are in 24?

24 ÷ 4 = _____

3. How many fives are in 15?

15 ÷ _____ = _____

4. How many fours are in 20?

20 ÷ _____ = _____

5. How many fives are in 30?

_____ ÷ _____ = _____

6. How many fours are in 28?

_____ ÷ _____ = _____

CRITICAL THINKING

7. Write how many

twos in 2. _____   threes in 3. _____   fours in 4. _____

2 ÷ 2 = _____   3 ÷ 3 = _____   4 ÷ 4 = _____

8. If this pattern continues, what comes next? _____   _____

Name _____

Put 10 🪙 in groups of 5. How many groups did you make?

Put 13 🪙 in groups of 5. How many groups did you make?

| 1st | Ring equal groups. |

| 2nd | Write how many are left over. | __2__ groups of 5 | __2__ groups of 5 |
| | | __0__ left over | __3__ left over |

1. How many groups of 3?

_12_ erasers in all

____ groups of 3

____ left over

2. How many groups of 3?

____ erasers in all

____ groups of 3

____ left over

3. How many groups of 4?

____ erasers in all

____ groups of 4

____ left over

4. How many groups of 4?

____ erasers in all

____ groups of 4

____ left over

SHARE YOUR THINKING

How are 1 and 2 alike? How are they different?
How are 3 and 4 alike? How are they different?

Model and draw. Find how many groups
and how many left over.

1. How many groups of 3 in 9?

Show __9__ in groups of __3__.

☐☐☐   ☐☐☐

☐☐☐

____ threes ____ left over

2. How many groups of 3 in 10?

Show ____ in groups of ____.

____ threes ____ left over

3. How many groups of 2 in 12?

Show ____ in groups of ____.

____ twos ____ left over

4. How many groups of 2 in 13?

Show ____ in groups of ____.

____ twos ____ left over

5. How many groups of 5 in 15?

Show ____ in groups of ____.

____ fives ____ left over

6. How many groups of 5 in 18?

Show ____ in groups of ____.

____ fives ____ left over

PROBLEM SOLVING

7. Fran has 26 🏝️. She puts 5 on
each page. How many pages
can she fill? How many 🏝️
are left over for the next page?

_____
pages filled

_____
cards left over

Name _____

Thirteen snow domes are shared equally among four friends. How many does each receive? Tally to solve.

|| ||| || ||

Each receives __3__ snow domes.

When you share you can have some things **left over**.

There is __1__ left over.

Share equally. Tally to solve.

1. 20 pretzels
   5 friends

   |||| |||| |||| |||| ||||

   Each receives __4__.

   There are __0__ left over.

2. 22 grapes
   5 friends

   Each receives ____.

   There are ____ left over.

3. 8 pennies
   2 friends

   Each receives ____.

   There are ____ left over.

4. 9 sheets of paper
   2 friends

   Each receives ____.

   There is ____ left over.

TALK IT OVER Explain why there are leftovers in 2 and 4.

10-10 Ask your child to show how to share 17 pennies with 5 friends and to explain the number left over.

three hundred ninety-three **393**

Model or draw. Show an equal number in each row. Complete.

1. 6 🚚 in all

   3 equal rows

   _____ in each row

   _____ left over

2. 8 🚗 in all

   3 equal rows

   _____ in each row

   _____ left over

3. 12 🎸 in all

   2 equal rows

   _____ in each row

   _____ left over

4. 13 🐱 in all

   2 equal rows

   _____ in each row

   _____ left over

5. 8 🎁 in all

   4 equal rows

   _____ in each row

   _____ left over

6. 10 🐴 in all

   4 equal rows

   _____ in each row

   _____ left over

FINDING TOGETHER

Use ▓. Write how many in each row.
Describe each pattern.

7. 21 ▓ in 2 rows

   _____ in each row

   _____ left over

   31 ▓ in 3 rows

   _____ in each row

   _____ left over

   41 ▓ in 4 rows

   _____ in each row

   _____ left over

8. 22 ▓ in 2 rows

   _____ in each row

   _____ left over

   33 ▓ in 3 rows

   _____ in each row

   _____ left over

   44 ▓ in 4 rows

   _____ in each row

   _____ left over

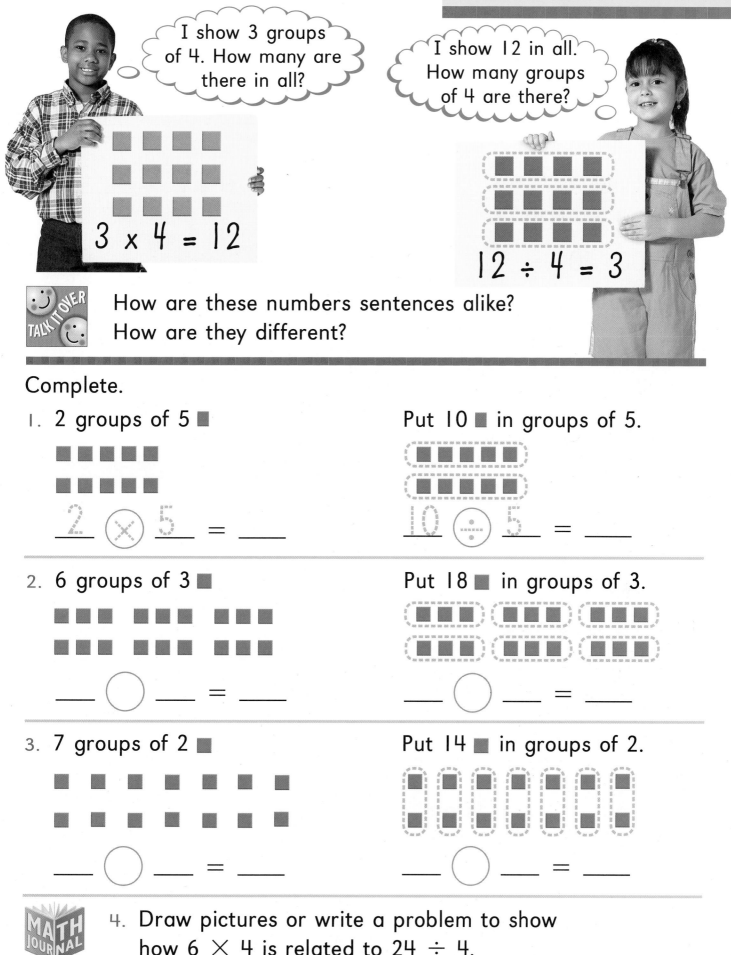

I show 3 groups of 4. How many are there in all?

I show 12 in all. How many groups of 4 are there?

$3 \times 4 = 12$

$12 \div 4 = 3$

TALK IT OVER

How are these numbers sentences alike?
How are they different?

Complete.

1. 2 groups of 5 ■

   2 ⊗ 5 = ___

   Put 10 ■ in groups of 5.

   10 ⊙ 5 = ___

2. 6 groups of 3 ■

   ___ ◯ ___ = ___

   Put 18 ■ in groups of 3.

   ___ ◯ ___ = ___

3. 7 groups of 2 ■

   ___ ◯ ___ = ___

   Put 14 ■ in groups of 2.

   ___ ◯ ___ = ___

MATH JOURNAL

4. Draw pictures or write a problem to show how 6×4 is related to $24 \div 4$.

10-11 Ask your child to show how 2 × 4 and 8 ÷ 4 are related.

three hundred ninety-five **395**

DO YOU REMEMBER ?

Solve the puzzle.

Across

1. 1 hundred 2 tens 3 ones

3. 420 + 100

5. 700, 701, _____

6. just before 200

8. 677   10. 465
   +126   +304

12. 600 + 80 + 2

14. 327   15. 283
    +227   + 84

17. 892 = 800 + _____ + 2

Down

1. 50 + 50 + 1

2. 192
   +197

3. 296
   +204

4. 162
   + 61

5. _____, 178, 278, 378

7. 115, 105, _____, 85

9. 106
   + 18

10. 594
    +119

11. 608
    +309

12. 260
    +390

13. ☐ + 111 = 196

16. 320 + 320

17. ☐ + 10 = 109

This page reviews the mathematical content presented in Chapter 9.

10

Multiplication and Division

You can use × and ÷ keys on a calculator to multiply and divide.

Press	Display	Press	Display
4 × 3 ▬	12.	8 ÷ 2 ▬	4.

Use a ▯ to find the products and quotients.

1. 1 × 10 =
1 × 1 0 ▬
[10.]

   2 × 10 = [ ]

   3 × 10 = [ ]

   4 × 10 = [ ]

   5 × 10 = [ ]

   6 × 10 = [ ]

   7 × 10 = [ ]

2. 45 ÷ 5 =
4 5 ÷ 5 ▬
[9.]

   40 ÷ 5 = [ ]

   35 ÷ 5 = [ ]

   30 ÷ 5 = [ ]

   25 ÷ 5 = [ ]

   20 ÷ 5 = [ ]

   15 ÷ 5 = [ ]

TALK IT OVER

Describe the patterns you see in 1 and 2.
Name the next two number sentences in each.

Use ✳ to multiply and / to divide on a computer.

Remember to use
Shift + [✳/8] to get ✳.

8 ✳ 5 means 8 × 5.

8/4 means 8 ÷ 4.

	Type	Press	Type	Press	Answer
1.	PRINT 2 ✳ 4	Enter	END	Enter	8
2.	PRINT 2 ✳ 5	Enter	END	Enter	_____
3.	PRINT 0 ✳ 7	Enter	END	Enter	_____
4.	PRINT 30/10	Enter	END	Enter	3
5.	PRINT 15/3	Enter	END	Enter	_____
6.	PRINT 18/2	Enter	END	Enter	_____

Use a 🖩 to find how many 3s you must subtract to get 0. Then complete.

7. $9 - 3 = 6$ ⟶ $6 - 3 = 3$ ⟶ $3 - 3 = 0$

There are __3__ threes in 9.

$9 ÷ 3 = $ ___

8. $12 - 3 = 9$ ⟶ _____

There are ___ threes in 12.

$12 ÷ 3 = $ ___

9. How is subtraction like division? How is it different?

Visit Sadlier on the Internet at
www.sadlier-oxford.com

Name _____

1. Read Plants grow 2 centimeters each day on Zork. On the first day, a plant was 2 cm tall. On what day will the plant be 10 cm tall?

Look for a pattern in each table you make.

Think
Write

Day	1st	2nd	3rd	4th	5th	6th	
Height in cm	2	4	6	8	10		

Check 10 centimeters tall on the __5th__ day

Model to check the answer.

2. Read A robot can make 5 spaceships each minute. How many minutes will it take a robot to make 35 of them?

Think
Write

Minutes	1	2	3					
Spaceships	5	10						

Check 35 spaceships in _____ minutes.

3. Read Og built 21 robots. Each day he sold 3 robots. On which day will there be 6 robots left?

$21 - 3 = 18$

Think
Write

Day	1st	2nd	3rd			
Number left	18	15				

Check 6 robots will be left on the _____ day.

10-13 Ask your child to describe the pattern she/he found in each table made to solve problems 1-3.

three hundred ninety-nine **399**

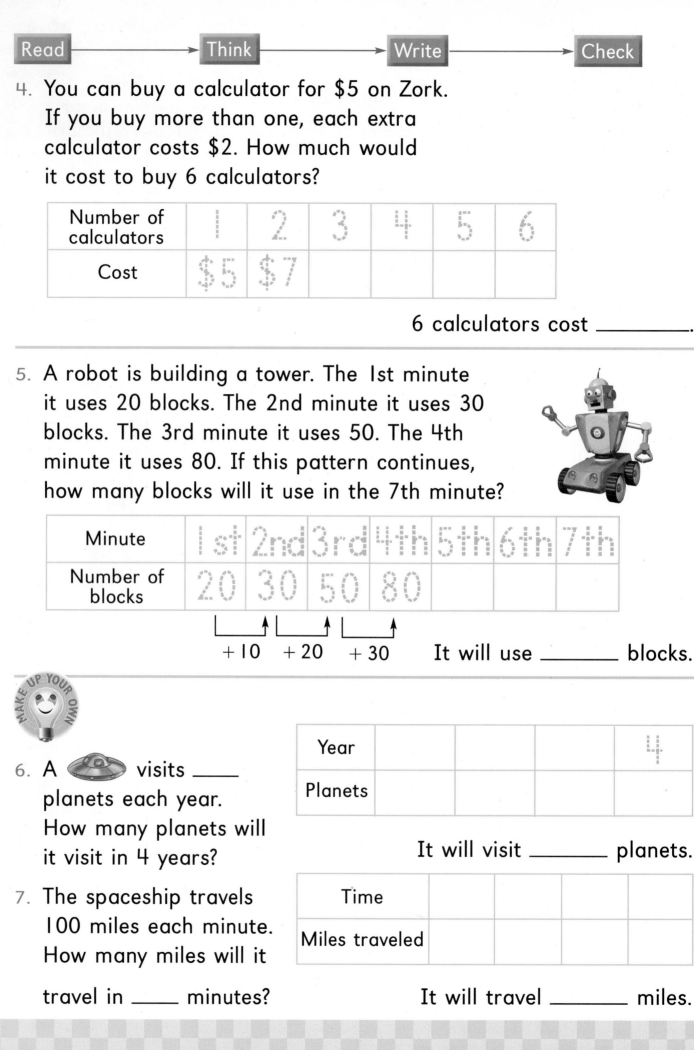

4. You can buy a calculator for $5 on Zork. If you buy more than one, each extra calculator costs $2. How much would it cost to buy 6 calculators?

Number of calculators	1	2	3	4	5	6
Cost	$5	$7				

6 calculators cost _____.

5. A robot is building a tower. The 1st minute it uses 20 blocks. The 2nd minute it uses 30 blocks. The 3rd minute it uses 50. The 4th minute it uses 80. If this pattern continues, how many blocks will it use in the 7th minute?

Minute	1st	2nd	3rd	4th	5th	6th	7th
Number of blocks	20	30	50	80			

+10   +20   +30

It will use _____ blocks.

MAKE UP YOUR OWN

6. A 🛸 visits _____ planets each year. How many planets will it visit in 4 years?

Year				4
Planets				

It will visit _____ planets.

7. The spaceship travels 100 miles each minute. How many miles will it travel in _____ minutes?

Time				
Miles traveled				

It will travel _____ miles.

PROBLEM SOLVING

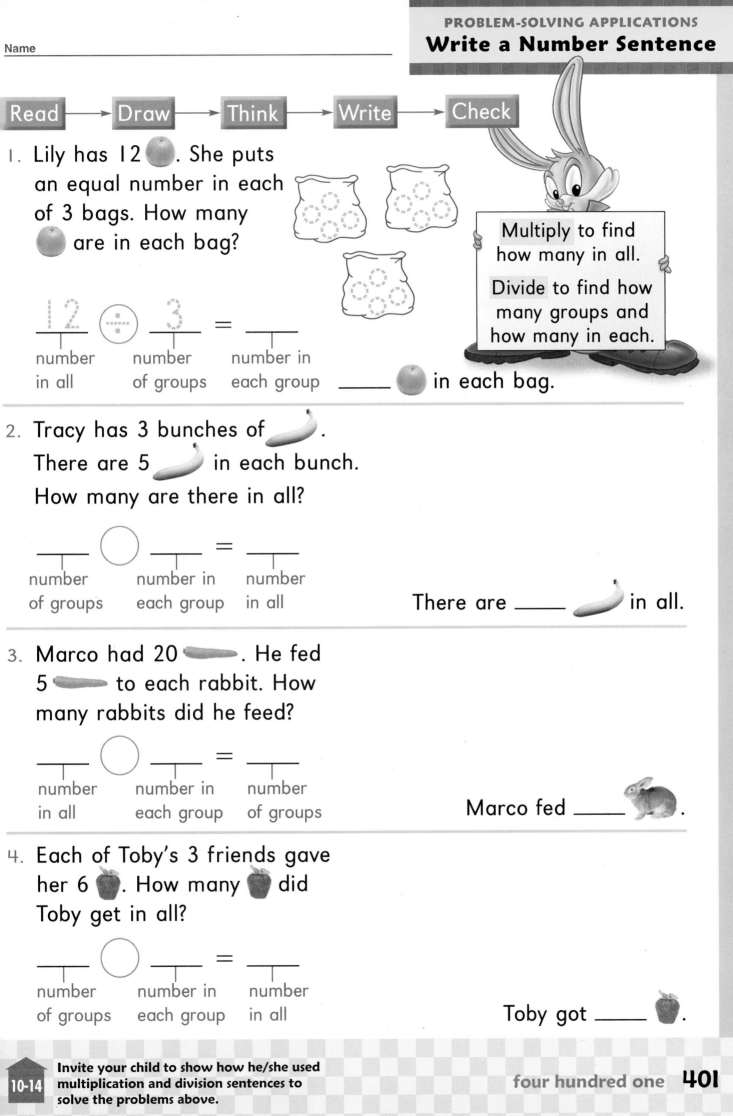

Name _____

PROBLEM-SOLVING APPLICATIONS
Write a Number Sentence

Read ⟶ Draw ⟶ Think ⟶ Write ⟶ Check

1. Lily has 12 🍊. She puts an equal number in each of 3 bags. How many 🍊 are in each bag?

   Multiply to find how many in all.

   Divide to find how many groups and how many in each.

   $$\underset{\substack{\text{number} \\ \text{in all}}}{12} \; \div \; \underset{\substack{\text{number} \\ \text{of groups}}}{3} \; = \; \underset{\substack{\text{number in} \\ \text{each group}}}{}$$

   _____ 🍊 in each bag.

2. Tracy has 3 bunches of 🍌. There are 5 🍌 in each bunch. How many are there in all?

   $$\underset{\substack{\text{number} \\ \text{of groups}}}{} \; \bigcirc \; \underset{\substack{\text{number in} \\ \text{each group}}}{} \; = \; \underset{\substack{\text{number} \\ \text{in all}}}{}$$

   There are _____ 🍌 in all.

3. Marco had 20 🥕. He fed 5 🥕 to each rabbit. How many rabbits did he feed?

   $$\underset{\substack{\text{number} \\ \text{in all}}}{} \; \bigcirc \; \underset{\substack{\text{number in} \\ \text{each group}}}{} \; = \; \underset{\substack{\text{number} \\ \text{of groups}}}{}$$

   Marco fed _____ 🐰.

4. Each of Toby's 3 friends gave her 6 🍎. How many 🍎 did Toby get in all?

   $$\underset{\substack{\text{number} \\ \text{of groups}}}{} \; \bigcirc \; \underset{\substack{\text{number in} \\ \text{each group}}}{} \; = \; \underset{\substack{\text{number} \\ \text{in all}}}{}$$

   Toby got _____ 🍎.

10-14 Invite your child to show how he/she used multiplication and division sentences to solve the problems above.

four hundred one **401**

Use a strategy you have learned.

STRATEGY FILE

Draw a Picture
Make a Table
Write a Number Sentence
Use a Graph

5. Bob has 4 vases. There are 2 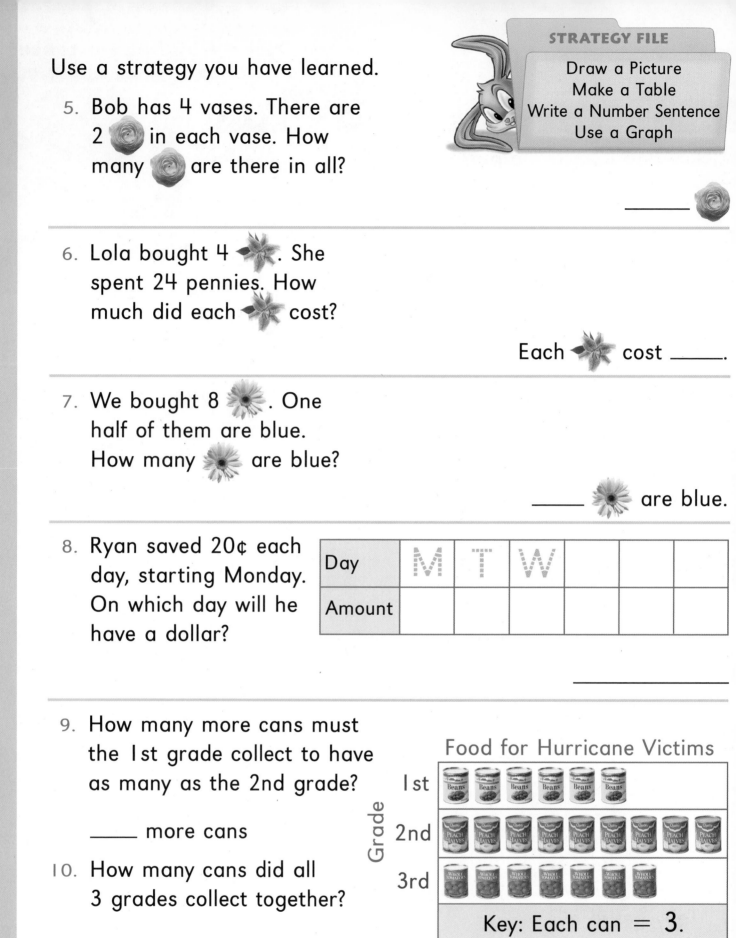 in each vase. How many are there in all?

_____

6. Lola bought 4 . She spent 24 pennies. How much did each cost?

Each cost _____.

7. We bought 8 . One half of them are blue. How many are blue?

_____ are blue.

8. Ryan saved 20¢ each day, starting Monday. On which day will he have a dollar?

Day	M	T	W			
Amount						

_____

9. How many more cans must the 1st grade collect to have as many as the 2nd grade?

_____ more cans

10. How many cans did all 3 grades collect together?

_____ cans

11. How many more cans must these grades collect to have a total of 100 cans?

_____ cans

Food for Hurricane Victims

Grade	
1st	Beans Beans Beans Beans Beans Beans
2nd	Peach Halves ×9
3rd	Whole Tomatoes ×7

Key: Each can = 3.

Name _____

Find the product. You can draw or model.

1. 3   2   3   4   3   5   2
   ×2   ×0   ×9   ×2   ×3   ×5   ×9
   ‾6

2. 5   2   5   4   2   5   4
   ×1   ×2   ×8   ×5   ×6   ×2   ×3

Multiply. Then change the order.

3. 7 × 2 = ____
   2 × ___ = ___

4. 6 × 3 = ____
   ___ × ___ = ___

5. 6 × 5 = ____
   ___ × ___ = ___

6. 3 × 5 = ____
   ___ × ___ = ___

7. 8 × 4 = ____
   ___ × ___ = ___

8. 7 × 3 = ____
   ___ × ___ = ___

Write how many. You can use models.

9. 16 in all
   groups of 4
   16 ÷ ___ = ___

10. 12 in all
    groups of 3
    12 ÷ ___ = ___

11. 20 in all
    groups of 5
    20 ÷ ___ = ___

PROBLEM SOLVING Use models.

12. You put 4 on each of 6 pages. How many in all?
    ____ in all

13. Three friends share 8 equally. How many does each one get? How many are left over?
    ____ each and ____ left over

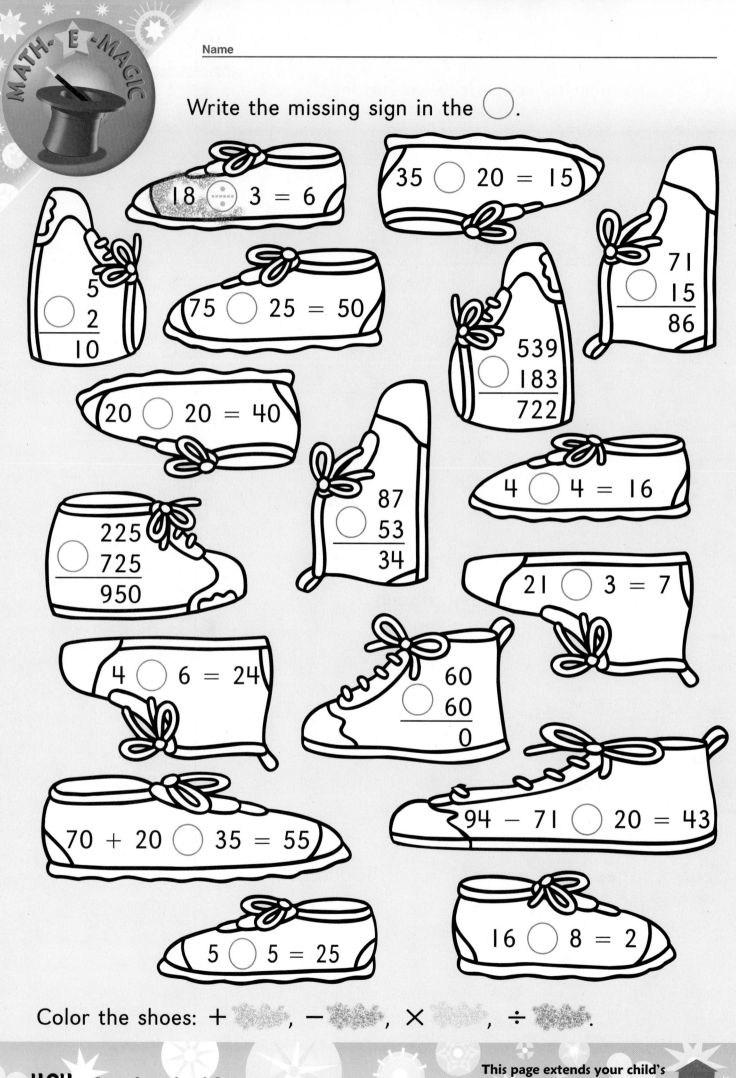

Write the missing sign in the ◯.

$18 \,\bigcirc\, 3 = 6$

$35 \,\bigcirc\, 20 = 15$

$\begin{array}{r} 5 \\ \bigcirc\ 2 \\ \hline 10 \end{array}$

$75 \,\bigcirc\, 25 = 50$

$\begin{array}{r} 71 \\ \bigcirc\ 15 \\ \hline 86 \end{array}$

$\begin{array}{r} 539 \\ \bigcirc\ 183 \\ \hline 722 \end{array}$

$20 \,\bigcirc\, 20 = 40$

$\begin{array}{r} 87 \\ \bigcirc\ 53 \\ \hline 34 \end{array}$

$4 \,\bigcirc\, 4 = 16$

$\begin{array}{r} 225 \\ \bigcirc\ 725 \\ \hline 950 \end{array}$

$21 \,\bigcirc\, 3 = 7$

$4 \,\bigcirc\, 6 = 24$

$\begin{array}{r} 60 \\ \bigcirc\ 60 \\ \hline 0 \end{array}$

$70 + 20 \,\bigcirc\, 35 = 55$

$94 - 71 \,\bigcirc\, 20 = 43$

$5 \,\bigcirc\, 5 = 25$

$16 \,\bigcirc\, 8 = 2$

Color the shoes: $+$ ░░░ , $-$ ░░░ , $\times$ ░░░ , $\div$ ░░░ .

This page extends your child's understanding of addition, subtraction, multiplication, and division.

10

1. Use models or drawings to show each of these. Explain what happens.

   • Change the order of the numbers you multiply.

   The products are _____.

   • Multiply 2, 3, 4, 5, or 10 by 1.

   The products are _____.

   • Multiply 2, 3, 4, 5, or 10 by 0.

   The products are _____.

2. Use a ten-frame and 🪙. Add, multiply, or divide to solve each.

   • Stack 4 🪙 in each of 5 □. How many 🪙 in all?

   • Share 18¢ equally with 3 □. How much is in each □?

   _____   _____

   Choose one of these projects. Use a separate sheet of paper.

3. Make a book about groups of 5. Make a table for each.

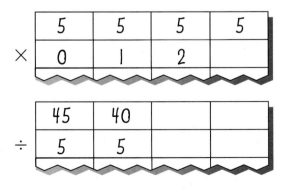

5	5	5	5
× 0	1	2	

45	40		
÷ 5	5		

4. Gwen predicts that the product of two even numbers is even. Give 3 examples to check her prediction. Then predict and check the following.

   • the product of two odd numbers

   • the product of an even and an odd number

10

This page provides a variety of informal assessment opportunities in order to measure your child's understanding of Chapter 10.

four hundred five **405**

Name _____

Find the sum. Then find the product.

1. $3 + 3 + 3 + 3 + 3 =$ ____ $2 + 2 + 2 + 2 =$ ____

   5 $\times$ ____ $=$ ____   ____ $\times$ ____ $=$ ____

Multiply. You can draw or model.

2.
$$\begin{array}{r} 4 \\ \times 1 \\ \hline \end{array} \quad \begin{array}{r} 2 \\ \times 3 \\ \hline \end{array} \quad \begin{array}{r} 3 \\ \times 4 \\ \hline \end{array} \quad \begin{array}{r} 2 \\ \times 4 \\ \hline \end{array} \quad \begin{array}{r} 5 \\ \times 9 \\ \hline \end{array} \quad \begin{array}{r} 5 \\ \times 0 \\ \hline \end{array} \quad \begin{array}{r} 2 \\ \times 8 \\ \hline \end{array}$$

3.
$$\begin{array}{r} 4 \\ \times 9 \\ \hline \end{array} \quad \begin{array}{r} 4 \\ \times 7 \\ \hline \end{array} \quad \begin{array}{r} 5 \\ \times 7 \\ \hline \end{array} \quad \begin{array}{r} 3 \\ \times 8 \\ \hline \end{array} \quad \begin{array}{r} 3 \\ \times 3 \\ \hline \end{array} \quad \begin{array}{r} 4 \\ \times 7 \\ \hline \end{array} \quad \begin{array}{r} 4 \\ \times 6 \\ \hline \end{array}$$

Multiply. Then change the order.

4. $8 \times 3 =$ ____ | 5. $7 \times 4 =$ ____ | 6. $9 \times 5 =$ ____

3 $\times$ ____ $=$ ____ | ____ $\times$ ____ $=$ ____ | ____ $\times$ ____ $=$ ____

Write how many.

7. 10 ⚃ in all | 8. 18 ⚃ in all | 9. 20 ⚃ in all
   groups of 5 | groups of 3 | groups of 4

10 $\div$ ____ $=$ ____ | ____ $\div$ ____ $=$ ____ | ____ $\div$ ____ $=$ ____

PROBLEM SOLVING   Use models.

10. There are 6 🍌 in a bunch. How many are in 2 bunches?

    ____ 🍌

11. Four friends share 30 🍎 equally. How many does each one get? How many are left over?

    ____ 🍎 each ____ left over

This page is a formal assessment of your child's understanding of the content presented in Chapter 10.

10

Place Value to 1000 and Subtraction

Bolts	20	20	5	5	10
Streaks	10	10	20	10	

CRITICAL THINKING

How many points in the fifth race do the Streaks need to score to win the tournament?

Dear Family,

Today your child began Chapter 11. As she/he studies subtraction of 3-digit numbers, you may want to read the poem below, which was read in class. Have your child talk about some of the math ideas pictured on page 407.

Look for the 🏠 at the bottom of each skills lesson. The suggestion on the page gives you an opportunity to improve your child's understanding of math and to reinforce her/his math language. You may want to have dollars, dimes, pennies and other countables available for your child to use throughout this chapter.

Home Activity

3-Digit Sack Race

Try this activity with your child. Prepare 3 sets of number cards (numbers 1–9) and place each set in a paper bag. Mark the bags and arrange them on a surface. (See diagram below.) Ask your child to draw two number cards from each bag and to place the cards in front of the bag from which they were drawn. Give him/her one minute to write two 3-digit numbers made from the numbers on the cards on a separate sheet of paper. Have your child draw a star beside the greater number.

Home Reading Connection

THE RACE IS ON

Just look at them go
In a quick scrambling flash,
Like lightning they're running
The hundred yard dash.

Two teams are competing,
(they've practiced for weeks—)
twenty feet go stampeding;
five *Bolts* and five *Streaks*.

Down lanes they go racing
(each mad dashing crew)
logging 1,000 yards
to capture the blue!

Rebecca Kai Dotlich

Name

The helpers for field day made beanbags.
Use the first container in each row to estimate.

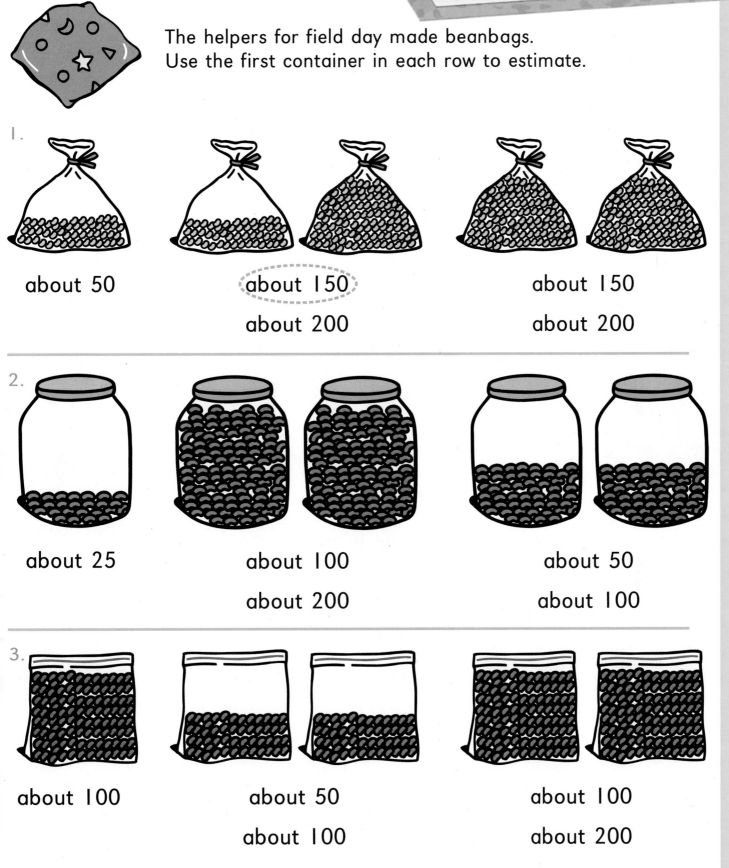

1.

about 50

about 150

about 200

about 150

about 200

2.

about 25

about 100

about 200

about 50

about 100

3.

about 100

about 50

about 100

about 100

about 200

Draw containers like the ones above. Explain how you estimate the number of beans in each.

CONNECTIONS

Field-Day Activities

Sack Race			Leaping Lizards		
Diana	Rosa	Greta	Ling	Jed	Asa
Ling	Asa	Pablo	Rosa	Tara	

Write the names on the diagram to sort the data.

1.

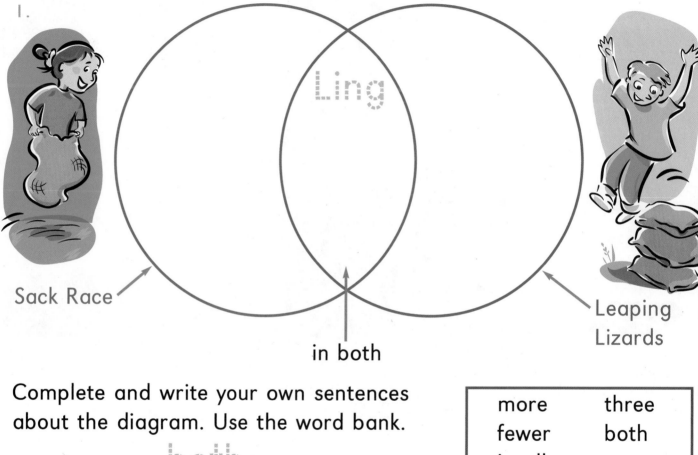

Sack Race → ↑ ← Leaping Lizards

Ling

in both

Complete and write your own sentences about the diagram. Use the word bank.

more	three
fewer	both
in all	

2. Ling was in ___both___ races.

_____

_____

_____

_____

 Survey 10 friends about two sports. Draw a diagram to show your data. Write about it.

You can put this in your
Math Portfolio. PORTFOLIO

Name _____

Use models to compare.

I hundred is less than 2 hundreds.

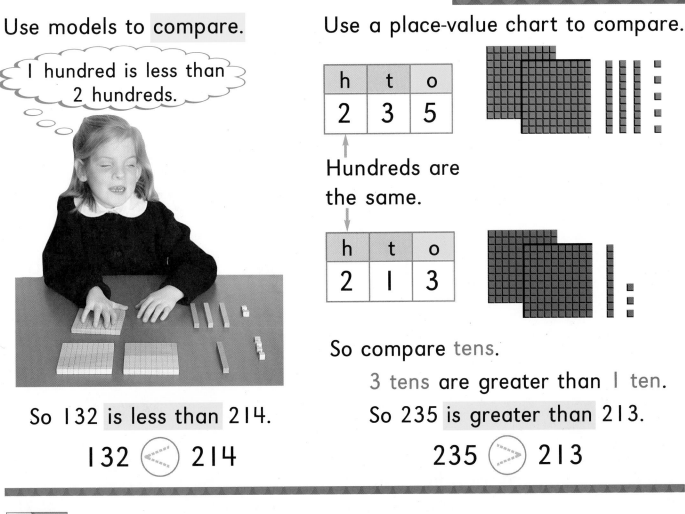

So 132 is less than 214.

132 ⟨<⟩ 214

Use a place-value chart to compare.

h	t	o
2	3	5

↑
Hundreds are the same.
↓

h	t	o
2	1	3

So compare tens.

3 tens are greater than 1 ten.

So 235 is greater than 213.

235 ⟨>⟩ 213

TALK IT OVER Which place should you compare first: hundreds, tens, or ones? Why?

Compare. Check using models.

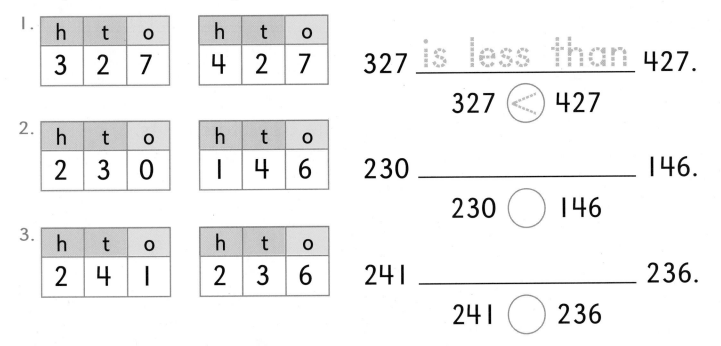

1.

h	t	o
3	2	7

h	t	o
4	2	7

327 __is less than__ 427.

327 ⟨<⟩ 427

2.

h	t	o
2	3	0

h	t	o
1	4	6

230 _____ 146.

230 ◯ 146

3.

h	t	o
2	4	1

h	t	o
2	3	6

241 _____ 236.

241 ◯ 236

11-1 Write two 3-digit numbers for your child and ask him/her to show which number is greater and which number is less.

four hundred eleven **411**

Compare. Write < or >.

1. 321 ◯< 421   493 ◯ 539   671 ◯ 598

2. 357 ◯ 306   587 ◯ 569   244 ◯ 255

3. 941 ◯ 914   653 ◯ 673   739 ◯ 722

4. 196 ◯ 241   892 ◯ 792   524 ◯ 349

5. 439 ◯ 437   284 ◯ 286   903 ◯ 905

6. 763 ◯ 637   736 ◯ 763   736 ◯ 737

PROBLEM SOLVING

7. P. J., Ita, and Feng each compared different numbers. P. J. and Feng used the tens place. Ita and Feng both compared numbers with less than 4 hundreds. Match each with their numbers.

_____ 347 ◯ 374

_____ 561 ◯ 570

_____ 216 ◯ 219

CHALLENGE Use 0 1 2 3 to write true statements.

8. 215 < 321   215 < _____   215 < _____

9. 215 < _____   215 > _____   215 > _____

10. 215 > _____   215 > _____   215 > _____

I count back to name the number just before.

I count on to name the number just after.

349 350 351 352 353 354 355 356 357 358

349 is just before 350. 357 is just after 356.

354 is between 353 and 355.

Complete the chart.

1. Just Before

_6 4 1_, 642

_____, 717

_____, 222

_____, 710

2. Between

399, _____, 401

804, _____, 806

111, _____, 113

460, _____, 462

3. Just After

902, _____

198, _____

329, _____

875, _____

Ring each number that is greater than the number on the ●.

4. 230

450   146

203   280

5. 791

789   801

798   699

Ring each number that is less than the number on the ⚾.

6. 403

480   398

401   510

7. 650

560   648

605   655

11-2 Name a number, then ask your child to tell you the number just before and after it.

four hundred thirteen 413

To list numbers in order from least to greatest, compare hundreds, then tens, then ones.

408   618
   280   448

2 hundreds are less than 4 hundreds and 6 hundreds.

280   408   448   618
least   0 tens is less than 4 tens   greatest

Write the numbers from least to greatest.

1.
138 342 703 243

138 , 243 , 342 , 703

2.
469 884 356 785

356 , ____ , ____ , ____

3.
177 117 777 770

____ , ____ , ____ , ____

4.
804 784 840 748

____ , ____ , ____ , ____

5.
511 251 501 295

____ , ____ , ____ , ____

6.
169 184 156 105

____ , ____ , ____ , ____

7. Which team scored the greatest number of points? _____

8. Which team scored the least number of points? _____

Team	Points Scored
Blue	707
White	607
Gold	670
Red	770

Name _____

Subtract 354 − 251 = __?__

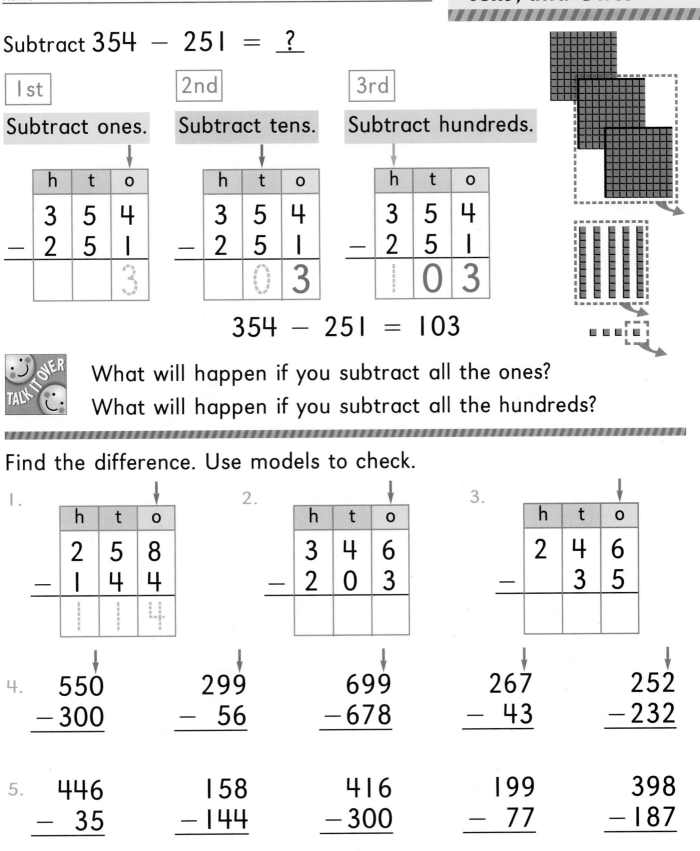

1st	2nd	3rd
Subtract ones.	Subtract tens.	Subtract hundreds.

h	t	o
3	5	4
− 2	5	1
		3

h	t	o
3	5	4
− 2	5	1
	0	3

h	t	o
3	5	4
− 2	5	1
	0	3

354 − 251 = 103

TALK IT OVER

What will happen if you subtract all the ones?

What will happen if you subtract all the hundreds?

Find the difference. Use models to check.

1.

h	t	o
2	5	8
− 1	4	4

2.

h	t	o
3	4	6
− 2	0	3

3.

h	t	o
2	4	6
−	3	5

4.
```
  550        299        699        267        252
− 300       −  56      − 678      −  43      − 232
```

5.
```
  446        158        416        199        398
−  35       − 144      − 300      −  77      − 187
```

MATH JOURNAL

6. Write what happens when you subtract
385 from 385 and 0 from 385.

11-3 Ask your child to subtract 566 − 333
and 566 − 303 and to subtract
the same amounts using money.

four hundred fifteen **415**

Subtract.

Subtracting dollars and cents is like subtracting 3-digit numbers.

1.
$7.32
− 0.21
$7.11

$6.25
− 0.13

$8.67
− 0.46

2.
$5.06
− 4.05

$8.75
− 2.31

$4.68
− 1.52

$7.89
− 2.45

3.
249
−119

199
− 99

756
−324

619
−317

985
− 74

4.
$8.30
− 1.10

$6.59
− 0.18

$2.96
− 1.11

$9.70
− 1.60

$4.56
− 1.23

5.
505
−404

674
−424

756
− 42

785
− 52

987
−221

PROBLEM SOLVING

6. There were 407 people at the park.
At dusk 205 people left.
How many stayed? _____ stayed.

7. Two trains carried 465 passengers.
153 passengers took the express.
How many took the local? _____ took the local.

MENTAL MATH

8. Subtract mentally. Write what comes next.

989 − 101   989 − 202   989 − 303   _____

= _____   = _____   = _____   = _____

Name _____

I count back to subtract mentally.

$280 - 1 =$ _279_

278   279   280   281

201	202	203	204	205	206	207	208	209	210
211	212	213	214	215	216	217	218	219	220
221	222	223	224	225	226	227	228	229	230
231	232	233	234	235	236	237	238	239	240
241	242	243	244	245	246	247	248	249	250
251	252	253	254	255	256	257	258	259	260
261	262	263	264	265	266	267	268	269	270
271	272	273	274	275	276	277	278	279	280
281	282	283	284	285	286	287	288	289	290
291	292	293	294	295	296	297	298	299	300

$280 - 10 =$ _270_

270 271 272 273 274 275 276 277 278 279 280

Write the number 1 less than each.

1. _325_ 326   ____ 452   ____ 613   ____ 807

Write the number 10 less than each.

2. _556_ 566   ____ 187   ____ 790   ____ 319

Write the number 100 less than each.

3. _576_ 676   ____ 505   ____ 400   ____ 467

Continue the pattern. Write the rule.

4. 627, 626, 625, ____   Rule: _____

5. 806, 706, 606, ____   Rule: _____

6. 714, 704, 694, ____   Rule: _____

Subtract.

7. $910 - 1 =$ ____ $587 - 10 =$ ____ $139 - 100 =$ ____

8. $233 - 1 =$ ____ $417 - 10 =$ ____ $112 - 100 =$ ____

Name _____

Use the rule to complete each table.
Color products green
and quotients blue.

Rule ÷ 2	
2	l
8	
16	
18	

Rule × 3	
0	0
4	
7	
8	

Rule ÷ 5	
10	
20	
35	
45	

Rule × 4	
1	
5	
4	
7	

Rule ÷ 3	
6	
9	
21	
27	

Rule × 5	
2	
4	
5	
7	

Rule × 5	
3	
6	
8	
9	

Rule ÷ 4	
12	
24	
32	
36	

Rule × 2	
3	
5	
6	
7	

This page reviews the mathematical
content presented in Chapter 10.
11

Name _____

Sometimes when you subtract, you do not have enough tens.

1 hundred 1 ten – 4 tens = __?__

> 1 hundred = 10 tens
> So 1 hundred 1 ten = 11 tens.

| 1st | Ask yourself, "Are there enough tens to subtract?" |

| 2nd | Regroup 1 hundred as 10 tens. |

| 3rd | Subtract. |

1 hundred 1 ten – 4 tens

= __11__ tens – 4 tens

= __7__ tens

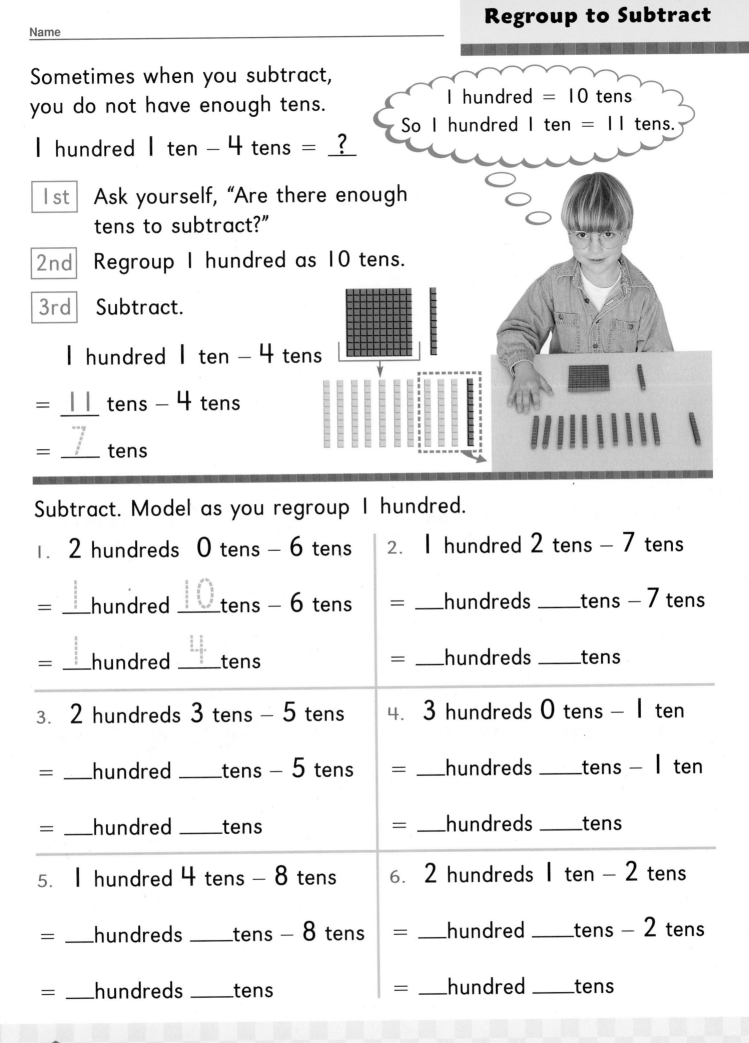

Subtract. Model as you regroup 1 hundred.

1. 2 hundreds 0 tens – 6 tens

= __1__ hundred __10__ tens – 6 tens

= __1__ hundred __4__ tens

2. 1 hundred 2 tens – 7 tens

= ___ hundreds ___ tens – 7 tens

= ___ hundreds ___ tens

3. 2 hundreds 3 tens – 5 tens

= ___ hundred ___ tens – 5 tens

= ___ hundred ___ tens

4. 3 hundreds 0 tens – 1 ten

= ___ hundreds ___ tens – 1 ten

= ___ hundreds ___ tens

5. 1 hundred 4 tens – 8 tens

= ___ hundreds ___ tens – 8 tens

= ___ hundreds ___ tens

6. 2 hundreds 1 ten – 2 tens

= ___ hundred ___ tens – 2 tens

= ___ hundred ___ tens

11-5 Provide hundreds number cards from 100–900 and tell your child to explain how to subtract by regrouping hundreds as tens.

When there are not enough ones, regroup 1 ten as 10 ones.

$$\begin{array}{r} \overset{5}{\cancel{6}} \quad \overset{15}{\cancel{5}} \\ 3 \text{ hundreds } \cancel{6} \text{ tens } \cancel{5} \text{ ones} \\ -\ 1 \text{ hundred } 3 \text{ tens } 8 \text{ ones} \\ \hline 2 \text{ hundreds } 2 \text{ tens } 7 \text{ ones} \end{array}$$

Why is regrouping not needed to subtract 1 hundred 2 tens 1 one from 5 hundreds 4 tens 3 ones?

Find the difference.

1.   4 hundreds 8 tens 2 ones
   − 1 hundred 4 tens 9 ones

   ____ hundreds ____ tens ____ ones

2.   7 hundreds 8 tens 3 ones
   − 5 hundreds 2 tens 6 ones

   ____ hundreds ____ tens ____ ones

3.   5 hundreds 5 tens 5 ones
   − 3 hundreds 0 tens 9 ones

   ____ hundreds ____ tens ____ ones

4.   9 hundreds 7 tens 0 ones
   − 4 hundreds 1 ten 7 ones

   ____ hundreds ____ tens ____ ones

5.   3 hundreds 6 tens 2 ones
   − 1 hundred 2 tens 6 ones

   ____ hundreds ____ tens ____ ones

6.   2 hundreds 4 tens 4 ones
   − 1 hundred 1 ten 9 ones

   ____ hundred ____ tens ____ ones

CHALLENGE Use models. Show how to regroup to subtract.

7.   3 hundreds 4 tens 5 ones
   − 1 hundred 6 tens 7 ones

8.   4 hundreds 3 tens 0 ones
   − 2 hundreds 8 tens 2 ones

Name

Subtract: $246 - 138 = \underline{\quad?\quad}$

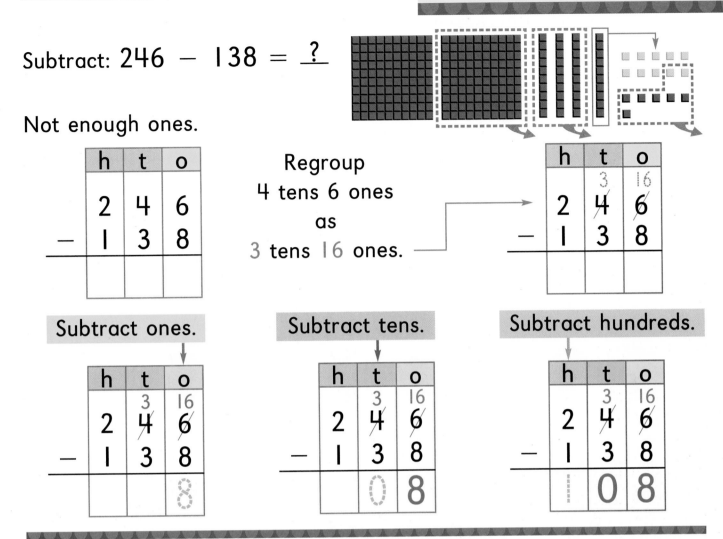

Not enough ones.

h	t	o
2	4	6
− 1	3	8

Regroup
4 tens 6 ones
as
3 tens 16 ones.

h	t	o
	³4̸	¹⁶6
2	4	6
− 1	3	8

Subtract ones.

h	t	o
	³4̸	¹⁶6̸
2	4	6
− 1	3	8
		8

Subtract tens.

h	t	o
	³4̸	¹⁶6̸
2	4	6
− 1	3	8
	0	8

Subtract hundreds.

h	t	o
	³4̸	¹⁶6̸
2	4	6
− 1	3	8
1	0	8

Subtract. Regroup tens as ones when needed.

1.

h	t	o
	⁷8̸	¹¹1̸
4	8	1
− 4	3	2
	4	9

2.

h	t	o
3	4	7
− 2	1	9

3.

h	t	o
1	8	0
−	4	4

4. $\begin{array}{r} 312 \\ -106 \\ \hline \end{array}$ $\qquad$ $\begin{array}{r} 242 \\ -\ 15 \\ \hline \end{array}$ $\qquad$ $\begin{array}{r} 361 \\ -228 \\ \hline \end{array}$ $\qquad$ $\begin{array}{r} 470 \\ -236 \\ \hline \end{array}$ $\qquad$ $\begin{array}{r} 193 \\ -176 \\ \hline \end{array}$

5. Check 4 by adding. Do you need to regroup when you check?

Subtract. ✔ when no regrouping is needed.

1.
$$\begin{array}{r} \overset{2\ 12}{4\cancel{3}2} \\ -\ \ 29 \\ \hline 403 \end{array}$$
$$\begin{array}{r} 263 \\ -\ 14 \\ \hline \end{array}$$
$$\begin{array}{r} 781 \\ -645 \\ \hline \end{array}$$
$$\begin{array}{r} 810 \\ -305 \\ \hline \end{array}$$
$$\begin{array}{r} 788 \\ -347 \\ \hline \end{array}$$

2.
$$\begin{array}{r} 651 \\ -239 \\ \hline \end{array}$$
$$\begin{array}{r} 392 \\ -111 \\ \hline \end{array}$$
$$\begin{array}{r} 267 \\ -\ 55 \\ \hline \end{array}$$
$$\begin{array}{r} 542 \\ -418 \\ \hline \end{array}$$
$$\begin{array}{r} 966 \\ -749 \\ \hline \end{array}$$

3.
$$\begin{array}{r} 992 \\ -517 \\ \hline \end{array}$$
$$\begin{array}{r} 783 \\ -526 \\ \hline \end{array}$$
$$\begin{array}{r} 977 \\ -847 \\ \hline \end{array}$$
$$\begin{array}{r} 851 \\ -534 \\ \hline \end{array}$$
$$\begin{array}{r} 530 \\ -122 \\ \hline \end{array}$$

4.
$$\begin{array}{r} 820 \\ -408 \\ \hline \end{array}$$
$$\begin{array}{r} 462 \\ -\ 49 \\ \hline \end{array}$$
$$\begin{array}{r} 555 \\ -337 \\ \hline \end{array}$$
$$\begin{array}{r} 281 \\ -158 \\ \hline \end{array}$$
$$\begin{array}{r} 379 \\ -\ 56 \\ \hline \end{array}$$

PROBLEM SOLVING

5. There are 444 people watching the soccer game. 229 people are watching the relay race. How many more people are watching the soccer game than the relay race?

_____ people

CRITICAL THINKING

6. I am less than 465 − 337.
I am more than 343 − 217.

I am _____.

7. I am more than 790 − 473.
I am less than 587 − 268.

I am _____.

Name

Subtract: 204 − 41 = ?

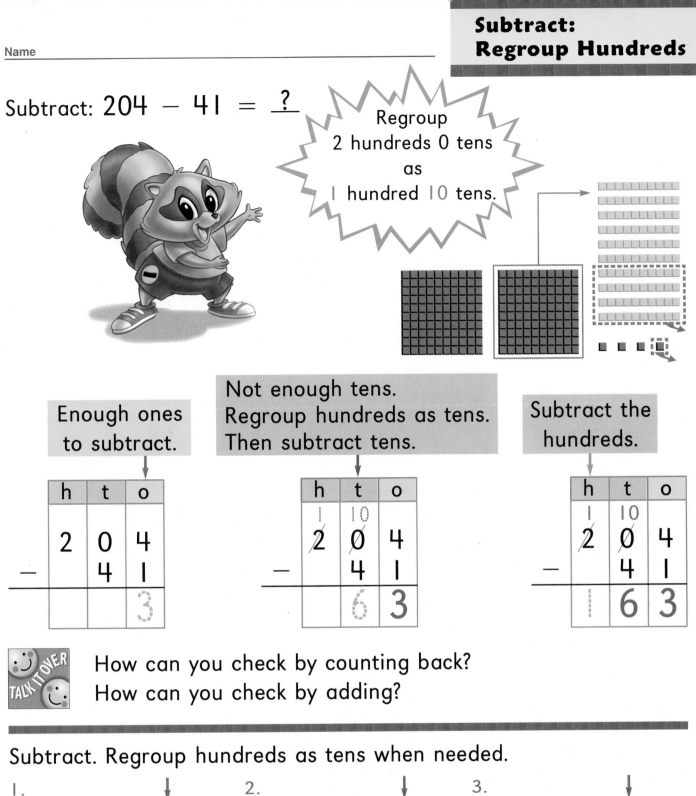

Regroup
2 hundreds 0 tens
as
1 hundred 10 tens.

Enough ones to subtract.

h	t	o
2	0	4
−	4	1
		3

Not enough tens. Regroup hundreds as tens. Then subtract tens.

h	t	o
2	0 (10)	4
−	4	1
	6	3

Subtract the hundreds.

h	t	o
2 (1)	0 (10)	4
−	4	1
1	6	3

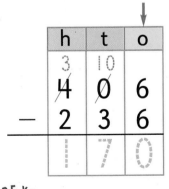 How can you check by counting back?
How can you check by adding?

Subtract. Regroup hundreds as tens when needed.

1.

h	t	o
4 (3)	0 (10)	6
− 2	3	6
1	7	0

2.

h	t	o
3	4	7
− 1	8	2

3.

h	t	o
1	0	8
−	3	8

 How is regrouping hundreds as tens
like regrouping tens as ones?

Subtract. Regroup if needed.

1.
$$\begin{array}{r} \overset{2\;11}{3\cancel{1}9} \\ -165 \\ \hline 154 \end{array}$$
$$\begin{array}{r} 874 \\ -291 \\ \hline \end{array}$$
$$\begin{array}{r} 548 \\ -377 \\ \hline \end{array}$$
$$\begin{array}{r} 454 \\ -170 \\ \hline \end{array}$$
$$\begin{array}{r} 680 \\ -170 \\ \hline \end{array}$$

2.
$$\begin{array}{r} 808 \\ -333 \\ \hline \end{array}$$
$$\begin{array}{r} 624 \\ -172 \\ \hline \end{array}$$
$$\begin{array}{r} 792 \\ -391 \\ \hline \end{array}$$
$$\begin{array}{r} 465 \\ -292 \\ \hline \end{array}$$
$$\begin{array}{r} 909 \\ -199 \\ \hline \end{array}$$

3.
$$\begin{array}{r} 107 \\ -\;\;55 \\ \hline \end{array}$$
$$\begin{array}{r} 666 \\ -\;\;76 \\ \hline \end{array}$$
$$\begin{array}{r} 337 \\ -\;\;94 \\ \hline \end{array}$$
$$\begin{array}{r} 469 \\ -\;\;29 \\ \hline \end{array}$$
$$\begin{array}{r} 843 \\ -\;\;83 \\ \hline \end{array}$$

4.
$$\begin{array}{r} 917 \\ -272 \\ \hline \end{array}$$
$$\begin{array}{r} 573 \\ -381 \\ \hline \end{array}$$
$$\begin{array}{r} 851 \\ -161 \\ \hline \end{array}$$
$$\begin{array}{r} 723 \\ -492 \\ \hline \end{array}$$
$$\begin{array}{r} 648 \\ -356 \\ \hline \end{array}$$

PROBLEM SOLVING

5. The Bolts scored 788 points. The Streaks scored 595 points. The team that scores more than 825 points wins a trophy. How many more points do the Streaks need to win?

_____ points

6. Arrange from greatest to least. Then subtract the least from the greatest.

685,   658,   746,   764,   693,   593

_____, _____, _____, _____, _____, _____

Subtract: $241 - 182 = \underline{\,?\,}$

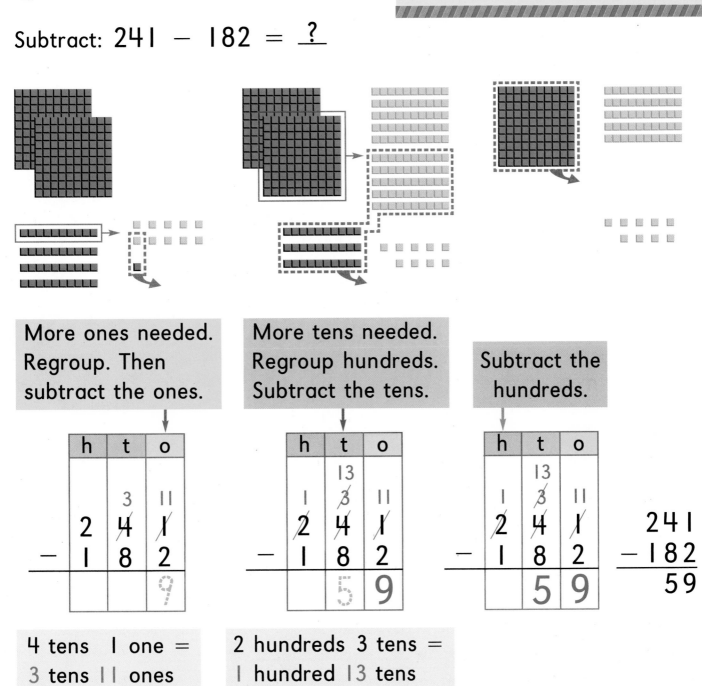

More ones needed. Regroup. Then subtract the ones.	More tens needed. Regroup hundreds. Subtract the tens.	Subtract the hundreds.

h	t	o
	3	11
2	4̸	1̸
− 1	8	2
		9

h	t	o
	13	
1	3̸	11
2̸	4̸	1̸
− 1	8	2
	5	9

h	t	o
	13	
1	3̸	11
2̸	4̸	1̸
− 1	8	2
	5	9

```
  241
− 182
   59
```

4 tens 1 one =
3 tens 11 ones

2 hundreds 3 tens =
1 hundred 13 tens

Subtract. Regroup twice.

1.
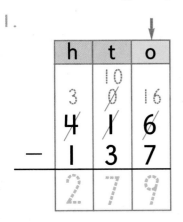

h	t	o
3	10	16
4̸	0̸ 1̸	6̸
− 1	3	7
2	7	9

2.

h	t	o
4	2	4
− 2	6	9

3.
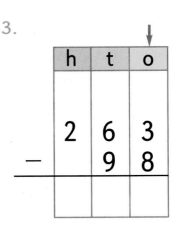

h	t	o
2	6	3
−	9	8

Subtract. Regroup twice if needed.

1.
$$\begin{array}{r} \overset{\overset{16}{8\;\not{9}\;15}}{\not{9}\not{7}5} \\ -398 \\ \hline 577 \end{array}$$
$$\begin{array}{r} 874 \\ -188 \\ \hline \end{array}$$
$$\begin{array}{r} 843 \\ -264 \\ \hline \end{array}$$
$$\begin{array}{r} 527 \\ -189 \\ \hline \end{array}$$
$$\begin{array}{r} 982 \\ -682 \\ \hline \end{array}$$

2.
$$\begin{array}{r} 474 \\ -195 \\ \hline \end{array}$$
$$\begin{array}{r} 523 \\ -234 \\ \hline \end{array}$$
$$\begin{array}{r} 615 \\ -\;\;36 \\ \hline \end{array}$$
$$\begin{array}{r} 798 \\ -265 \\ \hline \end{array}$$
$$\begin{array}{r} 843 \\ -357 \\ \hline \end{array}$$

3.
$$\begin{array}{r} 878 \\ -377 \\ \hline \end{array}$$
$$\begin{array}{r} 624 \\ -\;\;79 \\ \hline \end{array}$$
$$\begin{array}{r} 752 \\ -395 \\ \hline \end{array}$$
$$\begin{array}{r} 465 \\ -297 \\ \hline \end{array}$$
$$\begin{array}{r} 913 \\ -\;\;39 \\ \hline \end{array}$$

4.
$$\begin{array}{r} 621 \\ -268 \\ \hline \end{array}$$
$$\begin{array}{r} 842 \\ -744 \\ \hline \end{array}$$
$$\begin{array}{r} 630 \\ -163 \\ \hline \end{array}$$
$$\begin{array}{r} 712 \\ -153 \\ \hline \end{array}$$
$$\begin{array}{r} 931 \\ -857 \\ \hline \end{array}$$

PROBLEM SOLVING

5. Altogether Ms. Kane and Mr. Arnold donated 314 books to the library. Ms. Kane donated 192. How many books were donated by Mr. Arnold?

_____ books

CRITICAL THINKING

6. Subtract 99 from each. Describe the pattern you see. Write what comes next.

| 215 − 99 = | 225 − 99 = | |

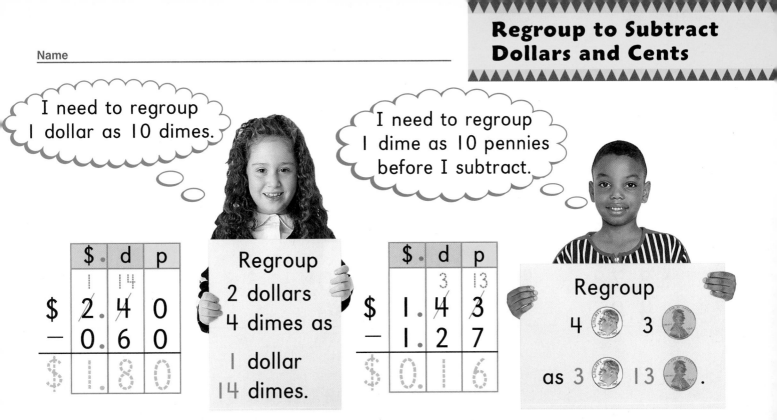

I need to regroup 1 dollar as 10 dimes.

I need to regroup 1 dime as 10 pennies before I subtract.

$.	d	p
¹2.⁴⁴0̸		0
− 0.	6	0
$1.	8	0

Regroup 2 dollars 4 dimes as 1 dollar 14 dimes.

$.	d	p
$ 1.	³4̸	¹³3̸
− 1.	2	7
$0.	1	6

Regroup 4 🪙 3 🪙 as 3 🪙 13 🪙.

 TALK IT OVER How is subtracting $2.40 − $0.60 like subtracting 240 − 60? How is it different?

Find the difference. Regroup dollars as dimes.

1.

$.	d	p
$ ⁸9̸.	¹⁵5̸	5
− 5.	6	4
$3.	9	1

2.

$.	d	p
$ 4.	6	6
− 0.	7	3
$.		

3.

$.	d	p
$ 5.	2	7
− 3.	5	0
$.		

Subtract. Regroup dimes as pennies.

4.

$.	d	p
$ 4.	⁶7̸	¹³3̸
− 0.	2	8
$4.	4	5

5.

$.	d	p
$ 5.	5	8
− 3.	4	9
$.		

6.

$.	d	p
$ 3.	9	2
− 1.	5	7
$.		

11-9 Ask your child to subtract $7.53 − $4.91 and to explain the regrouping.

four hundred twenty-seven **427**

Subtract. ✔ when you regroup dollars as dimes.

1.
$$\begin{array}{r} \overset{3\;\;13}{\$4.\cancel{3}6} \\ -\;1.56 \\ \hline \$2.80\;✔ \end{array}$$
$$\begin{array}{r} \$9.74 \\ -\;3.65 \\ \hline \end{array}$$
$$\begin{array}{r} \$5.29 \\ -\;2.79 \\ \hline \end{array}$$
$$\begin{array}{r} \$6.78 \\ -\;2.62 \\ \hline \end{array}$$
$$\begin{array}{r} \$7.56 \\ -\;0.39 \\ \hline \end{array}$$

2.
$$\begin{array}{r} \$8.78 \\ -\;4.97 \\ \hline \end{array}$$
$$\begin{array}{r} \$7.52 \\ -\;5.37 \\ \hline \end{array}$$
$$\begin{array}{r} \$4.50 \\ -\;2.90 \\ \hline \end{array}$$
$$\begin{array}{r} \$5.64 \\ -\;1.38 \\ \hline \end{array}$$
$$\begin{array}{r} \$3.96 \\ -\;0.65 \\ \hline \end{array}$$

3.
$$\begin{array}{r} \$9.57 \\ -\;4.18 \\ \hline \end{array}$$
$$\begin{array}{r} \$6.74 \\ -\;0.59 \\ \hline \end{array}$$
$$\begin{array}{r} \$7.25 \\ -\;2.45 \\ \hline \end{array}$$
$$\begin{array}{r} \$6.93 \\ -\;4.92 \\ \hline \end{array}$$
$$\begin{array}{r} \$4.36 \\ -\;1.93 \\ \hline \end{array}$$

PROBLEM SOLVING

4. Tai has $9.55. Rhonda has $5.64. How much more money does Tai have than Rhonda?

Tai has _____ more.

MENTAL MATH

5.

Rule: −$0.05	$2.26	$2.25	$2.24	$2.23	$2.22

6.

Rule: −$0.50	$3.60	$3.50	$3.40	$3.30	$3.20

Name _____

Subtract: **$3.23 − $1.75 = _?_**

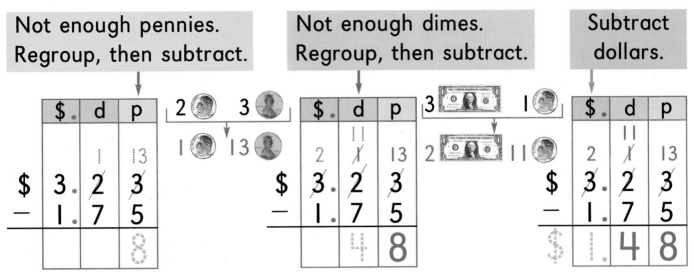

Not enough pennies. Regroup, then subtract.	Not enough dimes. Regroup, then subtract.	Subtract dollars.

 When do you need to regroup twice?

Find the difference. Regroup twice.

1.
$.	d	p
4	17 7̷	12
$ 5.	8 2̷	2̷
− 2.	9	7
$2.	8	5

2.
$.	d	p
$ 2.	5	6
− 0.	7	9
$		

3.
$.	d	p
$ 6.	7	0
− 4.	8	3
$		

4. $4.71 $3.35 $9.54 $6.40
  − 1.85   − 1.57   − 3.65   − 1.64

 Tell whether to use paper and pencil or mental math.

5. $6.87 − $4.80   6. $3.15 − $1.66

Find the difference. Regroup as needed.

1.
$$\begin{array}{r} \overset{10}{\underset{}{4\ \cancel{0}\ 16}} \\ \$5.\cancel{1}\cancel{6} \\ -\ \ 0.47 \\ \hline \$4.69 \end{array}$$

$$\begin{array}{r} \$8.75 \\ -\ 2.31 \\ \hline \end{array}$$

$$\begin{array}{r} \$4.60 \\ -\ 1.52 \\ \hline \end{array}$$

$$\begin{array}{r} \$6.77 \\ -\ 4.67 \\ \hline \end{array}$$

2.
$$\begin{array}{r} \$7.89 \\ -\ 2.45 \\ \hline \end{array}$$

$$\begin{array}{r} \$3.18 \\ -\ 1.87 \\ \hline \end{array}$$

$$\begin{array}{r} \$7.56 \\ -\ 3.24 \\ \hline \end{array}$$

$$\begin{array}{r} \$9.19 \\ -\ 5.24 \\ \hline \end{array}$$

3.
$$\begin{array}{r} \$4.39 \\ -\ 1.75 \\ \hline \end{array}$$

$$\begin{array}{r} \$6.20 \\ -\ 4.65 \\ \hline \end{array}$$

$$\begin{array}{r} \$2.32 \\ -\ 0.59 \\ \hline \end{array}$$

$$\begin{array}{r} \$3.24 \\ -\ 1.76 \\ \hline \end{array}$$

PROBLEM SOLVING

4. How much more does the plaque cost than the ribbon?

_____

Field Day Supplies
trophy $7.49
ribbon $0.98
plaque $4.89

5. Mrs. Connor has $5.50. How much more does she need to buy a trophy?

_____

CRITICAL THINKING

6. Subtract. Look for a pattern. Write what comes next.

$$\begin{array}{r} \$3.05 \\ -\ 0.95 \\ \hline \end{array}$$

$$\begin{array}{r} \$3.25 \\ -\ 0.90 \\ \hline \end{array}$$

$$\begin{array}{r} \$3.45 \\ -\ 0.85 \\ \hline \end{array}$$

$$\begin{array}{r} \$3.65 \\ -\ 0.80 \\ \hline \end{array}$$

$$-\ \rule{1cm}{0.4pt}$$

about

$3.93 is about $4.00.
Round $3.93 to $4.00.

LUNCH SPECIALS

$.95
Soup

$7.27
Pizza

$4.98
Steak Sandwich

$1.99
Salad

$5.37
Spaghetti

$2.79
Tuna Sandwich

Round to the nearest dollar. Write about how much.

1. Soup $1.00   Spaghetti _____   Steak Sandwich _____

   Salad _____   Pizza _____   Tuna Sandwich _____

About how much in all?

2. for soup
   and salad

$$\begin{array}{r} \$1.00 \\ +\ 2.00 \\ \hline \end{array}$$

3. for salad and
   spaghetti

---

4. About how much more does
   pizza cost than a steak sandwich?

   _____

---

5. Tia gave the cashier $9.00 for her
   soup, salad, and tuna sandwich.
   About how much change did Tia get?

   _____ change

First estimate. Then add or subtract.

1. 766 − 514	2. 679 − 328	3. 887 − 878
800 766 −500 −514 ——— ——— 300 252		
4. 234 + 98	5. 436 + 290	6. 482 + 178
7. 807 − 693	8. 532 + 422	9. 888 − 360

 PROBLEM SOLVING Choose the reasonable estimate.

10. About how much more is the book of western stories than the book of sports stories?

about $1.00   about $2.00

$1.25

$3.98

11. Mel bought books of fairy tales, sports stories, and horror tales. About how much did he spend?

about $6.00   about $7.00   about $8.00

$2.79

$4.98

Name _____

Read → Think → Write → Check

1. What is the sum of the numbers inside the circle?

   The sum is __612__.

   3 1 9
   + 2 9 3
   ‾‾‾‾‾
   6 1 2

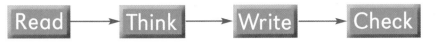

Look for the figure. Find the numbers inside or not inside.

2. What is the difference of the numbers inside the triangle?

   The difference is _____.

3. What is the sum of the numbers <u>not</u> inside the circle?

   The sum is _____.

516
319
293
387

4. What is the sum of the numbers <u>not</u> inside the triangle?

   The sum is _____.

5. What is the difference of the numbers inside the rectangle?

   The difference is _____.

6. What is the difference of the numbers inside the circle?

   The difference is _____.

317
368
125
492

7. What is the difference of the numbers <u>not</u> inside the circle?

   The difference is _____.

How did you check each problem?

Prompt your child to explain how he/she solved any of the problems above.

11-12

8. Which numbers have the greater sum, the numbers inside the circle or the numbers inside the triangle?

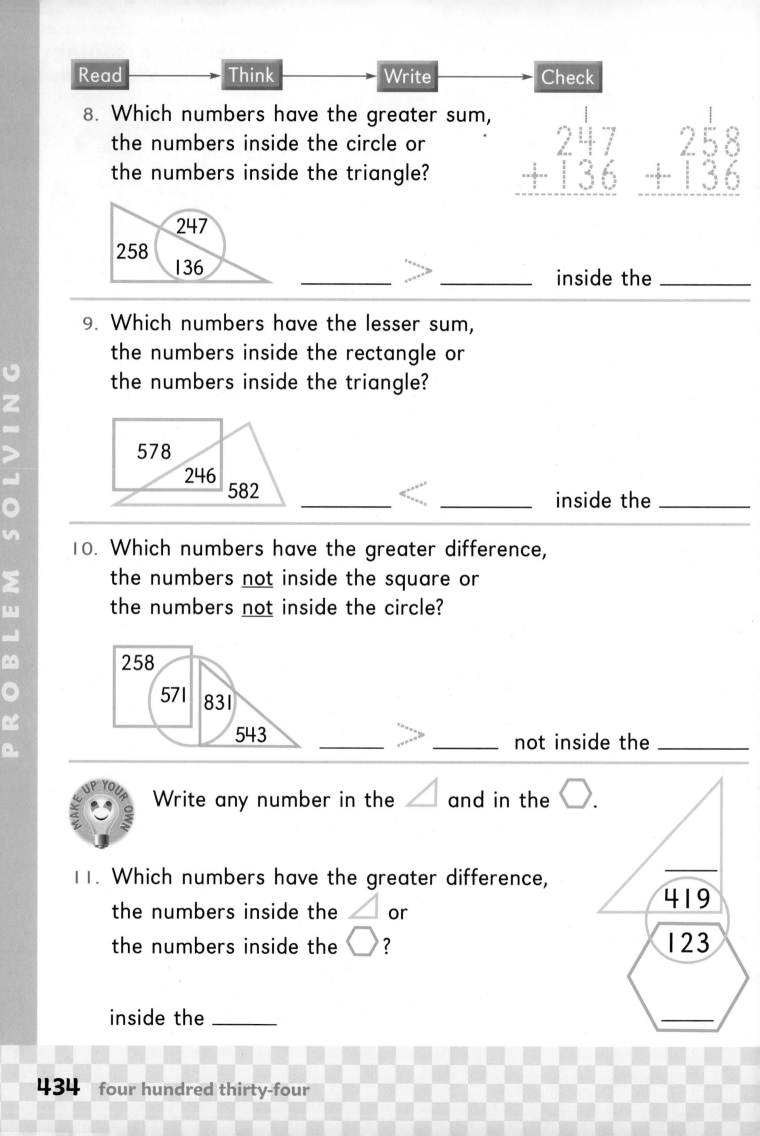

247
+136

258
+136

258 · 247 · 136

_____ > _____ inside the _____

9. Which numbers have the lesser sum, the numbers inside the rectangle or the numbers inside the triangle?

578 · 246 · 582

_____ < _____ inside the _____

10. Which numbers have the greater difference, the numbers <u>not</u> inside the square or the numbers <u>not</u> inside the circle?

258 · 571 · 831 · 543

_____ > _____ not inside the _____

Write any number in the △ and in the ⬡.

11. Which numbers have the greater difference, the numbers inside the △ or the numbers inside the ⬡?

419

123

inside the _____

Read → Think → Write → Check

Use the information in the table.

SPECIAL SALE!

T-Shirt $6.89	Juice....................$2.00
Sun Visor ...$3.79	Baseball Cap $5.95
Pinny $2.98	Game..................$6.19
Orange $0.85	Soccer Ball $7.29

1. Before the sale, a T-shirt cost $8.39. How much less does it cost on sale?

   It costs __$1.50__ less.

   $$\begin{array}{r} \$8.39 \\ -\ 6.89 \\ \hline \$1.50 \end{array}$$

2. Drew wants to buy 2 juices and an orange. How much money does he need?

   He needs _____.

3. Aram bought a sun visor and juice. Sandy bought a game. How much more did Sandy spend then Aram?

   Sandy spent _____ more.

4. How much more does a soccer ball cost than a pinny?

   It costs _____ more.

5. Jay spent $9.74. What 2 things did he buy?

   _____

6. Lori has $6.00. What 3 things can she buy?

   _____

PROBLEM SOLVING

11-13 Ask your child to explain how she/he solved problems 3 and 5 above and to name each step.

four hundred thirty-five **435**

Use a strategy you have learned.

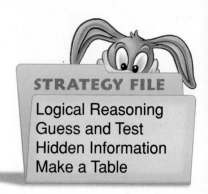

STRATEGY FILE

Logical Reasoning
Guess and Test
Hidden Information
Make a Table

7. Taisha had $2.84. She spent some money and has 3 quarters left. How much did she spend?

She spent _____.

---

8. Terri made 3-digit numbers using these cards. What is the difference between the greatest and the least number she can make?

The difference is _____.

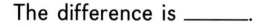

---

9. Our team scored 195 points in the first field-day game. In each of the following games, our team scored 12 fewer points. How many points did our team score in the fifth game?

_____ points

Game					
Points Scored					

---

10. What is the sum of the numbers <u>not</u> inside the rectangle? _____

11. What is the difference of the numbers inside the circle? _____

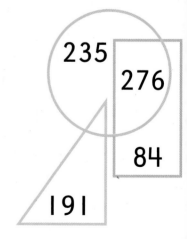

12. What is the difference of the numbers <u>not</u> inside the circle? _____

Name _____

Write the numbers after, before, or between.

1. 87, _____   _____, 914   234, _____, 236

Compare. Write < or >.

2. 426 ◯ 462   587 ◯ 758   362 ◯ 236

Write the numbers in order, least to greatest.

3. 416, 164, 260, 441 _____

Write the number.

4. 10 less than: 320 _____   406 _____   238 _____

5. 100 less than: 829 _____   759 _____   509 _____

Subtract. ✔ when you regroup dimes or tens.

6.
```
   749        418        586       $6.58       $9.50
 -  28      - 135      - 139      - 1.64      - 0.35
```

7.
```
   361        205        824       $4.19       $9.35
 - 207      -  81      - 168      - 0.89      - 2.79
```

Estimate. Round to the nearest hundred or dollar.

8.
```
   863        688       $2.71       $7.62
 - 217      + 136      + 0.89      - 3.49
```

PROBLEM SOLVING

162
228   316

9. What is the difference of the numbers inside the square?

_____

11   This page reviews the mathematical content presented in Chapter 11.

four hundred thirty-seven **437**

Name _____

Find the missing digits.

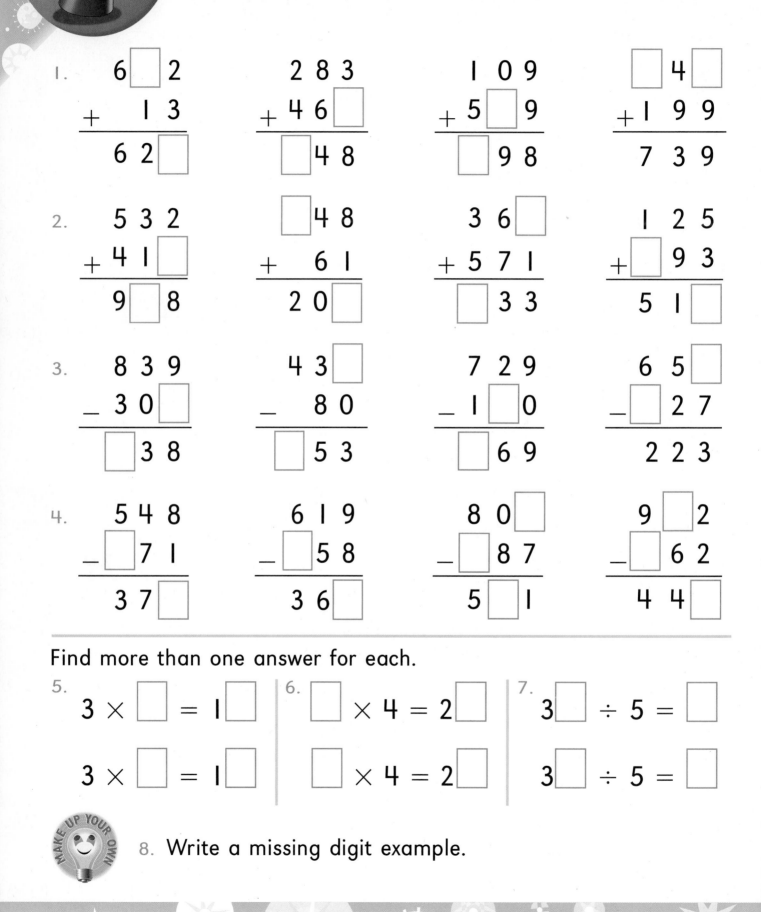

1.
6 □ 2
+   1 3
———
6 2 □

2 8 3
+ 4 6 □
———
□ 4 8

1 0 9
+ 5 □ 9
———
□ 9 8

□ 4 □
+ 1 9 9
———
7 3 9

2.
5 3 2
+ 4 1 □
———
9 □ 8

□ 4 8
+   6 1
———
2 0 □

3 6 □
+ 5 7 1
———
□ 3 3

1 2 5
+ □ 9 3
———
5 1 □

3.
8 3 9
− 3 0 □
———
□ 3 8

4 3 □
−   8 0
———
□ 5 3

7 2 9
− 1 □ 0
———
□ 6 9

6 5 □
− □ 2 7
———
2 2 3

4.
5 4 8
− □ 7 1
———
3 7 □

6 1 9
− □ 5 8
———
3 6 □

8 0 □
− □ 8 7
———
5 □ 1

9 □ 2
− □ 6 2
———
4 4 □

Find more than one answer for each.

5.   3 × □ = 1□

    3 × □ = 1□

6.   □ × 4 = 2□

    □ × 4 = 2□

7.   3□ ÷ 5 = □

    3□ ÷ 5 = □

8. Write a missing digit example.

1. Which can you can use, mental math or paper and pencil? Explain. Then find the difference.

$$256 - 184 \qquad 297 - 90 \qquad 254 - 204$$

— _____   — _____   — _____

2. Spin for an amount to subtract. Show how you can count back or regroup with dollars, dimes, and pennies.

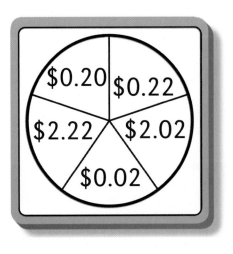

$0.20 $0.22
$2.22 $2.02
$0.02

$4.57 $3.90 $4.13
— _____ . _____   — _____ . _____   — _____ . _____

PORTFOLIO

Choose I of these projects.
Use a separate sheet of paper.

3. Use number cards from 0 to 9. Make two 3-digit numbers that have the same number of tens and two 3-digit numbers that have the same number of hundreds. Subtract them.

3 6 2 — 1 6 7

4. Pick 3 number cards from 0 to 9. Make two different 3-digit numbers. Write number sentences using <, =, and >.

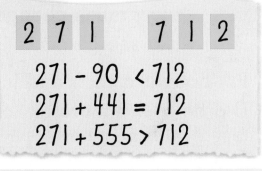

2 7 1   7 1 2

$$271 - 90 < 712$$
$$271 + 441 = 712$$
$$271 + 555 > 712$$

11 This page provides a variety of informal assessment opportunities in order to measure your child's understanding of Chapter II.

four hundred thirty-nine **439**

Write the numbers before, after, or between.

1. _____, 421   187, _____   449, _____, 451

Compare. Write < or >.

2. 368 ◯ 199   209 ◯ 229   612 ◯ 621

Write the numbers in order, least to greatest.

3. 187, 287, 180, 307 _____

Write the number.

4. 100 less than 218 _____   5. 10 less than 401 _____

Subtract. ✔ when you regroup dollars or hundreds.

6.
$$\begin{array}{r} 600 \\ -400 \\ \hline \end{array} \qquad \begin{array}{r} 537 \\ -212 \\ \hline \end{array} \qquad \begin{array}{r} 341 \\ -\ 28 \\ \hline \end{array} \qquad \begin{array}{r} \$8.45 \\ -\ 2.13 \\ \hline \end{array} \qquad \begin{array}{r} \$7.81 \\ -\ 0.67 \\ \hline \end{array}$$

7.
$$\begin{array}{r} 406 \\ -155 \\ \hline \end{array} \qquad \begin{array}{r} 728 \\ -\ 62 \\ \hline \end{array} \qquad \begin{array}{r} 750 \\ -381 \\ \hline \end{array} \qquad \begin{array}{r} \$5.74 \\ -\ 1.94 \\ \hline \end{array} \qquad \begin{array}{r} \$8.62 \\ -\ 6.86 \\ \hline \end{array}$$

Round each amount to the nearest dollar.

Item	Cost
taco	$2.46
fruit	$0.85
juice	$1.21

8. taco _____   fruit _____   juice _____

PROBLEM SOLVING

9. Anna paid $1.50 for juice. How much change did she get? _____

10. P.J. bought a taco and juice. Did he spend more or less than $3.50? _____

This page is a formal assessment of your child's understanding of the content presented in Chapter 11.

11

Name _____

Mark the ◯ for your answer.

Listening Section

A
◯ about $\frac{1}{2}$
◯ about $\frac{1}{4}$
◯ cannot tell

B
◯ inch
◯ pound
◯ gallon

1. Estimate.

    215
    +185

    ◯ about 300
    ◯ about 400
    ◯ about 500
    ◯ about 600

2. What is the amount?

    ◯ $1.76
    ◯ $1.81
    ◯ $1.86
    ◯ $1.96

3. Estimate.

    $6.40
    + 2.90

    ◯ about $4.00
    ◯ about $8.00
    ◯ about $9.00
    ◯ about $10.00

4. How many corners does this solid figure have?

    6   7   8   9
    ◯   ◯   ◯   ◯

5. $6.15
    + 0.50

    ◯ $7.45
    ◯ $6.55
    ◯ $6.65
    ◯ $6.45

6.   5
    ×4

    ◯ 9
    ◯ 54
    ◯ 16
    ◯ 20

7.   3
    ×1

    ◯ 2
    ◯ 1
    ◯ 3
    ◯ 4

8.   568
    −231

    ◯ 337
    ◯ 327
    ◯ 799
    ◯ 737

9. $8.78
    − 4.65

    ◯ $2.13
    ◯ $4.13
    ◯ $4.31
    ◯ $5.13

10.   5
    ×9

    ◯ 14
    ◯ 36
    ◯ 45
    ◯ 54

11. Order 387, 432, 519, and 423 from least to greatest.

    ◯ 519, 432, 423, 387
    ◯ 387, 423, 432, 519
    ◯ 423, 432, 519, 387
    ◯ 387, 432, 423, 519

Mark the ◯ for your answer.

12. Measure.
◯ 3 centimeters
◯ 4 centimeters
yellow
◯ 2 centimeters
◯ 5 centimeters

13. What comes next?
446, 447, _____
◯ 449   ◯ 448
◯ 445   ◯ 548

14. Compare.
◯ <
◯ >
146 ◯ 234
◯ =
◯ ×

15. 487
+304
◯ 183
◯ 791
◯ 783
◯ 781

16. What is the value of 6 in 632?
6   60   600
◯   ◯   ◯

17. 175
+ 37
◯ 202
◯ 212
◯ 222
◯ 232

18. 50
−18
◯ 32
◯ 42
◯ 38
◯ 48

19.
$2 \times 3 =$ _____
◯ 2+2+2   ◯ 3+3
◯ 3+2+2   ◯ 2+3+3

20. What part is shaded?
◯ $\frac{1}{5}$
◯ $\frac{1}{3}$
◯ $\frac{4}{5}$
◯ $\frac{1}{4}$

21. What part is shaded?
◯ $\frac{2}{3}$
◯ $\frac{1}{3}$
◯ $\frac{1}{4}$
◯ $\frac{3}{4}$

22. 3 friends share 10 crayons equally. How many crayons are left over?
0   1   2   3
◯   ◯   ◯   ◯

23. Tim has 2 🚗. Each 🚗 has 4 ⬤. How many ⬤ in all?
◯ 10
◯ 4
◯ 8
◯ 6

24. What time will it be in one half hour?
◯ 12:15   ◯ 1:15
◯ 12:45   ◯ 1:45

25.
15 in all
How many groups of 3?

$15 \div 3 =$ _____

◯ 4
◯ 12
◯ 5
◯ 3

CRITICAL THINKING

The treasure chest had 900 pieces of gold. Two divers each brought 400 pieces of gold to their ship. How many pieces are left in the chest?

For more information about Chapter 12, visit the Family Information Center at **www.sadlier-oxford.com**

Internet

Dear Family,

Today your child began Chapter 12. As she/he uses algebraic reasoning, you may want to read the poem below, which was read in class. Have your child talk about some of the math ideas pictured on page 443.

Look for the 🏠 at the bottom of each skills lesson. The suggestion on the page gives you an opportunity to improve your child's understanding of math and to reinforce her/his math language. You may want to have countables available for your child to use throughout this chapter.

Home Activity

Secrets of the Sea Stars

Try this activity with your child. On a sheet of light-colored construction paper, draw a sea star. Write a 2-digit number in the star's center and have this represent a sum or difference. Ask your child to write another name for the sum or difference on each arm of the star. Repeat the activity with a 3-digit number in the center.

174 – 144

30

15 + 15

Home Reading Connection

Mysteries of the Deep

I wonder what it would be like for me
To be swimming deep within the sea?
Would there be neon fish that brightly glow?
Would clown fish act in an ocean circus show?

I would ask sea creatures about hundreds of things—
Like "Angel fish, where did you get your wings?"
And, "Squids, how do you make your inky potion?"
And, "Sharks, where is your home within this ocean?"

I would ask thousands of fish passing my way,
"Do you get tired of being in school all day?
And do you ever wonder what it's like for me,
To live in the air *above* the sea?"

Christine Barrett

Name _____

Each of these tables has a different rule.
The rules use $+$, $-$, $\times$, or $\div$.

Write the rule and 1 more example.

1.

In	Out
20	4
35	7
15	3
25	5

Rule: $\div 5$

2.

In	Out
15	25
208	218
174	184
___	___

Rule: _____

3.

In	Out
17	7
165	155
32	22
___	___

Rule: _____

4.

In	Out
8	80
14	140
29	290
___	___

Rule: _____

5.

In	Out
26	16
354	344
159	149
___	___

Rule: _____

6.

In	Out
9	18
6	12
5	10
___	___

Rule: _____

7.

In	Out
9	15
29	35
49	55
___	___

Rule: _____

8.

In	Out
2	10
6	30
9	45
___	___

Rule: _____

9.

In	Out
20	10
18	9
12	6
___	___

Rule: _____

10. Write 2 tables.
Use a different rule for each.

MOVING ON: CONNECTIONS

MOVING ON: CONNECTIONS

Underwater explorers made a chart of the sea creatures they saw.

Sea Creature Count

Dolphins	Whales	Bluefish	Stingrays	Jellyfish
60	30	120	80	110

Make a pictograph for this data. Use the key.

1.   Sea Creature Count

Dolphins	☺
Whales	
Bluefish	
Stingrays	
Jellyfish	

Key: Each ☺ = 10 sea creatures.

Use your graph to write and answer questions about the data.

2. How many more _____ than

_____ were seen?   _____

3. How many _____ and

_____ were seen in all?   _____

4. Which creatures did they see the most?

_____ the fewest?   _____

5. How many more sea creatures need to
   be seen to have seen 500 altogether?   _____

You can put this in your
Math Portfolio. **PORTFOLIO**
12

Name

Does △ = 0?

Yes. In my multiplication sentence △ equals 0.

No. In my subtraction sentence △ does not equal 0.

△ × 5 = △
0 × 5 ? 0
0 does equal 0.

150 − △ = △
150 − 0 ? 0
150 does not equal 0.

Write Yes or No. Does ▢ = 2?

1. ▢ + ▢ = 4
2 + 2 = 4   Yes

2. ▢ × ▢ = 4
_____

Does ⬤ equal 10?

3. 476 + ⬤ = 596 − ⬤
_____

4. 16 − 5 + ⬤ = 31 − ⬤
_____

Does ▭ = 4?

5. 16 ÷ ▭ = 6 − ▭ + 2
_____

6. ▭ + ▭ = 13 − 4
_____

Does ▱ = 5?

7. 17 − ▱ + 8 = ▱ + ▱ + 10
_____

8. Can you write more than 1 answer for this number sentence? 1 × ⬭ = ⬭ − 0

MATH JOURNAL

Invite your child to make up a number sentence using the symbols above, and to solve it.
12-1
four hundred forty-seven   **447**

MOVING ON IN MATH

Hidden Numbers

In this number sentence Todd hid one number. Guess and test to find the hidden number.

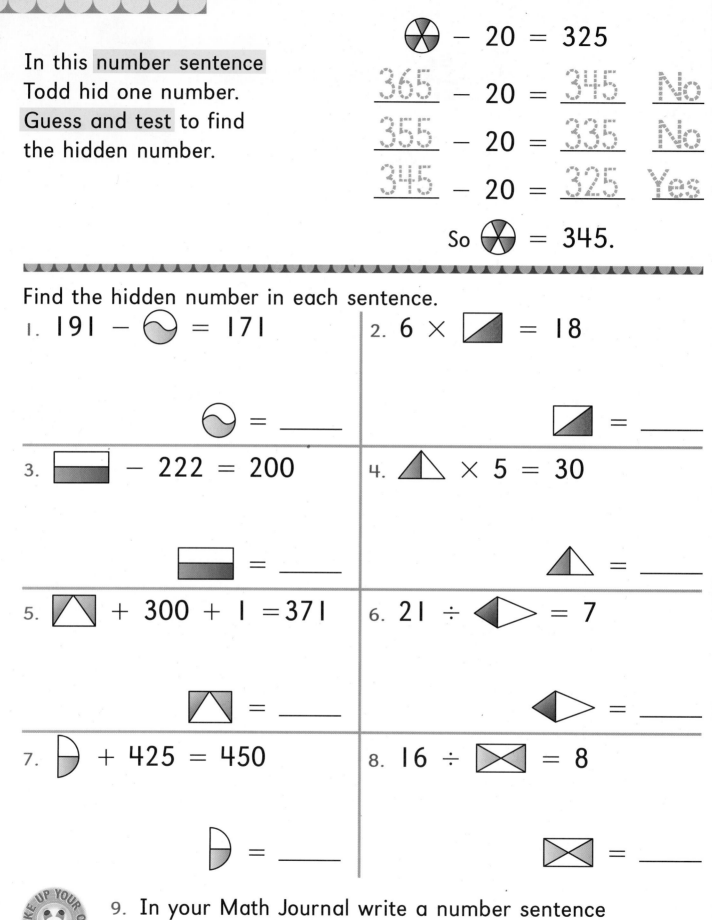

$$\bigotimes - 20 = 325$$

$365 - 20 = 345$ No

$355 - 20 = 335$ No

$345 - 20 = 325$ Yes

So $\bigotimes = 345$.

Find the hidden number in each sentence.

1. $191 - \bigcirc = 171$

   $\bigcirc =$ _____

2. $6 \times \square = 18$

   $\square =$ _____

3. $\square - 222 = 200$

   $\square =$ _____

4. $\triangle \times 5 = 30$

   $\triangle =$ _____

5. $\triangle + 300 + 1 = 371$

   $\triangle =$ _____

6. $21 \div \diamond = 7$

   $\diamond =$ _____

7. $\square + 425 = 450$

   $\square =$ _____

8. $16 \div \square = 8$

   $\square =$ _____

9. In your Math Journal write a number sentence with a hidden number. Have a classmate solve it.

Invite your child to find the hidden number in $333 - \bigcirc = 33$.

 12-2

Name _____

There are 2 ways to make the sides equal.

$$29 + 10 - 3 = 9 \bigcirc \square$$

$$29 + 10 - 3 = 9 \oplus \boxed{27}$$

$$36 = 36$$

$$29 + 10 - 3 = 9 \otimes \boxed{4}$$

$$36 = 36$$

Find two ways to make the sides equal.

1. $45 - 20 - 1 = 6 \bigcirc \square$

$\underline{24} = 6 \oplus \boxed{18}$

and $\underline{24} = 6 \otimes \boxed{4}$

2. $312 - 311 = 5 \bigcirc \square$

$\underline{\quad} = 5 \bigcirc \square$

and $\underline{\quad} = 5 \bigcirc \square$

3. $405 - 400 = 25 \bigcirc \square$

$\underline{\quad} = 25 \bigcirc \square$

and $\underline{\quad} = 25 \bigcirc \square$

4. $17 + 3 - 12 = 4 \bigcirc \square$

$\underline{\quad} = 4 \bigcirc \square$

and $\underline{\quad} = 4 \bigcirc \square$

5. $10 + 20 + 5 = 7 \bigcirc \square$

$\underline{\quad} = 7 \bigcirc \square$

and $\underline{\quad} = 7 \bigcirc \square$

6. $172 - 140 = 8 \bigcirc \square$

$\underline{\quad} = 8 \bigcirc \square$

and $\underline{\quad} = 8 \bigcirc \square$

7. $717 - 710 + 2 = 18 \bigcirc \square$

$\underline{\quad} = 18 \bigcirc \square$

and $\underline{\quad} = 18 \bigcirc \square$

8. $625 - 615 = 50 \bigcirc \square$

$\underline{\quad} = 50 \bigcirc \square$

and $\underline{\quad} = 50 \bigcirc \square$

MOVING ON IN MATH

12-3 **Ask your child to explain how she/he solved exercises 5–8.**

You can find more than 1 answer for
number sentences that have < or >.

$100 + \boxed{} < 103$

$100 + \underline{1} < 103$

$ 101 < 103 \quad \text{true}$

Guess
and
test.

$10 > 4 \times \boxed{}$

$10 > 4 \times \underline{0}$

$10 > 0 \quad \text{true}$

and $100 + \underline{2} < 103$

$ 102 < 103 \quad \text{true}$

and $10 > 4 \times \underline{1}$

$ 10 > 4 \quad \text{true}$

TALK IT OVER Is there any other number you can
use to make each sentence true?

Find 2 answers for each number sentence.

1. $258 - \boxed{} < 200 + 56$

$258 - \boxed{3} < \underline{256}$

$\underline{255} < \underline{256}$

and $258 - \boxed{} < \underline{256}$

$\underline{} < \underline{}$

2. $9 - 5 > \boxed{} \div 3$

$\underline{4} > \boxed{9} \div 3$

$\underline{} > \underline{}$

and $\underline{} > \boxed{} \div 3$

$\underline{} > \underline{}$

3. $172 + \boxed{} < 178 - 3$

$172 + \boxed{} < \underline{}$

$\underline{} < \underline{}$

and $172 + \boxed{} < \underline{}$

$\underline{} < \underline{}$

4. $28 \div 4 > \boxed{} - 700$

$\underline{} > \boxed{} - 700$

$\underline{} > \underline{}$

and $\underline{} > \boxed{} - 700$

$\underline{} > \underline{}$

5. In your Math Journal write 2 number sentences
like those above. Find 2 answers for each.

Name _____

Each letter has a different value.
Find the value of the word ocean.

O	C	E	A	N
↓	↓	↓	↓	↓
300 +	10 +	200 +	100 +	30

A = 100	I = 20	R = 4
B = 1	K = 12	S = 40
C = 10	L = 3	T = 13
E = 200	M = 400	W = 500
F = 11	N = 30	Y = 5
H = 2	O = 300	

Count on: 310, 510, 610, 640   OCEAN = 640

Find the value of each word.

1. SHARK = __40__ + __2__ + __100__ + __4__ + __12__

   SHARK = _____

2. WHALE = ____ + ____ + ____ + ____ + ____

   WHALE = _____

3. FISH = ____ + ____ + ____ + ____

   FISH = _____

4. CORAL = ____ + ____ + ____ + ____ + ____

   CORAL = _____

5. CLAM = ____ + ____ + ____ + ____

   CLAM = _____

CHALLENGE   Ring the word with the greatest value.

6. EEL   CRAB   SEA

12-4   Spell a word, assign a value to each letter, and tell your child to find the value of the word.

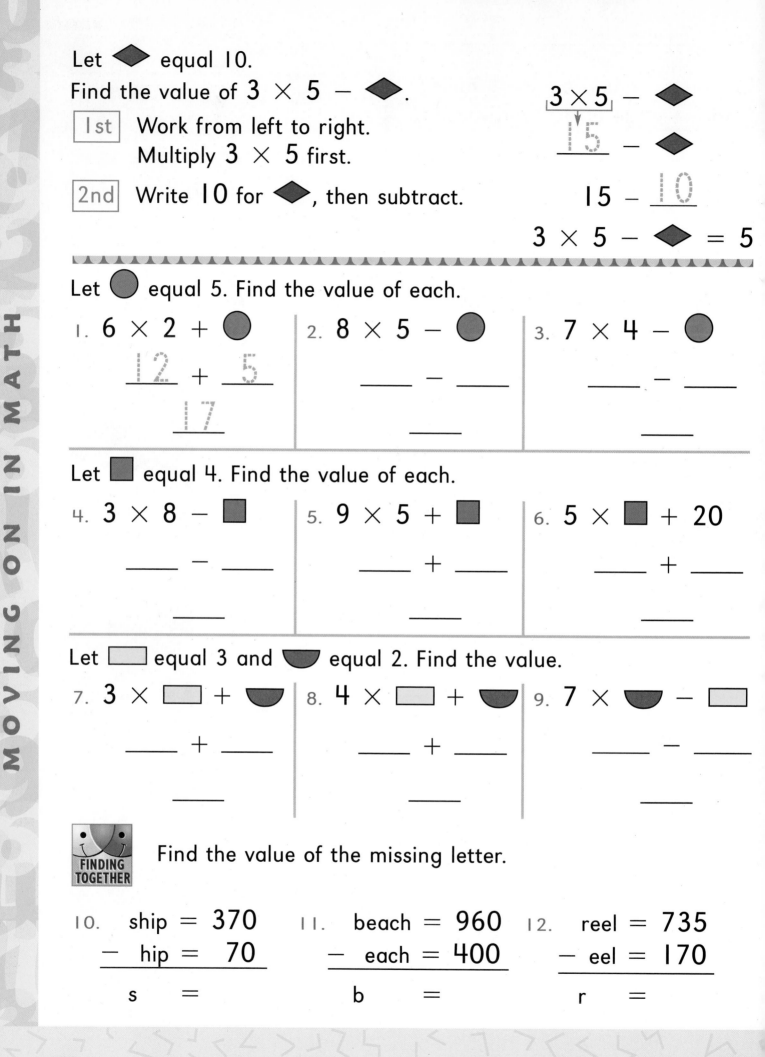

Let ◆ equal 10.

Find the value of $3 \times 5 - ◆$.

| 1st | Work from left to right. Multiply 3×5 first. |

| 2nd | Write 10 for ◆, then subtract. |

$$3 \times 5 - ◆$$
$$15 - ◆$$
$$15 - 10$$
$$3 \times 5 - ◆ = 5$$

Let ● equal 5. Find the value of each.

1. $6 \times 2 + ●$

$$12 + 5$$
$$17$$

2. $8 \times 5 - ●$

___ − ___

___

3. $7 \times 4 - ●$

___ − ___

___

Let ■ equal 4. Find the value of each.

4. $3 \times 8 - ■$

___ − ___

___

5. $9 \times 5 + ■$

___ + ___

___

6. $5 \times ■ + 20$

___ + ___

___

Let ▭ equal 3 and ◡ equal 2. Find the value.

7. $3 \times ▭ + ◡$

___ + ___

___

8. $4 \times ▭ + ◡$

___ + ___

___

9. $7 \times ◡ - ▭$

___ − ___

___

FINDING TOGETHER   Find the value of the missing letter.

10.
$$\begin{array}{r} \text{ship} = 370 \\ - \text{ hip} = 70 \\ \hline \text{s} = \end{array}$$

11.
$$\begin{array}{r} \text{beach} = 960 \\ - \text{ each} = 400 \\ \hline \text{b} = \end{array}$$

12.
$$\begin{array}{r} \text{reel} = 735 \\ - \text{ eel} = 170 \\ \hline \text{r} = \end{array}$$

Name _____

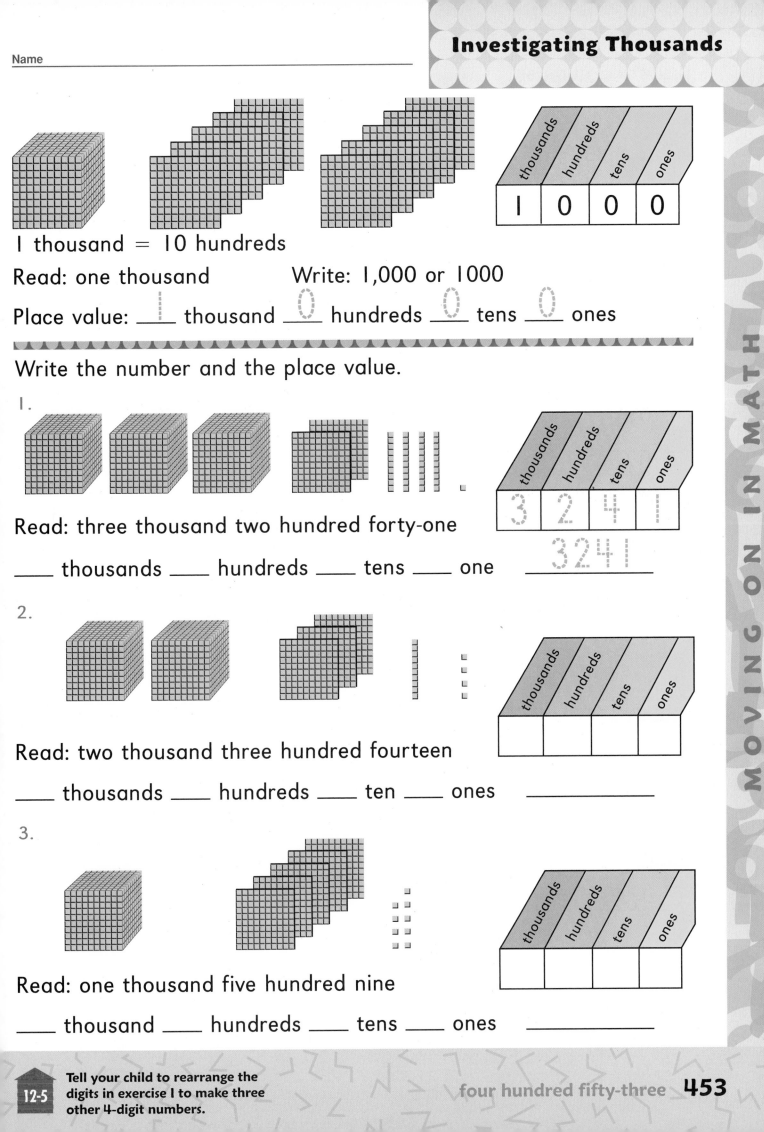

1 thousand = 10 hundreds

Read: one thousand   Write: 1,000 or 1000

Place value: __1__ thousand __0__ hundreds __0__ tens __0__ ones

Write the number and the place value.

1.

Read: three thousand two hundred forty-one

____ thousands ____ hundreds ____ tens ____ one   _____

2.

Read: two thousand three hundred fourteen

____ thousands ____ hundreds ____ ten ____ ones   _____

3.

Read: one thousand five hundred nine

____ thousand ____ hundreds ____ tens ____ ones   _____

12-5 Tell your child to rearrange the digits in exercise 1 to make three other 4-digit numbers.

Write the number.

1. 3 thousands 4 hundreds 7 tens 8 ones 3478

2. 6 thousands 7 hundreds 0 tens 4 ones _____

3. 8 thousands 0 hundreds 5 tens 2 ones _____

4. 5 thousands 2 hundreds 6 tens 0 ones _____

Ring what the 8 means in each number.

5. 2148	8 hundreds	8 tens	(8 ones)
6. 8406	8 thousands	8 hundreds	8 tens
7. 5781	8 hundreds	8 tens	8 ones
8. 1892	8 thousands	8 hundreds	8 tens

Write the number for each number word.

9. seven thousand five hundred forty-six _____

10. five thousand seven hundred _____

11. nine thousand five hundred twenty-nine _____

12. four thousand seventy-five _____

CHALLENGE Write the missing numbers.

13. $4670 = 4000 + \underline{\quad} + 70 + 0$

14. $8125 = \underline{\quad} + 100 + 20 + \underline{\quad}$

15. $2904 = \underline{\quad} + \underline{\quad} + \underline{\quad} + \underline{\quad}$

Add: 1824 + 1143 = ?

| 1st | Estimate about how many thousands. |

1824 + 1143

about   2000 + 1000 = 3000

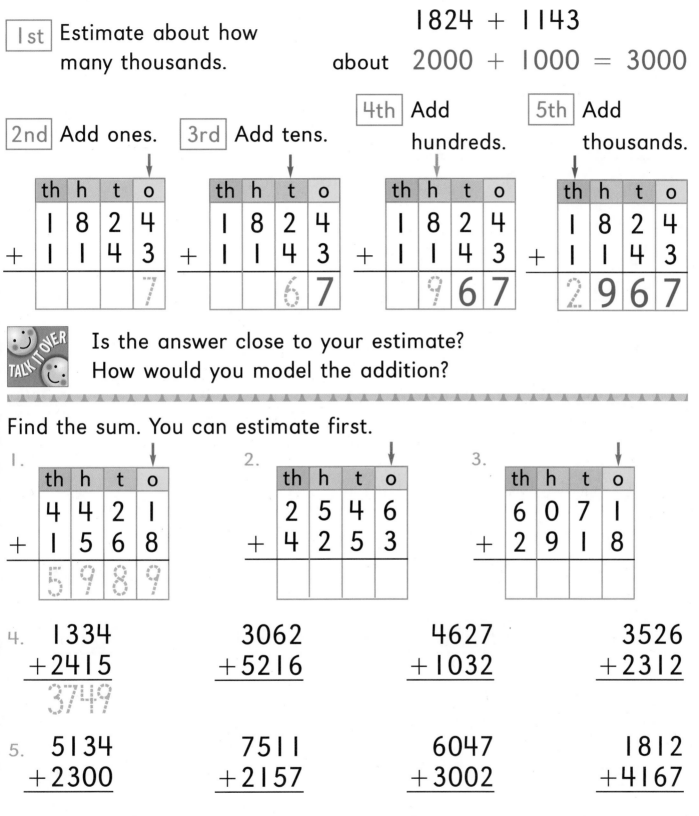

| 2nd | Add ones. | | 3rd | Add tens. | | 4th | Add hundreds. | | 5th | Add thousands. |

th	h	t	o
1	8	2	4
+ 1	1	4	3
			7

th	h	t	o
1	8	2	4
+ 1	1	4	3
		6	7

th	h	t	o
1	8	2	4
+ 1	1	4	3
	9	6	7

th	h	t	o
1	8	2	4
+ 1	1	4	3
2	9	6	7

TALK IT OVER Is the answer close to your estimate? How would you model the addition?

Find the sum. You can estimate first.

1.
th	h	t	o
4	4	2	1
+ 1	5	6	8
5	9	8	9

2.
th	h	t	o
2	5	4	6
+ 4	2	5	3

3.
th	h	t	o
6	0	7	1
+ 2	9	1	8

4.
```
  1334        3062        4627        3526
+ 2415      + 5216      + 1032      + 2312
  3749
```

5.
```
  5134        7511        6047        1812
+ 2300      + 2157      + 3002      + 4167
```

MATH JOURNAL

6. How is adding 5000 + 2000 like adding 500 + 200? How is it different?

12-6 Invite your child to explain how to find the sum of 2240 + 3125.

four hundred fifty-five   **455**

I use paper and pencil.

$$\begin{array}{r} 3412 \\ +2185 \\ \hline 5597 \end{array}$$

TALK IT OVER

Explain how to estimate each sum.

I use mental math by counting on.

$$\begin{array}{r} 6140 \\ +2000 \\ \hline 8140 \end{array}$$

Add. ✔ when you use mental math.

1. $\begin{array}{r} 8614 \\ +134 \\ \hline 8748 \end{array}$ $\begin{array}{r} 1702 \\ +263 \\ \hline \end{array}$ $\begin{array}{r} 3020 \\ +600 \\ \hline \end{array}$ $\begin{array}{r} 6430 \\ +400 \\ \hline \end{array}$

2. $\begin{array}{r} 7132 \\ +1010 \\ \hline \end{array}$ $\begin{array}{r} 8276 \\ +1223 \\ \hline \end{array}$ $\begin{array}{r} 6093 \\ +1706 \\ \hline \end{array}$ $\begin{array}{r} 5499 \\ +2300 \\ \hline \end{array}$

CRITICAL THINKING

Write each number that is:

3. 2000 more	5176	1642	679	4083	951

4. 200 more	5176	1642	679	7083	951

Write the missing addends.

5. $1528 + \boxed{} = 1558$

6. $2875 + \boxed{} = 5875$

7. $6119 = 6115 + \boxed{}$

8. $9298 = \boxed{} + 60$

9. $4753 = \boxed{} + 300$

10. $9384 = \boxed{} + 3000$

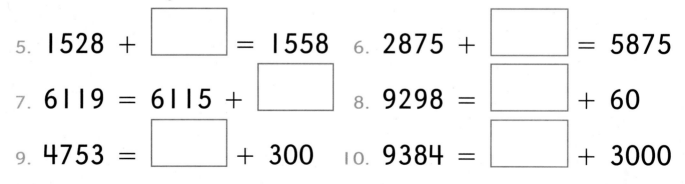

Name _____

Subtract: 5694 − 2173 = ?

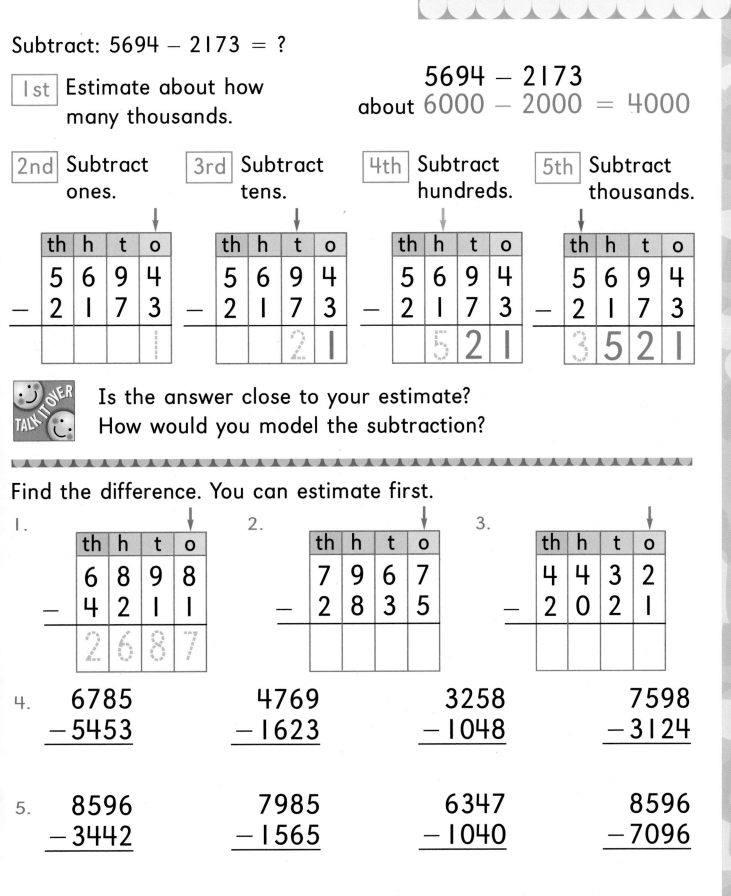

| 1st | Estimate about how many thousands. |

5694 − 2173
about 6000 − 2000 = 4000

| 2nd | Subtract ones. |

th	h	t	o
5	6	9	4
2	1	7	3
			1

| 3rd | Subtract tens. |

th	h	t	o
5	6	9	4
2	1	7	3
		2	1

| 4th | Subtract hundreds. |

th	h	t	o
5	6	9	4
2	1	7	3
	5	2	1

| 5th | Subtract thousands. |

th	h	t	o
5	6	9	4
2	1	7	3
3	5	2	1

TALK IT OVER Is the answer close to your estimate?
How would you model the subtraction?

Find the difference. You can estimate first.

1.

th	h	t	o
6	8	9	8
4	2	1	1
2	6	8	7

2.

th	h	t	o
7	9	6	7
2	8	3	5

3.

th	h	t	o
4	4	3	2
2	0	2	1

4.
```
  6785          4769          3258          7598
− 5453        − 1623        − 1048        − 3124
```

5.
```
  8596          7985          6347          8596
− 3442        − 1565        − 1040        − 7096
```

6. How can you check your answers for 5?

Invite your child to explain how to subtract 9578 − 4216.

MOVING ON IN MATH

I use paper and pencil.

$$\begin{array}{r} 2658 \\ -1235 \\ \hline 1423 \end{array}$$

I use mental math by counting back.

$$\begin{array}{r} 4314 \\ -2000 \\ \hline 2314 \end{array}$$

TALK IT OVER Explain how to estimate each difference.

Subtract. ✔ when you use mental math.

1.
$$\begin{array}{r} 4669 \\ -3035 \\ \hline 1634 \end{array}$$
$$\begin{array}{r} 6897 \\ -1005 \\ \hline \end{array}$$
$$\begin{array}{r} 5782 \\ -2781 \\ \hline \end{array}$$
$$\begin{array}{r} 9347 \\ -6300 \\ \hline \end{array}$$

2.
$$\begin{array}{r} 6580 \\ -300 \\ \hline \end{array}$$
$$\begin{array}{r} 7241 \\ -1010 \\ \hline \end{array}$$
$$\begin{array}{r} 9665 \\ -243 \\ \hline \end{array}$$
$$\begin{array}{r} 5627 \\ -1400 \\ \hline \end{array}$$

CRITICAL THINKING

Write each number that is:

3.
2000 less	2261	6604	9215	7800	8602

4.
200 less	2261	6604	1550	7800	1150

Write the missing numbers.

5. $6404 - \boxed{} = 3404$ 6. $\boxed{} - 30 = 7262$

7. $5030 = 5630 - \boxed{}$ 8. $1120 = \boxed{} - 60$

9. $4379 = \boxed{} - 2000$ 10. $\boxed{} - 200 = 5788$

Name _____

Read → Think → Test → Write → Check

1. Two divers find coins. Adam finds 600 coins, Carlos finds 900. How many must Carlos give Adam so each has the same number?

Guess the answer.
Test your answer.
Guess again if needed.

1st guess: Try __100__
$$900 - 100 = 800$$
$$600 + 100 = 700$$

2nd guess: Try __200__
$$900 - 200 = 700$$
$$600 + 200 = 800$$

3rd guess: Try __150__
$$900 - 150 = 750$$
$$600 + 150 = 750$$

Carlos must give Adam __150__ coins.

2. Marta has $4.95. Which two fish can she buy and have 41¢ change?

$2.39

$2.49

$2.15

She can buy fish that cost _____ and _____.

3. A deep-sea diving team saw a total of 4580 sea creatures in 2 days. The difference in the team's daily totals was 320. What days did the team dive?

Sea Creatures

Monday	2130
Tuesday	2140
Thursday	2440
Friday	2450

They dove on _____ and _____.

12-8
Invite your child to tell you how she/he solved problems using the Guess and Test strategy.

four hundred fifty-nine **459**

MOVING ON: PROBLEM SOLVING

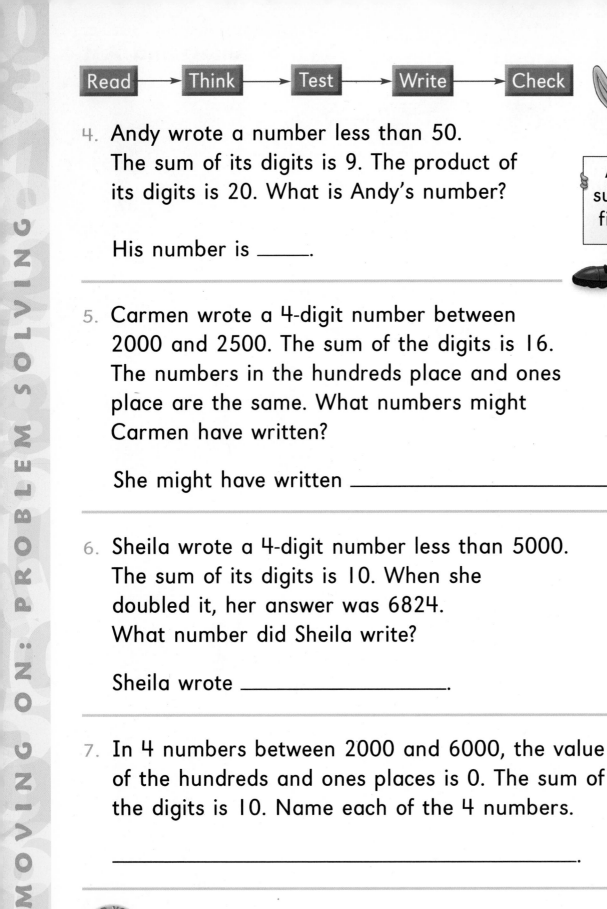

Add to find a sum. Multiply to find a product.

4. Andy wrote a number less than 50. The sum of its digits is 9. The product of its digits is 20. What is Andy's number?

His number is _____.

5. Carmen wrote a 4-digit number between 2000 and 2500. The sum of the digits is 16. The numbers in the hundreds place and ones place are the same. What numbers might Carmen have written?

She might have written _____.

6. Sheila wrote a 4-digit number less than 5000. The sum of its digits is 10. When she doubled it, her answer was 6824. What number did Sheila write?

Sheila wrote _____.

7. In 4 numbers between 2000 and 6000, the value of the hundreds and ones places is 0. The sum of the digits is 10. Name each of the 4 numbers.

_____.

8. My 2-digit number is less than _____.

The sum of its digits is _____. The product

of its digits is _____. What is my number?

Name _____

Read ⟶ Think ⟶ Write ⟶ Check

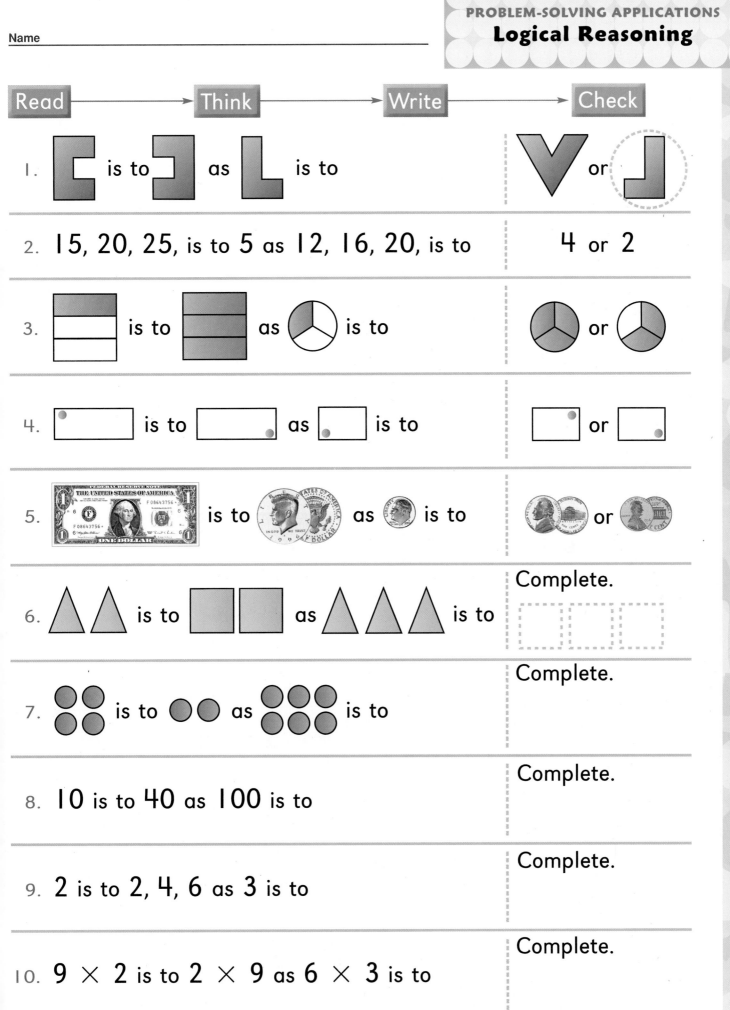

1. ⊏ is to ⊐ as ∟ is to   ⋁ or ⌐

2. 15, 20, 25, is to 5 as 12, 16, 20, is to   4 or 2

3. ▭ is to ▭ as ◐ is to   ◓ or ◔

4. ▭ is to ▭ as ▭ is to   ▭ or ▭

5. $1 is to ½ dollar as dime is to   nickels or penny

6. △△ is to ▢▢ as △△△ is to   Complete.
   ⬚ ⬚ ⬚

7. ⦾⦾ is to ◯◯ as ⦾⦾⦾ is to   Complete.

8. 10 is to 40 as 100 is to   Complete.

9. 2 is to 2, 4, 6 as 3 is to   Complete.

10. 9 × 2 is to 2 × 9 as 6 × 3 is to   Complete.

MOVING ON: PROBLEM SOLVING

Use a strategy you have learned.

11. Alonzo has $6.45. Which two key rings can he buy and have 67¢ change?

STRATEGY FILE

Logical Reasoning
Guess and Test
Draw a Picture
Choose the Operation

A 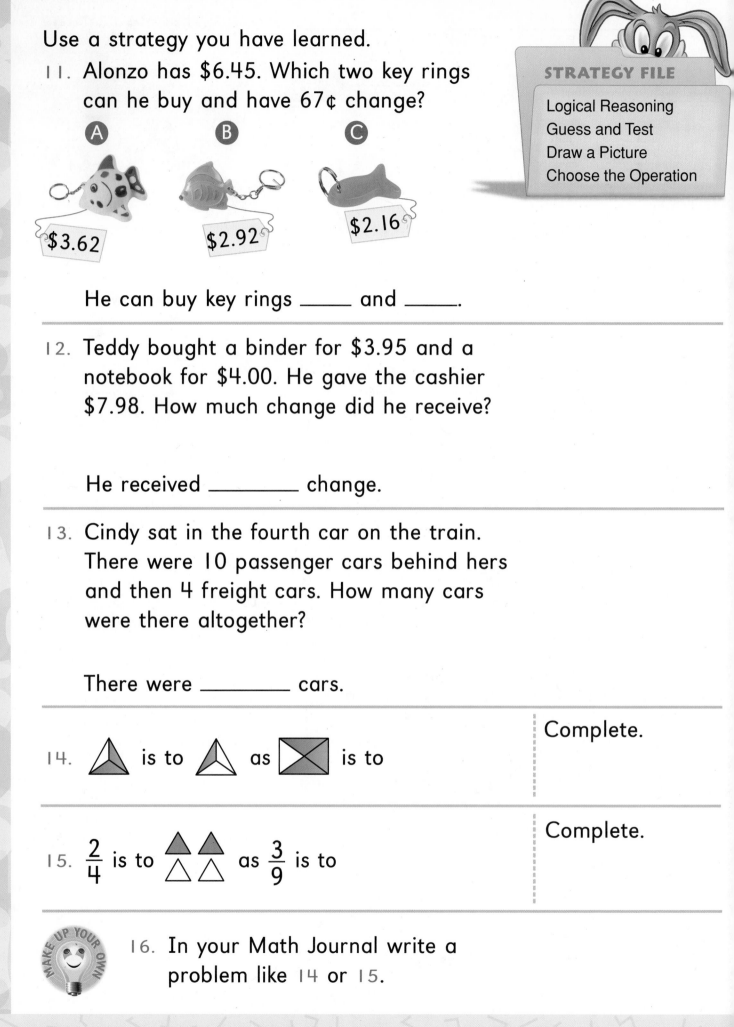 $3.62

B $2.92

C $2.16

He can buy key rings _____ and _____.

12. Teddy bought a binder for $3.95 and a notebook for $4.00. He gave the cashier $7.98. How much change did he receive?

He received _____ change.

13. Cindy sat in the fourth car on the train. There were 10 passenger cars behind hers and then 4 freight cars. How many cars were there altogether?

There were _____ cars.

Complete.

14. △ is to △ as ⊠ is to

Complete.

15. $\frac{2}{4}$ is to △△ / △△ as $\frac{3}{9}$ is to

16. In your Math Journal write a problem like 14 or 15.

Write Yes or No. Does ▲ = 12?

1. 153 − ▲ = 141

2. 100 + ▲ − 50 = 162

_____   _____

Find the hidden number.

3. 18 ÷ ◡ = 9

   ◡ = ___

4. ▪ + 11 = 775

   ▪ = ___

Find two answers for each number sentence.

5. 234 − 230 = 16 ◯ ▢

   ___ = 16 ◯ ▢

   ___ = 16 ◯ ▢

6. 182 − 172 > 2 ◯ ▢

   ___ > 2 ◯ ▢

   ___ > 2 ◯ ▢

Let ◆ equal 5. Find the value of each.

7. 4 × ◆ − 10

8. 3 × 9 + ◆

_____   _____

9. What is the value of 7 in: 8723 _____   7415 _____

Add or subtract.

10. 6542
    +1027

11. 7763
    +2123

12. 6154
    −1042

13. 7468
    −2431

PROBLEM SOLVING

14. ⎮⎮ is to 20 as ▦ ▦ is to _____.

12   This page reviews the mathematical content presented in Chapter 12.

four hundred sixty-three   **463**

MOVING ON: REINFORCEMENT

MATH·E·MAGIC

Add, subtract, or multiply. Use the code to answer the question.

Why are fish so smart?

Code:

A	B	C	E	H	I	L	M	N	O	S	T	U	W	Y
0	24	15	9	10	12	14	16	18	20	25	29	33	360	40

4×6	$11 - 2$	$24 - 9$	4×0	$52 - 19$	$19 + 6$	3×3
24						
B						

$25 + 4$	5×2	$6 + 3$	5×8

$67 - 67$	2×7	$280 + 80$	5×0	$19 + 21$	$77 - 52$

$85 - 60$	$407 - 47$	4×3	$143 - 127$

3×4	$9 + 9$

$18 + 7$	3×5	$760 - 750$	$75 - 55$	$13 + 7$	$8 + 6$	5×5

This page extends your child's understanding of addition, subtraction, and multiplication.

12

1. Unscramble the place values.
   Then model and describe the number.

   6 tens 2 thousands 0 ones 3 hundreds

   _____

   The number just before it is _____.

   The number just after it is _____.

2. Add or subtract. Describe the pattern.
   Write what comes next.

2121	2222	2323	
+2121	+2222	+2323	+ _____

2944	2948	2952	
−1111	−1111	−1111	− _____

PORTFOLIO Choose 1 of these projects.
Use a separate sheet of paper.

3. Make a number line for each
   counting pattern. Complete
   and then write the rule.

   • 1030, 1035, 1040, __?__

   • 998, 1000, 1002, __?__

   • 1003, 1006, 1009, __?__

   • 1015, 1025, 1035, __?__

4. Use the last 4 digits of
   your telephone number to
   write a riddle. Give clues
   in 10 sentences.

   Number Please
   My phone number is
   close to 7000. It is an
   even number. The 9
   means 900.

12 This page provides a variety of informal
assessment opportunities in order to measure
your child's understanding of Chapter 12.   four hundred sixty-five **465**

Write Yes or No. Does 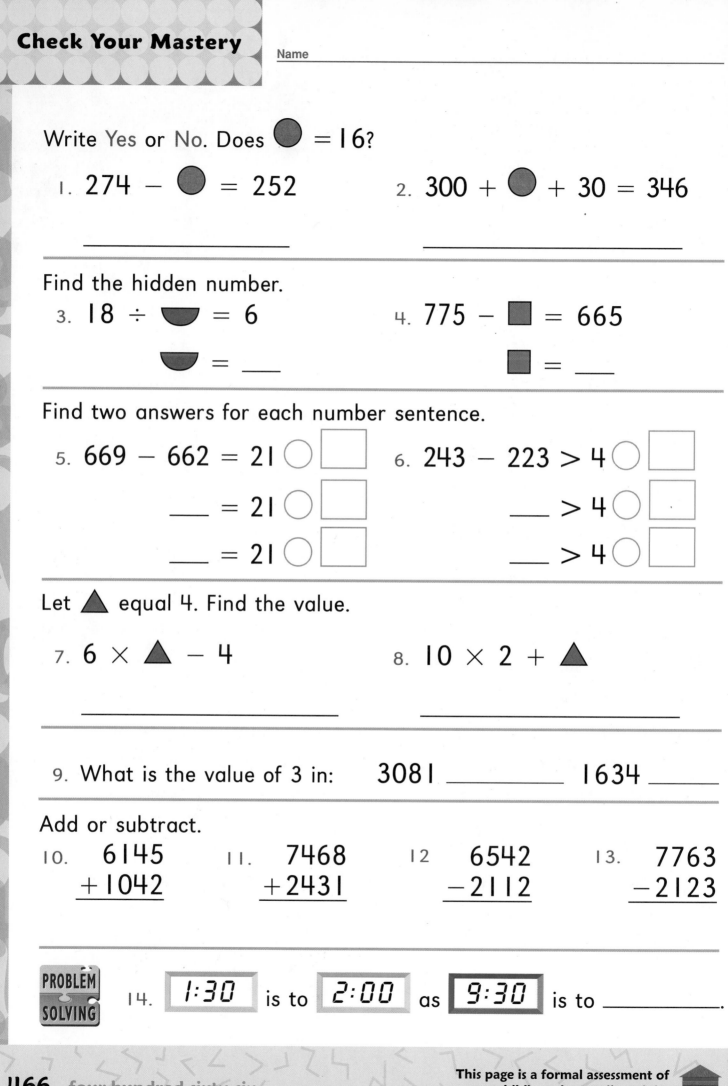 = 16?

1. 274 − ● = 252

2. 300 + ● + 30 = 346

_____   _____

Find the hidden number.

3. 18 ÷ ⌣ = 6

 ⌣ = ___

4. 775 − ■ = 665

 ■ = ___

Find two answers for each number sentence.

5. 669 − 662 = 21 ◯ ☐

 ___ = 21 ◯ ☐

 ___ = 21 ◯ ☐

6. 243 − 223 > 4 ◯ ☐

 ___ > 4 ◯ ☐

 ___ > 4 ◯ ☐

Let ▲ equal 4. Find the value.

7. 6 × ▲ − 4

_____

8. 10 × 2 + ▲

_____

9. What is the value of 3 in: 3081 _____   1634 _____

Add or subtract.

10. 6145
   +1042

11. 7468
   +2431

12 6542
   −2112

13. 7763
   −2123

PROBLEM SOLVING

14. 1:30 is to 2:00 as 9:30 is to _____.

This page is a formal assessment of your child's understanding of the content presented in Chapter 12.

12

Name _____

Write the number word.

1. _____

Ring the even numbers.

2. | 7 | 8 |

3. | 14 | 19 |

Compare. Write < or >.

4. 14 ◯ 18   5. 20 ◯ 16   6. 7 ◯ 17

Write the missing numbers.

7. ____, 15   8. 12, ____   9. 16, ____, 18

Find the sum or difference. Watch for + and −.

10. 5 + 3 = ____   7 + 2 = ____   8 + 3 = ____

11. 10 − 2 = ____   9 − 5 = ____   12 − 3 = ____

12.
$$\begin{array}{r} 8 \\ +1 \\ \hline \end{array} \quad \begin{array}{r} 7 \\ +3 \\ \hline \end{array} \quad \begin{array}{r} 4 \\ +4 \\ \hline \end{array} \quad \begin{array}{r} 11 \\ -4 \\ \hline \end{array} \quad \begin{array}{r} 9 \\ -9 \\ \hline \end{array} \quad \begin{array}{r} 9 \\ -3 \\ \hline \end{array}$$

Find the missing addend.

13. 2 + ☐ = 8

14. 7 + ☐ = 10

Add.

15. 3 + 4 + 4 = ____

16. 7 + 2 + 3 = ____

PROBLEM SOLVING

17. Cody had 6 🐟. He bought 4 more. How many 🐟 did he have in all?

____

18. ☀️ ☀️ ☀️ ☀️ ☀️ ☀️

☀️ means 2 sunny days. How many sunny days were there?

____ sunny days

REINFORCEMENT

Write the number.

1. 4 tens 3 ones	2. 60 + 5	3. eighty-nine	4.

	tens	ones
	9	1

____

Compare. Write < or >.

5. 23 ◯ 39   74 ◯ 67   86 ◯ 68

Write the number before, after, and between.

6. ___, 48   89, ___   67, ___, 69

Write in order from least to greatest.

7. 26, 31, 9, 19, _____

Count by 4. Write the missing numbers.

8. 60, 64, ____, ____, ____, ____, 84

9. Color these squares.

🌸 51st   🌸 56th
🌸 52nd   🌸 54th

50th   55th

10. Round 74 to the nearest ten.   74 rounds to ____.

⑦⓪ 71 72 73 **74** 75 76 77 78 79 ⑧⓪

PROBLEM SOLVING

11. Celia read 43 books. Dagmar read 39 books. Who read fewer books?

_____ read fewer books.

12. Eduardo baked 16 🥧. He put 2 in each bag. How many bags did he use?

He used ____ bags.

Name

Find the sum or difference.

1.
```
   9        6        7        9        8        7
  +5       +7       +5       +9       +9       +7
```

2.
```
  18       13       16       17       15       16
 − 9      − 5      − 7      − 8      − 9      − 8
```

3. 4 + 3 + 6 = ___   2 + 3 + 3 + 5 = ___

Add or subtract. Ring sums of 16.

4.
```
   9        9        7        5        8        9
  +7       +6       +9       +8       +8       +8
```

5.
```
  13       14       16       15       17       13
 − 9      − 7      − 9      − 7      − 9      − 8
```

Write the missing addends.

6. ☐ + 6 = 15   4 + ☐ = 13

PROBLEM SOLVING

7. There are 14 🐰 and 9 🥕. How many fewer 🥕 are there?

_____ fewer 🥕

8. Mai picked four numbers. They are 6, 8, 13, and 14. Which three of these numbers make up a fact family?

____ ____ ____

REINFORCEMENT

Name _____

1. Write how much.

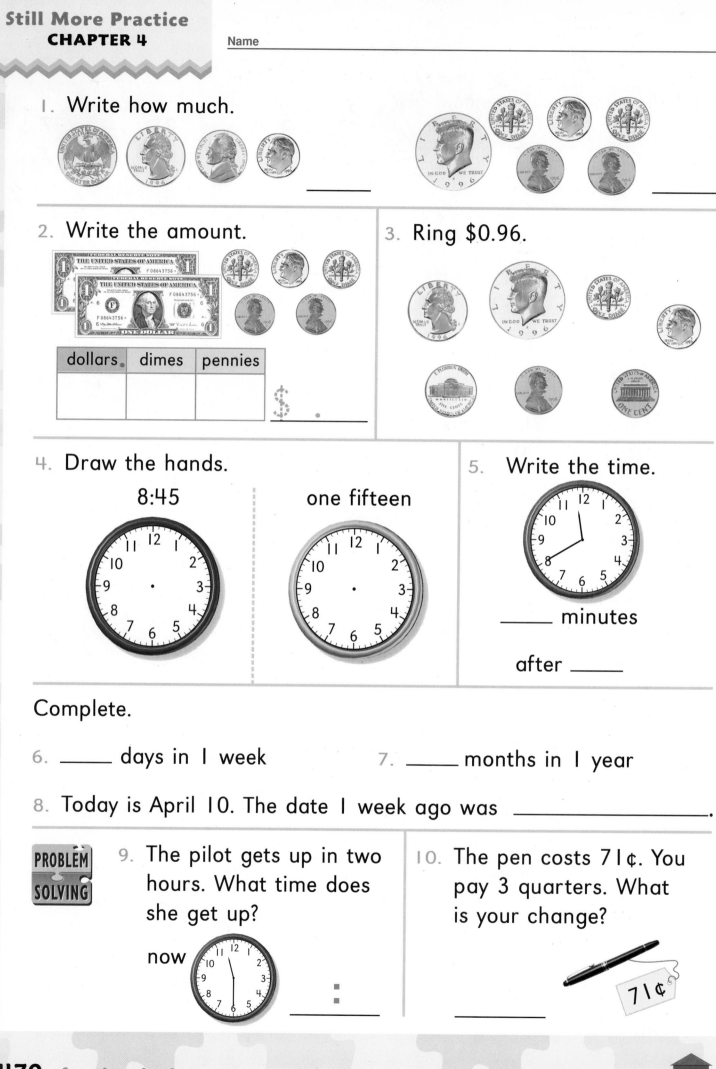

_____   _____

2. Write the amount.

dollars	dimes	pennies

$ ____ . ____

3. Ring $0.96.

4. Draw the hands.

8:45   one fifteen

5. Write the time.

_____ minutes

after _____

Complete.

6. _____ days in 1 week   7. _____ months in 1 year

8. Today is April 10. The date 1 week ago was _____.

PROBLEM SOLVING

9. The pilot gets up in two hours. What time does she get up?

now   __:__

10. The pen costs 71¢. You pay 3 quarters. What is your change?

_____

71¢

REINFORCEMENT

Name

Add. Use models.

1. 2 tens 7 ones + 5 ones   1 ten 1 one + 9 ones

    ____ tens ____ ones   ____ tens ____ ones

Add. ✔ the greatest sum in each row.

2.
30	71	17	24¢	56¢
+40	+21	+31	+44¢	+ 3¢

3.
65	49	46	83¢	39¢
+23	+30	+18	+ 7¢	+33¢

4.
18	25	21	72¢	37¢
+24	+66	+58	+ 6¢	+25¢

5.
34	56	37	27¢	69¢
+56	+16	+ 7	+36¢	+19¢

6.
28	11	47	14¢	23¢
20	34	11	25¢	12¢
+31	+12	+18	+35¢	+ 6¢

PROBLEM SOLVING

7. Lupe gave 59¢ to the nature fund.
    Then she gave 3 nickels more. How
    much is that in all?

    _____

REINFORCEMENT

Subtract. Use models to check.

1. 2 tens 0 ones − 4 ones =

   _____ ten _____ ones

2. 3 tens 6 ones − 9 ones

   _____ tens _____ ones

3.
```
  60        95        48        55¢       89¢
- 30      - 40      - 13      - 52¢     - 36¢
```

4.
```
  78        89        64        91¢       82¢
- 32      - 35      - 39      -  2¢     - 43¢
```

Subtract. Add to check.

5.
```
  76        59        82        43¢       62¢
- 54      -  9      - 67      - 28¢     - 36¢
```

6.
```
  84        51        95        60¢       90¢
- 27      -  6      - 87      - 47¢     - 55¢
```

PROBLEM SOLVING

7. Gloria has 47 stickers. Ed has 28 stickers. About how many do they have altogether?

   _____ ⟶
   _____ ⟶ _____

   about _____ stickers

8. The pet store had 63 🐱. 44 🐱 were sold. How many 🐱 were left?

   _____ 🐱

Name _____

Make each a closed figure. Complete.

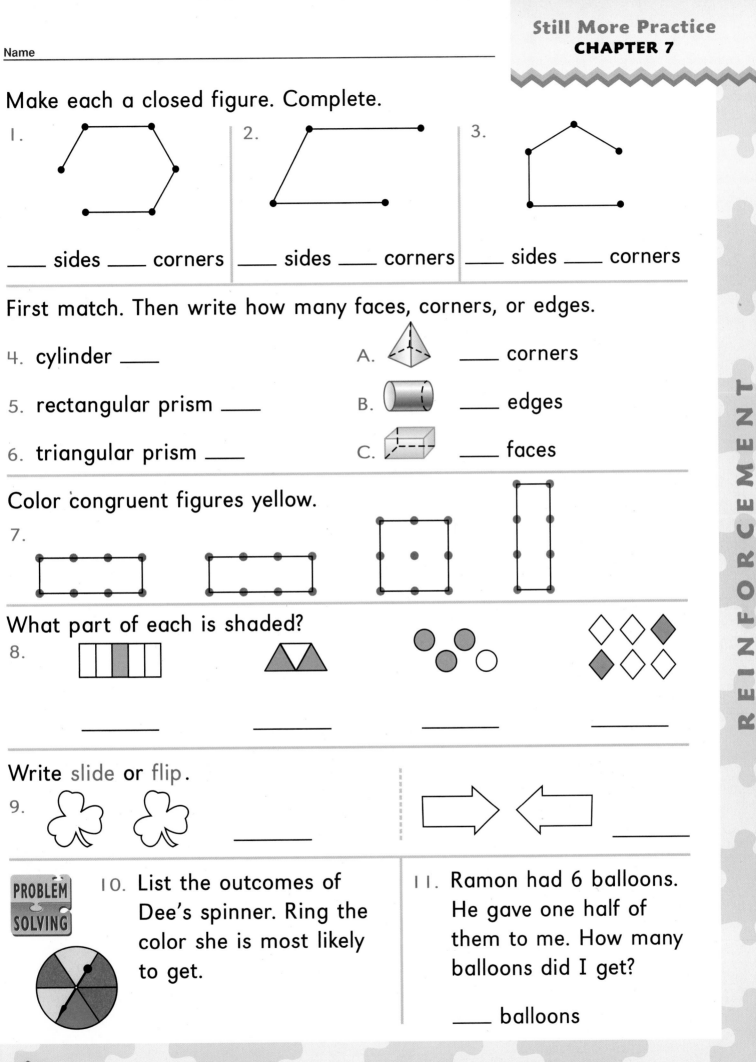

1. ____ sides ____ corners

2. ____ sides ____ corners

3. ____ sides ____ corners

First match. Then write how many faces, corners, or edges.

4. cylinder ____

5. rectangular prism ____

6. triangular prism ____

A. ____ corners

B. ____ edges

C. ____ faces

Color congruent figures yellow.

7.

What part of each is shaded?

8. _____   _____   _____   _____

Write slide or flip.

9. _____   _____

PROBLEM SOLVING

10. List the outcomes of Dee's spinner. Ring the color she is most likely to get.

11. Ramon had 6 balloons. He gave one half of them to me. How many balloons did I get?

____ balloons

Use your ruler to measure.

1. 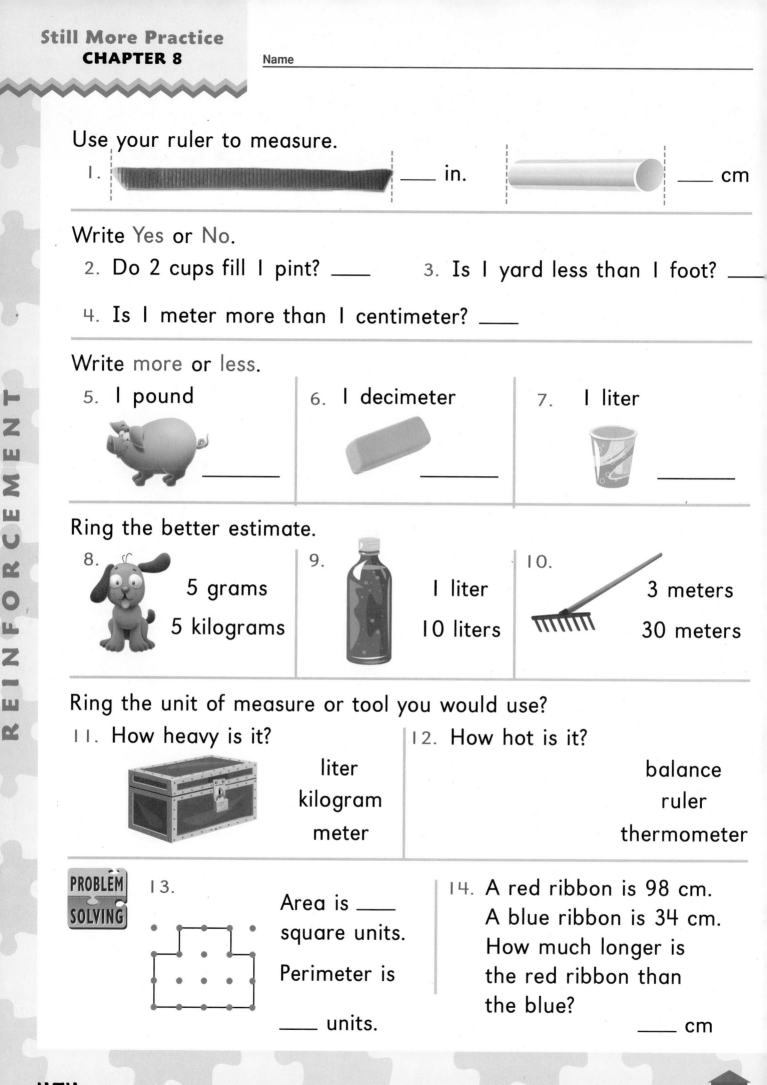 ___ in.   ___ cm

Write Yes or No.

2. Do 2 cups fill 1 pint? ___   3. Is 1 yard less than 1 foot? ___

4. Is 1 meter more than 1 centimeter? ___

Write more or less.

5. 1 pound

_____

6. 1 decimeter

_____

7. 1 liter

_____

Ring the better estimate.

8.   5 grams
      5 kilograms

9.   1 liter
      10 liters

10.   3 meters
      30 meters

Ring the unit of measure or tool you would use?

11. How heavy is it?

liter
kilogram
meter

12. How hot is it?

balance
ruler
thermometer

PROBLEM SOLVING

13.

Area is ___ square units.

Perimeter is

___ units.

14. A red ribbon is 98 cm. A blue ribbon is 34 cm. How much longer is the red ribbon than the blue?

___ cm

Complete.

1. 387 = ___ hundreds ___ tens ___ ones

2. 801 = ___ hundreds ___ tens ___ one

3. 100 + 40 = ___   4. 200 + 70 + 1 = ___

5. 178 = _____   6. 302 = _____

7. 100 more than   8. 10 more than

  106 ___   577 ___   217 ___   892 ___

Add. Regroup where needed.

9.
237	632	404	$8.12	$2.46
+ 42	+ 56	+265	+ 1.73	+ 1.15

10.
652	368	237	$1.44	$6.41
+ 29	+122	+290	+ 3.64	+ 0.85

11.
583	277	379	$6.37	$6.51
+ 53	+289	+242	+0.65	+ 1.59

PROBLEM SOLVING

12. How many 3-digit numbers between 800 and 900 have the same number of hundreds and ones?

_____ numbers

13. Mina spends $2.82 for a sandwich and $0.95 for juice. She also spends $1.00 for cherries. How much does she spend in all?

_____

REINFORCEMENT

Find the sum. Then multiply.

1. $5 + 5 =$ ___ $3 + 3 + 3 + 3 =$ ___ $2 + 2 + 2 =$ ___

   2 fives $=$ ___   4 _____ $=$ ___   3 _____ $=$ ___

   $2 \times$ ___ $=$ ___ $4 \times$ ___ $=$ ___ $3 \times$ ___ $=$ ___

Find the product.

2. $\begin{array}{r} 3 \\ \times 1 \end{array}$ $\begin{array}{r} 5 \\ \times 6 \end{array}$ $\begin{array}{r} 3 \\ \times 7 \end{array}$ $\begin{array}{r} 4 \\ \times 4 \end{array}$ $\begin{array}{r} 5 \\ \times 8 \end{array}$ $\begin{array}{r} 4 \\ \times 9 \end{array}$ $\begin{array}{r} 3 \\ \times 8 \end{array}$

3. $\begin{array}{r} 4 \\ \times 0 \end{array}$ $\begin{array}{r} 2 \\ \times 6 \end{array}$ $\begin{array}{r} 3 \\ \times 3 \end{array}$ $\begin{array}{r} 5 \\ \times 5 \end{array}$ $\begin{array}{r} 2 \\ \times 5 \end{array}$ $\begin{array}{r} 5 \\ \times 9 \end{array}$ $\begin{array}{r} 2 \\ \times 9 \end{array}$

Multiply. Then change the order.

4. $6 \times 4 =$ ___ $9 \times 3 =$ ___ $8 \times 2 =$ ___

   ___ $\times$ ___ $=$ ___   ___ $\times$ ___ $=$ ___   ___ $\times$ ___ $=$ ___

Complete.

5. There are 8 in all. How many groups of 2?

   There are 15 in all. How many groups of 3?

   There are 20 in all. How many groups of 5?

   ___ $\div$ ___ $=$ ___   ___ $\div$ ___ $=$ ___   ___ $\div$ ___ $=$ ___

PROBLEM SOLVING

6. Asa puts cards into groups of 4. Write how many.

   ___ groups of 4

   ___ left over

7. There are 18 🌑. 3 girls share them equally. How many does each girl get?

   ___ ◯ ___ $=$ ___

   Each girl gets ___ 🌑.

Compare. Write < or >.

1. 236 ◯ 275   509 ◯ 648   889 ◯ 799

---

2. Write 323, 113, 449, and 368 in order from least to greatest.

_____

3. Write 323, 432, 129, 340 in order from least to greatest.

_____

Subtract. Regroup where needed.

4. 5 hundreds 3 tens − 4 tens

= ____ hundreds ____ tens

2 hundreds 2 tens − 5 tens

= ____ hundred ____ tens

5.
819	766	341	$9.78	$6.46
−409	− 38	−225	− 1.75	− 0.05

6.
987	532	635	$3.20	$9.43
−579	− 72	−349	− 1.12	− 0.75

Estimate. Round to the nearest hundred or dollar.

7.
234	$1.98	648	$5.27
+183	+ 6.89	− 97	− 3.66

PROBLEM SOLVING

8. What is the difference of the numbers

324   562   172

inside the triangle? _____

inside the circle? _____

not inside both figures? _____

MAINTENANCE

Set 1	
1 to 10	ten to one
5 to 14	six to fifteen
8 to 17	nine to twenty
11 to 20	four to twelve
12 to 3	eleven to zero
16 to 10	thirteen to two
20 to 8	nineteen to ten
18 to 7	sixteen to six

Set 2		
4	3	6
8	5	9
11	17	12
19	14	10
twelve, _____, fourteen		
nine, _____, eleven		
ten, _____, twelve		
eighteen, _____, twenty		

Set 3			
4	8	11	20
7	5	16	6
10	1	18	15
12	14	13	19
3	13	7	12
0	10	14	16
20	2	17	8
15	14	9	1

Set 4	
1	0, 2, 4, 6, 8, 10, 12
1	1, 3, 5, 7, 9, 11, 13
2	0, 1, 2, 3, 4, 5, 6, 7, 8, 9, 10
3	0, 1, 2, 3, 4, 5, 6, 7, 8, 9, 10
4	0, 1, 2, 3, 4, 5, 6, 7, 8, 10
5	0, 1, 2, 3, 4, 5, 6, 7, 10
6	0, 1, 2, 3, 4, 5, 6, 10

Set 5	
1	12, 10, 8, 6, 4, 2
1	11, 9, 7, 5, 3, 1
2	12, 11, 10, 9, 8, 7, 6, 5
3	12, 11, 10, 9, 8, 7, 6, 5
4	12, 11, 10, 9, 8, 7, 6, 5
5	12, 11, 10, 9, 8, 7, 6, 5
6	12, 11, 10, 9, 8, 7, 6

Set 6	
3, 6, 9	2, 8, 10
1, 7, 8	8, 4, 12
2, 4, 6	3, 8, 11
3, 4, 7	3, 7, 10
4, 5, 9	6, 4, 10
3, 5, 8	4, 7, 11
2, 7, 9	5, 7, 12

Set 7		
tens	ones	
8	5	70 + 1
2	4	20 + 7
1	9	30 + 8
6	0	50 + 9
26	93	14
37	82	56
41	70	98

Set 8		
Count by	from	to
1	73	81
2	42	60
5	20	70
10	50	100
3	3	24
4	4	24
5	5	40

Set 9		
Compare.		
58	_____	85
96	_____	51
66	_____	38
32	_____	43
17	_____	29
72	_____	43
46	_____	64

Set 10		Set 11		Set 12	
Double...	**then...**	**Add...**	**to**	**Subtract...**	**from**
6	Add:	3	9, 8, 7, 6, 5, 4	9	19, 18, 17, 16, 15, 14, ...
5	0, 1, 2, 3, 10	4	4, 5, 6, 7, 8, 9	8	18, 17, 16, 15, 14, 13, ...
8		5	9, 8, 7, 6, 5, 4	7	17, 16, 15, 14, 13, 12, ...
9	Subtract:	6	4, 5, 6, 7, 8, 9	6	16, 15, 14, 13, 12, 11, ...
7	0, 1, 2, 3, 10	7	9, 8, 7, 6, 5, 4	5	15, 14, 13, 12, 11, 10, ...
10		8	4, 5, 6, 7, 8, 9	4	14, 13, 12, 11, 10, 9, ...
20		9	9, 8, 7, 6, 5, 4	3	13, 12, 11, 10, 9, 8, ...

Set 13

- 13 🐤, 9 fly away
- 8 🍌, 5 more 🍌
- 13 🐟, 6 swim away
- 14 ✈, 7 fly away
- 5 ⚾, 9 more ⚾
- 9 🐦, 4 more 🐦
- 16 🎁, lose 7 🎁

Set 14

nickels: 4, 6, 5, 10
dimes: 3, 5, 7, 9
quarters: 2, 4, 1, 3

1 🪙	2 🪙	1 🪙
2 🪙	3 🪙	2 🪙
1 🪙	1 🪙	4 🪙
1 🪙	2 🪙	3 🪙

Set 15

Set 16			Set 17			Set 18	
1:30	2:30	3:30	10 +60	40 +40	20 +70	20 + 6	10 + 18
9:15	8:15	7:15				40 + 9	60 + 17
10:45	11:45	12:45	10 +80	30 +50	60 +20	50 + 3	70 + 15
9:00	9:10	9:20				80 + 6	20 + 14

Sun.	Mon.	Tues.	Wed.
	1	2	?
7	?	?	10
?	15	16	?
21	?	?	24

1, 11, 21, 31, ...
3, 13, 23, 33, ...
7, 17, 27, 37, ...
9, 19, 29, 39, ...
5, 15, 25, 35, ...

30¢ + 8¢	40¢ + 11¢
60¢ + 2¢	30¢ + 12¢
90¢ + 4¢	50¢ + 13¢

Set 19

Add... to

3	93, 83, 73, 63, 53, 43
4	4, 14, 24, 34, 44, 54
5	85, 75, 65, 55, 45, 35
6	6, 16, 26, 36, 46, 56
7	87, 77, 67, 57, 47, 37
8	8, 18, 28, 38, 48, 58
9	89, 79, 69, 59, 49, 39

Set 20

7¢ + 3¢ + 15¢

12¢ + 6¢ + 4¢

1¢ + 9¢ + 32¢

27¢ + 2¢ + 2¢

5¢ + 49¢ + 5¢

3¢ + 85¢ + 7¢

9¢ + 41¢ + 1¢

Set 21

70	80	90
−20	−50	−80

50	90	60
−20	−40	−30

98, 88, 78, 68, ...

92, 82, 72, 62, ...

96, 86, 76, 66, ...

94, 84, 74, 64, ...

Set 22

37 − 7	35 − 11
51 − 1	88 − 11
45 − 5	79 − 11
89 − 9	46 − 11
97¢ − 7¢	91¢ − 11¢
62¢ − 2¢	62¢ − 11¢
74¢ − 4¢	53¢ − 11¢

Set 23

Subtract... from

3	99, 88, 77, 66, 55
4	64, 65, 66, 67, 68
5	37, 47, 57, 67, 77
6	76, 77, 78, 79, 80
7	59, 48, 37, 29, 18, 7
8	88, 89, 90, 78, 79, 80
9	99, 88, 77, 66, 55

Set 24

Set 25

Set 26

Set 27

1 inch	1 foot
1 inch	$\frac{1}{2}$ inch
1 foot	1 yard
1 cup	1 quart
1 quart	1 pint
1 centimeter	1 meter
1 pound	1 gallon
1 kilogram	1 liter
10°F	80°F

Set 28

Have 11 inches of .
Buy 6 inches more.

Had 18 liters of .
Used 5 liters.

Have 9 pounds of .
Buy 7 pounds more.

Had 14 pints of .
8 pints spilled.

Set 29

scale   ruler

measuring cup

balance

thermometer

---

inches   pounds

liters   meters

degrees   gallons

Set 30

hundreds	tens	ones
1	3	2
2	8	7
4	5	0
283	436	271
592	742	853
411	329	647

Set 31

from 320 to 720

from 112 to 312

from 210 to 510

from 799 to 999

---

from 410 to 110

from 905 to 605

from 368 to 168

Set 32

Compare.

132 _____ 321

254 _____ 146

456 _____ 478

952 _____ 925

367 _____ 309

187 _____ 245

Set 33

100 + 80	300 + 40 + 8
200 + 60	400 + 50 + 1
500 + 30	600 + 70 + 5
700 + 20	800 + 90 + 6
900 + 10	200 + 50 + 2
600 + 40	400 + 80 + 7

Set 34

Add... to

2	117, 127, 137, 147
3	912, 812, 712, 612
4	64, 164, 264, 364
5	483, 473, 463, 453
9	151, 161, 171, 181
10	718, 618, 518, 418
11	104, 114, 124, 134

Set 35

Multiply.

2, 3, 5, 1, 7, 8, 9, 6, 4	2
4, 9, 8, 2, 5, 3, 6, 1, 7	3
6, 1, 8, 3, 7, 5, 9, 2, 4	4
8, 2, 5, 7, 3, 1, 4, 6, 9	5

Set 36

Groups	of
3	2, 3, 4, 5, 10
5	10, 5, 4, 3, 2
4	2, 3, 4, 5, 10
2	10, 5, 4, 3, 2
6	2, 3, 4, 5, 10
7	2, 3, 4, 5, 10
8	2, 3, 4, 5, 10

MAINTENANCE

Set 37	Set 38	Set 39
How many in each?	805 508	800 700 900
twos in: 8, 2, 4, 10, 12, 6, 18, 14	533 335	−200 −500 −300
threes in: 3, 12, 21, 9, 15, 6, 27, 18	423 432	680 580 480
fours in: 8, 16, 20, 32, 4, 12, 24, 36	124 214	−500 −400 −200
	787 687 778	212 316 518
fives in: 30, 25, 20, 15, 10, 5, 35	634 643 534	−100 −200 −400
	108 808 118	325 176 485
		− 25 −170 −480

Set 40	Set 41	Set 42
Subtract... from	**Add... to**	**Subtract... from**
3 | 127, 137, 147, 157	2 | 9, 19, 29, 39, 49, 59, 69, 79, 89, 99	2 | 11, 21, 31, 41, 51, 61, 71, 81, 91, 101
4 | 919, 819, 719, 619	3 | 7, 17, 27, 37, 47 57, 67, 77, 87, 97	3 | 10, 110, 210, 310, 410, 510, 610, 710, 810
5 | 208, 218, 228, 238		
6 | 847, 837, 827, 817	4 | 19, 119, 219, 319 419, 519, 619, 719	5 | 13, 23, 33, 43, 53 63, 73, 83, 93, 103
7 | 119, 219, 319, 419		
8 | 988, 978, 968, 958	5 | 925, 935, 945, 955, 965, 975	9 | 18, 27, 36, 45, 54, 118, 127, 136, 145
9 | 119, 219, 319, 419		

Set 43	Set 44	Set 45
▲ = 1¢ ■ = 10¢ ● = $1.00	$117 − \square = 117$	$2 + 6 = 18 - \square$
85¢ + ▲ | ● + 16¢	$\square + 10 = 215$	$35 - 10 = 5 \times \square$
■ + 16¢ | ● + 25¢	$123 + \square = 124$	$2 \times 5 = \square - 110$
75¢ − ▲ | $8.00 − ●	$131 − \square = 31$	$125 - 100 = 25 + \square$
18¢ − ■ | $2.11 + ●	$250 − \square = 0$	$400 - 300 = \square + 90$
$3.00 − ● + ■	$145 − \square = 135$	$30 + 20 = 250 - \square$
$5.50 + ● − ▲	$102 − \square = 101$	$95 - 20 = 85 - \square$
$1.25 − ● + ▲		

GLOSSARY

A

This magazine belongs to ___

addend $3 + 4 = 7$

bar graph
Number of Sides

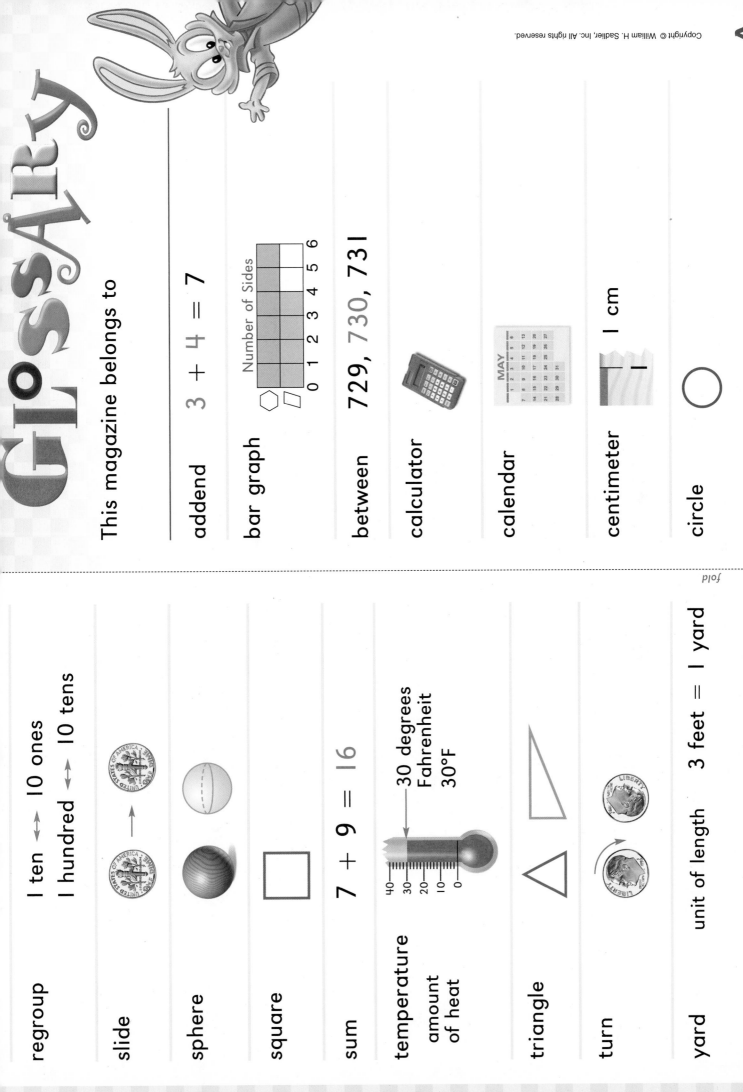

between   729, **730**, 731

calculator

calendar
MAY

centimeter   1 cm

circle

fold

regroup   1 ten ⟷ 10 ones
1 hundred ⟷ 10 tens

slide

sphere

square

sum $7 + 9 = 16$

temperature   30 degrees
Fahrenheit
30°F
amount
of heat

triangle

turn

yard   unit of length   3 feet = 1 yard

cone

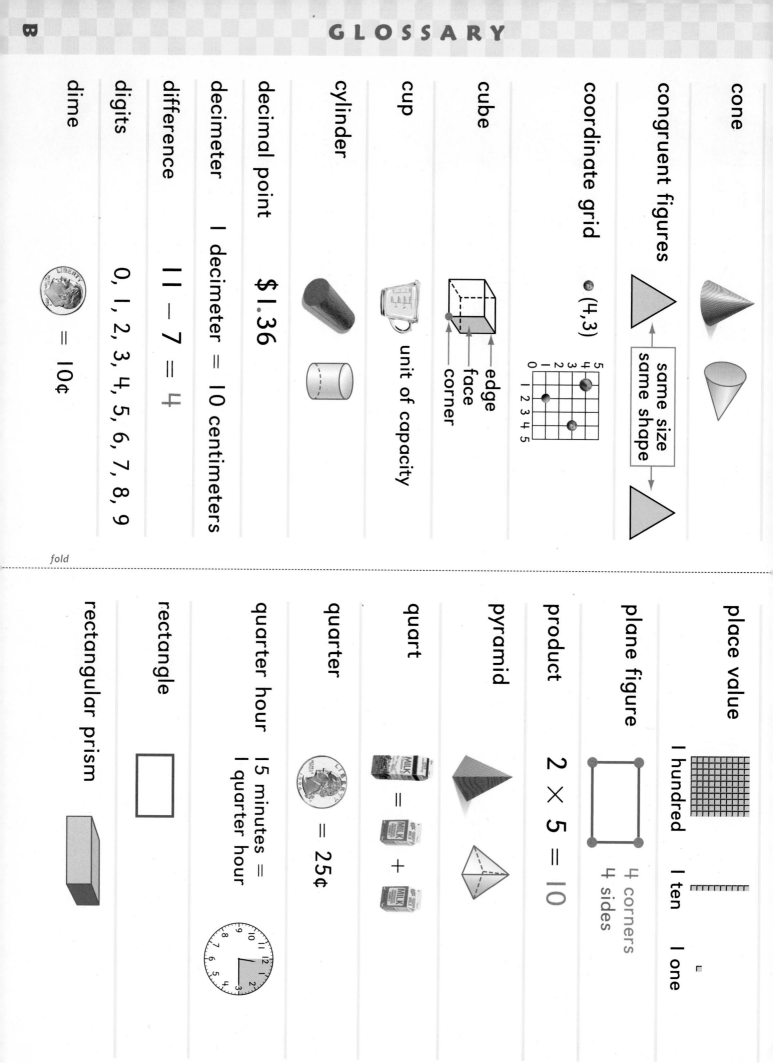

congruent figures

same size
same shape

coordinate grid

(4,3)

cube

edge
face
corner

cylinder

cup

unit of capacity

decimal point $1.36

decimeter   1 decimeter = 10 centimeters

difference   11 − 7 = 4

digits   0, 1, 2, 3, 4, 5, 6, 7, 8, 9

dime   = 10¢

fold

place value   1 hundred   1 ten   1 one

plane figure   4 corners
4 sides

product   2 × 5 = 10

pyramid   =   +

quart

quarter   = 25¢

quarter hour   15 minutes =
1 quarter hour

rectangle

rectangular prism

division $6 = \underline{?}$ groups of 2
$6 = 3$ groups of $\underline{?}$

division sentence $15 \div 5 = 3$
divided by (division sign)

dollar $100¢$
$\$1.00$

dollar sign $\$2.25$

dozen   12 things =
1 dozen

even numbers   0, 2, 4, 6, 8,...

fact family $8 + 4 = 12$ $12 - 4 = 8$
$4 + 8 = 12$ $12 - 8 = 4$

flip

foot   unit of length   12 inches = 1 foot

fractions $\frac{1}{2}$ one half $\frac{2}{3}$ two thirds

fold

---

odd numbers   1, 3, 5, 7, 9,...

order property $7 + 2 = 9$ $2 \times 5 = 10$
$2 + 7 = 9$ $5 \times 2 = 10$

ordinal numbers   1st, 2nd, 3rd, 4th, 5th,...

patterns   △○○, △○○, △○○,...
1, 4, 7, 10, 13,...

penny   = 1¢

perimeter   distance around
$3 + 4 + 5 = 12$

pictograph

Colors of Bicycles

Blue					
Purple					
Silver					

Key: Each = 2 friends.

pint   = +

gallon

= + + + + +

half dollar

= 50¢

half gallon

= +

half past the hour   30 minutes

1:30

30 minutes
after the hour

half hour   30 minutes

hour

60 minutes = 1 hour

inch

1 in.

is greater than

8 > 2

is less than

16 < 20

fold

kilogram   unit of mass

line of symmetry

liter   unit of capacity

meter   unit of length

100 centimeters = 1 meter

multiplication

3 twos = 6

3 × 2 = 6

multiplication sentence   4 × 2 = 8

times (multiplication sign)

nickel

= 5¢

number line

0 1 2 3 4 5

number sentence

8 + 7 = 15   3 × 4 = 12
15 − 7 = 8   12 ÷ 4 = 3